AF508659

# Clinical Features and Long-Term Outcomes of Systemic Lupus Erythematosus

# Clinical Features and Long-Term Outcomes of Systemic Lupus Erythematosus

Editors

**Christopher Sjöwall**
**Ioannis Parodis**

MDPI • Basel • Beijing • Wuhan • Barcelona • Belgrade • Manchester • Tokyo • Cluj • Tianjin

*Editors*
Christopher Sjöwall
Department of Biomedical
and Clinical Sciences (BKV)
Linköping University
Linköping
Sweden

Ioannis Parodis
Department of Medicine Solna
Division of Rheumatology
Karolinska Institutet
Stockholm
Sweden

*Editorial Office*
MDPI
St. Alban-Anlage 66
4052 Basel, Switzerland

This is a reprint of articles from the Special Issue published online in the open access journal *Journal of Clinical Medicine* (ISSN 2077-0383) (available at: www.mdpi.com/journal/jcm/special_issues/lupus_erythematosus).

For citation purposes, cite each article independently as indicated on the article page online and as indicated below:

LastName, A.A.; LastName, B.B.; LastName, C.C. Article Title. *Journal Name* **Year**, *Volume Number*, Page Range.

**ISBN 978-3-0365-6110-3 (Hbk)**
**ISBN 978-3-0365-6109-7 (PDF)**

# Contents

# Preface to "Clinical Features and Long-Term Outcomes of Systemic Lupus Erythematosus"

The clinical spectrum of systemic lupus erythematosus (SLE) is highly heterogeneous, ranging from mild disease, which can be limited to skin and joint involvement, to life-threatening conditions with renal impairment, severe cytopenias, central nervous system disease, and thromboembolic events. Apart from the host genetics, several environmental factors, such as sunlight, infections, drugs, and probably hormonal factors, can trigger the onset of symptoms related to SLE. Despite significant advances in our understanding of the pathophysiology and optimization of medical care, patients with SLE still have significant rates of premature mortality and many patients experience severe disease with increased risk of sustaining organ damage and having a reduced health-related quality of life. The development of effective drugs that can induce remission or low disease activity, the unanimous use of definitions of remission and low or high disease activity, flare, and response to therapy, the identification of non-invasive biomarkers of disease activity and long-term outcomes, and the implementation of SLE patients' perspectives as an integral part of the clinical assessment constitute only a few of the many unmet needs in the field of SLE.

**Christopher Sjöwall and Ioannis Parodis**

*Editors*

Journal of
*Clinical Medicine*

*Editorial*

# Clinical Heterogeneity, Unmet Needs and Long-Term Outcomes in Patients with Systemic Lupus Erythematosus

Christopher Sjöwall [1,*,†] and Ioannis Parodis [2,3]

1    Department of Biomedical and Clinical Sciences, Division of Inflammation and Infection/Rheumatology, Linköping University, SE-581 85 Linköping, Sweden
2    Department of Medicine Solna, Division of Rheumatology, Karolinska Institutet and Karolinska University Hospital, SE-171 76 Stockholm, Sweden
3    Department of Rheumatology, Faculty of Medicine and Health, Örebro University, SE-701 82 Örebro, Sweden
*    Correspondence: christopher.sjowall@liu.se; Tel.: +46-10-1032416
†    Current address: Rheumatology Unit, University Hospital in Linköping, SE-581 85 Linköping, Sweden.

**Citation:** Sjöwall, C.; Parodis, I. Clinical Heterogeneity, Unmet Needs and Long-Term Outcomes in Patients with Systemic Lupus Erythematosus. *J. Clin. Med.* **2022**, *11*, 6869. https://doi.org/10.3390/jcm11226869

Received: 15 November 2022
Accepted: 17 November 2022
Published: 21 November 2022

**Publisher's Note:** MDPI stays neutral with regard to jurisdictional claims in published maps and institutional affiliations.

## 1. Introduction

The clinical presentation of systemic lupus erythematosus (SLE) is highly heterogeneous, ranging from mild disease limited to skin and joint involvement to life-threatening conditions with renal impairment, severe cytopenias, central nervous system disease, and thromboembolic events [1]. Despite significant advances in our understanding of the pathophysiology and optimization of medical care, SLE populations still exhibit premature mortality. Many patients with SLE experience poor health-related quality of life (HRQoL), even after successful treatment in terms of clinical and laboratory parameters [2], as well as severe disease flares with an increased risk of organ damage [3]. The development of effective drugs for SLE—which can induce remission or lower disease activity—the unanimous use of definitions of remission and low or high disease activity, flare, and response to therapy, the identification of non-invasive biomarkers of disease activity and long-term outcomes, and the implementation of the patient perspective as an integral part of the clinical assessment constitute only a few of the many unmet needs in the field of SLE.

In this Special Issue hosted by the Journal of Clinical Medicine, we selected a series of articles that highlight current and contribute new knowledge related to aspects such as clinical heterogeneity, autoantibodies, and long-term outcomes in SLE. Several of the contributions focus on patients' perspectives and SLE patients' unmet needs, for instance, fatigue, poor HRQoL experience, and non-adherence to medications. In this Editorial, we provide an overview of challenges and opportunities in the management of SLE, as presented by authors who contributed to the collection, and we hope that this will prove valuable both for clinicians and people living with SLE.

## 2. Clinical Heterogeneity

In a study by Jung and co-authors, hierarchical clustering was performed to gain insights into the clinical heterogeneity of SLE [4]. Three distinct clusters of patients with different manifestations and antibody profiles were identified among 389 patients through the combination of laboratory test results at SLE onset and linear discriminant analysis, utilized to construct prediction models. In a comprehensive review, Mahler et al. summarized the history and future directions regarding anti-Ki/SL antibodies in SLE and Sjögren's syndrome, which were first described in 1981 [5].

Register data can be used to evaluate changes over time. Moreno-Torres et al. used the Spanish national registry to evaluate trends in hospital admissions and causes of death from the late 1990s until 2015 using ICD codes [6]. The authors concluded that the improved control of SLE over the past two decades has led to a decrease in early admissions to hospital and disease chronification. In line with data from other groups [7,8], cardiovascular disease, infections, malignancies, and thromboembolic events were among the most common causes of death.

Similarly, ICD codes and reliable national patient register data were used during the coronavirus disease 2019 (COVID-19) pandemic to improve our understanding and patient care. In this Special Issue, Cordtz and colleagues demonstrated that Danish patients with SLE were at an approximately threefold increased risk of hospitalization due to COVID-19 compared with age- and sex-matched comparators from the general population [9]. Interestingly, no obvious impact of the use of glucocorticoids or hydroxychloroquine was seen on the risk of hospitalization.

Diagnosis of autoimmune liver diseases (AILD) among individuals with an already established diagnosis of SLE is challenging since liver enzyme test abnormalities and hyper-gammaglobulinemia are common laboratory findings in SLE, and antinuclear antibodies (ANA) constitute a prerequisite. Heijke et al. demonstrated why the autoimmune hepatitis criteria [10] are less useful in SLE populations and why a liver biopsy should be performed with the measurement of other AILD-associated autoantibodies [11].

## 3. Patient-Reported Experience

Several contributions in this Special Issue dealt with patient-reported outcome measures (PROMs), mainly PROMs capturing SLE patients' HRQoL.

Nguyen and colleagues reviewed the literature for the use of PROMs to assess HRQoL, both in research settings and clinical practice, and described the characteristics of commonly used PROMs [12]. The authors advocate that the increased use of PROMs may help alleviate the discordance of health perception between patients and clinicians, which would be especially useful for patient populations with a high comorbidity burden.

Lai et al. compared the well-established SLE disease activity index 2000 (SLEDAI-2K) with the more recent SLE disease activity score (SLE-DAS) in terms of their correlation with the Lupus Quality of Life questionnaire (LupusQoL) in Taiwanese SLE patients and found that both activity indices perform fairly well in capturing SLE patients' health experience, with no substantial differences [13].

Fatigue is a common and multifaceted phenomenon in SLE, oftentimes neglected by clinicians. One of the reasons for this is due to the scarceness of interventions that have shown effectiveness in improving fatigue. Kawka and colleagues reviewed the literature to shed light on the impact, determinants, and management of fatigue in patients with SLE [14]. Some pharmaceuticals have demonstrated ability to alleviate fatigue, as has non-pharmacological management such as psychosocial interventions and lifestyle improvements. The authors suggest that the management of fatigue in SLE should rely on person-centered approaches and targeted interventions. Fatigue was also addressed in a review by Dey et al., which demonstrated how fatigue is manifested and managed in patients with SLE and rheumatoid arthritis (RA) [15]. While the two diseases differ in terms of clinical manifestations, fatigue is commonly reported in both patient populations. It is associated with pain, depression, and anxiety, and affects function, work capacity, and quality of life. Comorbidities contribute to fatigue, further complicating its management. Collectively, fatigue should be managed in a holistic manner, along with the management of comorbidities and management of factors that augment its impact on patients' lives.

As a tool to fight fatigue, a study from Sweden by Skoglund and colleagues revived an old mechanism that has shown promise, namely the adrenal hormone dehydroepiandrosterone (DHEA) [16]. The authors studied DHEA dosages in relation to SLE activity, glucocorticoid use, concomitant immunosuppressants, and patient-reported pain, fatigue, well-being, HRQoL, and functional disability. DHEA treatment was safe but did not alter disease activity or organ damage progression over time. Some improvement was seen regarding fatigue; however, this did not reach statistical significance. The authors nevertheless suggested that the determination of DHEA blood concentrations should be performed prior to treatment commencement, along with the exclusion of comorbidities that may require other therapeutic approaches.

Undoubtedly, one of the major challenges in SLE management is posed by non-adherence to therapy. Emamikia and colleagues interviewed patients with SLE from two Swedish centers in a qualitative study aiming at identifying influenceable contributors to non-adherence and suggesting interventions to alleviate this phenomenon [17]. The reasons for non-adherence were complex and multifaceted, both intentional and unintentional, related to the relationship between patients and caregivers, lack of information about the disease and medications, and influence from family and friends. Increased communication between patients and caregivers, patient education, psychosocial support, and the involvement of family members in the patients' journey through their disease were some of the potential contributors that the authors suggested for the increased adherence of SLE patients to their medications.

Anxiety and depression are major concerns in patients with SLE. Nikoloudaki et al. examined longitudinal trends in anxiety, depression, and SLE activity and showed that a high mental disease burden persists despite disease control in some patients [18]. Based on these findings, the authors suggested that socioeconomic facets should be a part of comprehensive patient evaluations. Importantly, and to make the connection with the study by Emamikia et al. [17], anxiety and depression were associated with non-adherence to medications.

## 4. Long-Term Outcomes

Longitudinal follow-up of patients with SLE using validated tools to assess disease activity and organ damage is crucial for understanding the true nature and burden of the disease. Gerosa et al. reported data from the Milan SLE consortium cohort (SMiLE) with an impressively long follow-up and data on the attainment of remission and lupus low-disease-activity state (LLDAS) [19]. In line with observations of other groups [20], the authors demonstrated that the attainment of remission or LLDAS was associated with less organ damage. Furthermore, this study showed that patients with a longer distance from disease onset are at a higher risk of developing disease flares, which in turn constitutes a risk factor for late damage accrual.

The potential genetic background underlying damage development in distinct organ domains was the focus of a study by Ceccarelli and colleagues [21]. The authors provided new insights into the genetic susceptibility for damage accrual in the renal and neuropsychiatric domains in particular, based on gene polymorphisms.

In a systematic literature review, Reppe Moe et al. compiled data on mortality, end-stage kidney disease (ESKD), and cancer from population-based studies, and found that cardiovascular disease was the most frequent cause of death over 15 years of follow-up. Moreover, 5–11% of patients developed ESKD, and no evidence for increased cancer incidence was found [22].

Last but not least, Suzon et al. summarized long-term population-based data in Afro-descendant patients with lupus nephritis (LN) from Martinique [23]. The main purpose of this study was to determine the rates of ESKD and mortality in an Afro-descendant LN population with a generally high income as well as easy and free access to healthcare. Unlike the stale notion that patients of African descent overall have worse ESKD and mortality rates compared to Caucasians, this study reported overall favorable rates, comparable to those seen in Caucasians, estimated at 21.3% for ESKD and 7.9% for mortality at 20 years of follow-up. These results underscore the importance of optimizing modifiable contributors to poor outcomes, especially socioeconomic factors.

## 5. Perspective

Herein, we present a rich collection of important contributions by esteemed colleagues in the field of SLE and autoimmunity, ranging from disease classification and genetic susceptibility to disease evolution to facilitators towards improved long-term outcomes and patient-reported health experience. We firmly believe that this article collection contributes to novel knowledge and substantiates older, well-established notions, all presented in a manner that provides direct clinical implications and emphasizes the importance of incorporating the patient perspective in a holistic, patient-centered, and tailored management of people living with SLE.

The articles of this Special Issue will also be made available in the form of an electronic book. We would like to express our gratitude to all colleagues who contributed works to this Special Issue and look forward to seeing the findings reported herein being discussed and implemented in clinical practice, making an impact and ultimately a difference in SLE patients' lives.

**Author Contributions:** C.S. and I.P. made a substantial, direct, and intellectual contribution to the work, and approved it for publication. All authors have read and agreed to the published version of the manuscript.

**Funding:** This work was supported by grants from the Swedish Rheumatism Association (R-969696; R-939149), King Gustaf V's 80-year Foundation (FAI-2020-0741; FAI-2020-0663), King Gustaf V and Queen Victoria's Freemasons foundation (2021-22), Swedish Society of Medicine (SLS-974449), Nyckelfonden (OLL-974804); Professor Nanna Svartz Foundation (2020-00368), Ulla and Roland Gustafsson Foundation (2021-26), Region Stockholm (FoUI-955483), and Karolinska Institutet.

**Conflicts of Interest:** C.S. declares no conflict of interest related to this work. I.P. has received research funding and/or honoraria from Amgen, AstraZeneca, Aurinia Pharmaceuticals, Elli Lilly and Company, Gilead Sciences, GlaxoSmithKline, Janssen Pharmaceuticals, Novartis and F. Hoffmann-La Roche AG. The funders had no role in the design of the study, the analyses or interpretation of data, or the writing of the manuscript.

## References

1. Kaul, A.; Gordon, C.; Crow, M.K.; Touma, Z.; Urowitz, M.B.; van Vollenhoven, R.; Ruiz-Irastorza, G.; Hughes, G. Systemic lupus erythematosus. *Nat. Rev. Dis. Prim.* **2016**, *2*, 16039. [CrossRef] [PubMed]
2. Gomez, A.; Qiu, V.; Cederlund, A.; Borg, A.; Lindblom, J.; Emamikia, S.; Enman, Y.; Lampa, J.; Parodis, I. Adverse Health-Related Quality of Life Outcome Despite Adequate Clinical Response to Treatment in Systemic Lupus Erythematosus. *Front. Med.* **2021**, *8*, 651249. [CrossRef] [PubMed]
3. Björk, M.; Dahlström, Ö.; Wetterö, J.; Sjöwall, C. Quality of life and acquired organ damage are intimately related to activity limitations in patients with systemic lupus erythematosus. *BMC Musculoskelet. Disord.* **2015**, *16*, 188. [CrossRef] [PubMed]
4. Jung, J.-Y.; Lee, H.-Y.; Lee, E.; Kim, H.-A.; Yoon, D.; Suh, C.-H. Three Clinical Clusters Identified through Hierarchical Cluster Analysis Using Initial Laboratory Findings in Korean Patients with Systemic Lupus Erythematosus. *J. Clin. Med.* **2022**, *11*, 2406. [CrossRef]
5. Mahler, M.; Bentow, C.; Aure, M.-A.; Fritzler, M.J.; Satoh, M. Significance of Autoantibodies to Ki/SL as Biomarkers for Systemic Lupus Erythematosus and Sicca Syndrome. *J. Clin. Med.* **2022**, *11*, 3529. [CrossRef]
6. Moreno-Torres, V.; Tarín, C.; Ruiz-Irastorza, G.; Castejón, R.; Gutiérrez-Rojas, Á.; Royuela, A.; Durán-del Campo, P.; Mellor-Pita, S.; Tutor, P.; Rosado, S.; et al. Trends in Hospital Admissions and Death Causes in Patients with Systemic Lupus Erythematosus: Spanish National Registry. *J. Clin. Med.* **2021**, *10*, 5749. [CrossRef]
7. Chambers, S.A.; Allen, E.; Rahman, A.; Isenberg, D. Damage and mortality in a group of British patients with systemic lupus erythematosus followed up for over 10 years. *Rheumatology* **2009**, *48*, 673–675. [CrossRef]
8. Frodlund, M.; Reid, S.; Wetterö, J.; Dahlström, Ö.; Sjöwall, C.; Leonard, D. The majority of Swedish systemic lupus erythematosus patients are still affected by irreversible organ impairment: Factors related to damage accrual in two regional cohorts. *Lupus* **2019**, *28*, 1261–1272. [CrossRef]
9. Cordtz, R.; Kristensen, S.; Dalgaard, L.P.H.; Westermann, R.; Duch, K.; Lindhardsen, J.; Torp-Pedersen, C.; Dreyer, L. Incidence of COVID-19 Hospitalisation in Patients with Systemic Lupus Erythematosus: A Nationwide Cohort Study from Denmark. *J. Clin. Med.* **2021**, *10*, 3842. [CrossRef]
10. Hennes, E.M.; Zeniya, M.; Czaja, A.J.; Parés, A.; Dalekos, G.N.; Krawitt, E.L.; Bittencourt, P.L.; Porta, G.; Boberg, K.M.; Hofer, H.; et al. Simplified criteria for the diagnosis of autoimmune hepatitis. *Hepatology* **2008**, *48*, 169–176. [CrossRef]
11. Heijke, R.; Ahmad, A.; Frodlund, M.; Wirestam, L.; Dahlström, Ö.; Dahle, C.; Kechagias, S.; Sjöwall, C. Usefulness of Clinical and Laboratory Criteria for Diagnosing Autoimmune Liver Disease among Patients with Systemic Lupus Erythematosus: An Observational Study. *J. Clin. Med.* **2021**, *10*, 3820. [CrossRef]
12. Nguyen, M.H.; Huang, F.F.; O'Neill, S.G. Patient-Reported Outcomes for Quality of Life in SLE: Essential in Clinical Trials and Ready for Routine Care. *J. Clin. Med.* **2021**, *10*, 3754. [CrossRef]
13. Lai, N.-S.; Lu, M.-C.; Chang, H.-H.; Lo, H.-C.; Hsu, C.-W.; Huang, K.-Y.; Tung, C.-H.; Hsu, B.-B.; Wu, C.-H.; Koo, M. A Comparison of the Correlation of Systemic Lupus Erythematosus Disease Activity Index 2000 (SLEDAI-2K) and Systemic Lupus Erythematosus Disease Activity Score (SLE-DAS) with Health-Related Quality of Life. *J. Clin. Med.* **2021**, *10*, 2137. [CrossRef]

14. Kawka, L.; Schlencker, A.; Mertz, P.; Martin, T.; Arnaud, L. Fatigue in Systemic Lupus Erythematosus: An Update on Its Impact, Determinants and Therapeutic Management. *J. Clin. Med.* **2021**, *10*, 3996. [CrossRef]
15. Dey, M.; Parodis, I.; Nikiphorou, E. Fatigue in Systemic Lupus Erythematosus and Rheumatoid Arthritis: A Comparison of Mechanisms, Measures and Management. *J. Clin. Med.* **2021**, *10*, 3566. [CrossRef]
16. Skoglund, O.; Walhelm, T.; Thyberg, I.; Eriksson, P.; Sjöwall, C. Fighting Fatigue in Systemic Lupus Erythematosus: Experience of Dehydroepiandrosterone on Clinical Parameters and Patient-Reported Outcomes. *J. Clin. Med.* **2022**, *11*, 5300. [CrossRef]
17. Emamikia, S.; Gentline, C.; Enman, Y.; Parodis, I. How Can We Enhance Adherence to Medications in Patients with Systemic Lupus Erythematosus? Results from a Qualitative Study. *J. Clin. Med.* **2022**, *11*, 1857. [CrossRef]
18. Nikoloudaki, M.; Repa, A.; Pitsigavdaki, S.; Molla Ismail Sali, A.; Sidiropoulos, P.; Lionis, C.; Bertsias, G. Persistence of Depression and Anxiety despite Short-Term Disease Activity Improvement in Patients with Systemic Lupus Erythematosus: A Single-Centre, Prospective Study. *J. Clin. Med.* **2022**, *11*, 4316. [CrossRef]
19. Gerosa, M.; Beretta, L.; Ramirez, G.A.; Bozzolo, E.; Cornalba, M.; Bellocchi, C.; Argolini, L.M.; Moroni, L.; Farina, N.; Segatto, G.; et al. Long-Term Clinical Outcome in Systemic Lupus Erythematosus Patients Followed for More Than 20 Years: The Milan Systemic Lupus Erythematosus Consortium (SMiLE) Cohort. *J. Clin. Med.* **2022**, *11*, 3587. [CrossRef]
20. Ugarte-Gil, M.F.; Hanly, J.; Urowitz, M.; Gordon, C.; Bae, S.C.; Romero-Diaz, J.; Sanchez-Guerrero, J.; Bernatsky, S.; Clarke, A.E.; Wallace, D.J.; et al. Remission and low disease activity (LDA) prevent damage accrual in patients with systemic lupus erythematosus: Results from the Systemic Lupus International Collaborating Clinics (SLICC) inception cohort. *Ann. Rheum. Dis.* **2022**, *81*, 1541–1548. [CrossRef]
21. Ceccarelli, F.; Olivieri, G.; Pirone, C.; Ciccacci, C.; Picciariello, L.; Natalucci, F.; Perricone, C.; Spinelli, F.R.; Alessandri, C.; Borgiani, P.; et al. The Impacts of the Clinical and Genetic Factors on Chronic Damage in Caucasian Systemic Lupus Erythematosus Patients. *J. Clin. Med.* **2022**, *11*, 3368. [CrossRef] [PubMed]
22. Reppe Moe, S.; Haukeland, H.; Molberg, Ø.; Lerang, K. Long-Term Outcome in Systemic Lupus Erythematosus; Knowledge from Population-Based Cohorts. *J. Clin. Med.* **2021**, *10*, 4306. [CrossRef] [PubMed]
23. Suzon, B.; Louis-Sidney, F.; Aglaé, C.; Henry, K.; Bagoée, C.; Wolff, S.; Moinet, F.; Emal-Aglaé, V.; Polomat, K.; DeBandt, M.; et al. Good Long-Term Prognosis of Lupus Nephritis in the High-Income Afro-Caribbean Population of Martinique with Free Access to Healthcare. *J. Clin. Med.* **2022**, *11*, 4860. [CrossRef] [PubMed]

Journal of
*Clinical Medicine*

*Article*

# Three Clinical Clusters Identified through Hierarchical Cluster Analysis Using Initial Laboratory Findings in Korean Patients with Systemic Lupus Erythematosus

Ju-Yang Jung [1,†], Hyun-Young Lee [2,†], Eunyoung Lee [2], Hyoun-Ah Kim [1], Dukyong Yoon [3] and Chang-Hee Suh [1,4,*]

1   Department of Rheumatology, Ajou University School of Medicine, Suwon 16499, Korea;
    serinne20@hanmail.net (J.-Y.J.); nakhada@naver.com (H.-A.K.)
2   Department of Biomedical Informatics, Ajou University School of Medicine, Suwon 16499, Korea;
    ajoustat@gmail.com (H.-Y.L.); e.angie.lee@gmail.com (E.L.)
3   Center for Digital Health, Yongin Severance Hospital, Yonsei University Health System,
    Yongin 16995, Korea; yoon8302@gmail.com
4   Department of Molecular Science and Technology, Ajou University, Suwon 16499, Korea
*   Correspondence: chsuh@ajou.ac.kr; Tel.: +82-31-219-5118
†   These authors contributed equally to this work.

**Abstract:** Systemic lupus erythematosus (SLE) is a heterogeneous disorder with diverse clinical manifestations. This study classified patients by combining laboratory values at SLE diagnosis via hierarchical cluster analysis. Linear discriminant analysis was performed to construct a model for predicting clusters. Cluster analysis using data from 389 patients with SLE yielded three clusters with different laboratory characteristics. Cluster 1 had the youngest age at diagnosis and showed significantly lower lymphocyte and platelet counts and hemoglobin and complement levels and the highest erythrocyte sedimentation rate (ESR) and anti-double-stranded DNA (dsDNA) antibody level. Cluster 2 showed higher white blood cell (WBC), lymphocyte, and platelet counts and lower ESR and anti-dsDNA antibody level. Cluster 3 showed the highest anti-nuclear antibody titer and lower WBC and lymphocyte counts. Within approximately 171 months, Cluster 1 showed higher SLE Disease Activity Index scores and number of cumulative manifestations, including malar rash, alopecia, arthritis, and renal disease, than did Clusters 2 and 3. However, the damage index and mortality rate did not differ significantly between them. In conclusion, the cluster analysis using the initial laboratory findings of the patients with SLE identified three clusters. While disease activities, organ involvements, and management patterns differed between the clusters, damages and mortalities did not.

**Keywords:** classification; cluster analysis; laboratory; linear discriminant analysis; systemic lupus erythematosus

**Citation:** Jung, J.-Y.; Lee, H.-Y.; Lee, E.; Kim, H.-A.; Yoon, D.; Suh, C.-H. Three Clinical Clusters Identified through Hierarchical Cluster Analysis Using Initial Laboratory Findings in Korean Patients with Systemic Lupus Erythematosus. *J. Clin. Med.* **2022**, *11*, 2406. https://doi.org/10.3390/jcm11092406

Academic Editors: Christopher Sjöwall and Ioannis Parodis

Received: 17 March 2022
Accepted: 22 April 2022
Published: 25 April 2022

**Publisher's Note:** MDPI stays neutral with regard to jurisdictional claims in published maps and institutional affiliations.

## 1. Introduction

Systemic lupus erythematosus (SLE) is a chronic autoimmune disease characterized by inflammatory responses in diverse organs due to an abnormal immune system, including autoantibody production or hyperactive immune cells [1]. It shows various manifestations depending on the organ in which the inflammatory response occurs, with varying severity, and treatment is determined by such manifestations. When patients have only a skin rash or mild arthritis, cytotoxic drugs are not indicated; however, when they have nephritis or vasculitis, aggressive treatment, including glucocorticoids and immunosuppressants, is necessary [2,3]. In addition, if inflammation is not well-controlled, sustained hyperactive immune responses can lead to organ damage, such as renal failure. Some patients have mild symptoms continuously, while other patients have recurrent episodes of flare-up or active disease.

The mortality and morbidity of SLE are still remarkable despite the fact that management of this disease has advanced over the past two decades [4,5]. The causes of mortality are serious infection, atherosclerosis, and active disease, and poor outcomes are associated with high disease activity and renal damage in patients with SLE [5–7]. Patients with higher disease activity are vulnerable to permanent organ damage owing to the need for glucocorticoids and immunosuppressants. These drugs play an essential role in controlling disease activity but result in complications, including infection and atherosclerosis, in patients with SLE [8,9]. Monitoring the current disease status and modifying treatment are essential to minimize organ damage and drug complications in the management of SLE [10]. Both clinical manifestations and laboratory findings should be used to monitor the disease status of patients within SLE. Several disease activity indices used to represent disease severity include the Systemic Lupus Erythematosus Disease Activity Index (SLEDAI) and British Isles Lupus Assessment Group Index (BILAG), which have been designed on the basis of clinical symptoms and laboratory findings [11–14]. These scoring systems have been used in clinical trials or research and are recommended to identify the degree of severity among different subsets of patients with SLE in clinical practice. Anti-double-stranded DNA (dsDNA) antibody and complement levels are used as disease activity biomarkers, with high anti-dsDNA antibody titers or low complement levels indicating high disease activity [15]. In addition, the Systemic Lupus International Collaborating Clinics (SLICC)/American College of Rheumatology (ACR) Damage Index could predict organ damage and mortality as a tool for evaluating the long-term outcomes of SLE [16]. These tools that identify disease activity could be used to classify patients with SLE, and such differentiation could help modify further treatment to prevent clinical worsening. Several attempts have been made to classify patients based on clinical or immunological data, suggesting different phenotypes of SLE [17–19]. The formulation of classified subtypes through more sophisticated statistical methods can help assess and manage the disease and educate patients with mixed symptoms.

Herein, we classified the phenotypic clusters of Korean patients with SLE using their initial laboratory findings at the time of SLE diagnosis. Classifying clusters of SLE aimed to analyze subgroup characteristics, including their disease presentation and activities, management patterns, organ damage, and mortality.

## 2. Materials and Methods

### 2.1. Study Design

A total of 389 patients who met the Systemic Lupus International Collaborating Clinics (SLICC) classification criteria and the revised ACR classification criteria for SLE receiving standard-of-care treatment for SLE were enrolled [20,21]. The laboratory findings obtained at the time of SLE diagnosis, including complete blood count, erythrocyte sedimentation rate (ESR), C-reactive protein (CRP) level, complement 3 (C3) and 4 (C4) levels, anti-nuclear antibody (ANA) titer, and anti-dsDNA antibody level, were collected. The ANA titer was measured via immunofluorescence assay using ANA HEp-2 Plus (GA Generic Assays, Dahlewitz, Germany) and categorized by the International Consensus on ANA Patterns (ICAP) as follows: homogenous (AC-1), speckled (AC4,5), nucleolar (AC-8,9,10), or cytoplasmic (AC-15 to AC-23) [22]. The anti-dsDNA antibody level was also measured via enzyme immunoassay (GA Generic Assays), with a cut-off value of 7 IU/mL. The duration was defined as the period from the time when the initial tests were performed to the time when the cumulative data were collected. Data on cumulative manifestations, including oral ulcer, malar rash, alopecia, arthritis, and renal disease, were obtained. Disease activity and disease-related damage were assessed using the SLEDAI score and SLICC/ACR damage index at the time of data collection [16,23]. Comprehensive medication histories, including the use of glucocorticoids and immunosuppressants, were obtained. For medication data for immunosuppressants, taking the drugs for more than 1 month was considered as "use". Data were collected from the medical records within 9 years (2006–2015) of patients with SLE managed at Ajou University Hospital using

MS-SQL 2012 (Microsoft, Redmond, WA, USA). This study was conducted in accordance with the principles of the Declaration of Helsinki. The study protocol was reviewed and approved by the institutional review board of Ajou University Hospital (AJIRB-MED-MDB-17-147); the need for informed consent was waived because of the retrospective nature of the study.

### 2.2. Statistical Analysis

Hierarchical cluster analysis was performed on 389 patients with SLE with 10 different laboratory values. The laboratory values, including the white blood cell (WBC), lymphocyte, and platelet counts; hemoglobin, CRP, C3, C4, and anti-dsDNA antibody levels; ESR; and ANA titer, were transformed into Z-scores for hierarchical clustering [24]. To classify the patients with SLE according to laboratory values, we applied Ward's method as an agglomeration method applied with Spearman correlation as a distance metric [25]. The clinical characteristics among clusters were examined using ANOVA with Tukey's and Fisher's exact tests (SPSS version 23.0; IBM SPSS Statistics, IBM Corporation, Chicago, IL, USA). To predict the classified SLE clusters, we constructed a discriminant model using Fisher's linear discriminant (LD) functions with six variables, including the WBC count, ESR, C3 level, C4 level, anti-dsDNA antibody level, and ANA titer. Statistical significance was set at $p < 0.05$.

### 3. Results

### 3.1. Three Clusters with Different Characteristics Were Identified among the Patients with SLE

A total of 389 patients with SLE were divided into three clusters via hierarchical cluster analysis based on the 10 laboratory values (Figure 1). The analysis showed different patterns between the following two laboratory sets: positive sign set for SLE, including the ESR, CRP level, anti-dsDNA antibody level, and ANA titer, and negative sign set, including the C3 level, C4 level, hemoglobin level, platelet count, WBC count, and lymphocyte counts. In contrast, Cluster 2 showed a tendency to deviate from Cluster 3 in terms of the ANA titer, platelet count, WBC count, and lymphocyte count.

Table 1 shows the differences between the initial laboratory test results. The patients in Cluster 1 were significantly younger than those in the other clusters (1 vs. 2, $p = 0.044$ and 1 vs. 3, $p < 0.001$). Sex and disease duration did not differ between the SLE clusters. Cluster 1 had significantly higher anti-dsDNA antibody titers and lower platelet counts, hemoglobin levels, and C3/4 levels than Clusters 2 and 3 ($p < 0.001$, $p = 0.004$, $p < 0.001$, and $p < 0.001$, respectively). Cluster 2 had a significantly higher lymphocyte count and a lower ESR ($p < 0.001$ and $p < 0.001$, respectively). The WBC count was significantly lower in Cluster 3 than in the other clusters (3 vs. 1, $p = 0.029$ and 3 vs. 2, $p < 0.001$, respectively). The ANA titers were significantly different between all SLE clusters ($p < 0.001$), with the highest values in Cluster 3, followed by those in Clusters 1 and 2. The proportion of the homogeneous type (AC-1) was significantly higher in Clusters 1 and 2 (46/131, 35.4% and 64/183, 35%, respectively, $p = 0.002$) than in Cluster 3. The proportion of the nucleolar (AC-8,9,10) and cytoplasmic types (AC-15 to AC-23) was significantly higher in Cluster 2 (15/183, 8.2% and 22/183, 12.0%, respectively, $p < 0.001$) than in Clusters 1 and 3. The proportion of the speckled type (AC-4,5) was significantly different between all SLE clusters, with the highest proportion in Cluster 3 (54/75, 72%), followed by those in Clusters 1 and 2 (46/131, 44.6% and 56/183, 30.6%, respectively, $p < 0.001$). There was no significant association between the CRP level and SLE clusters.

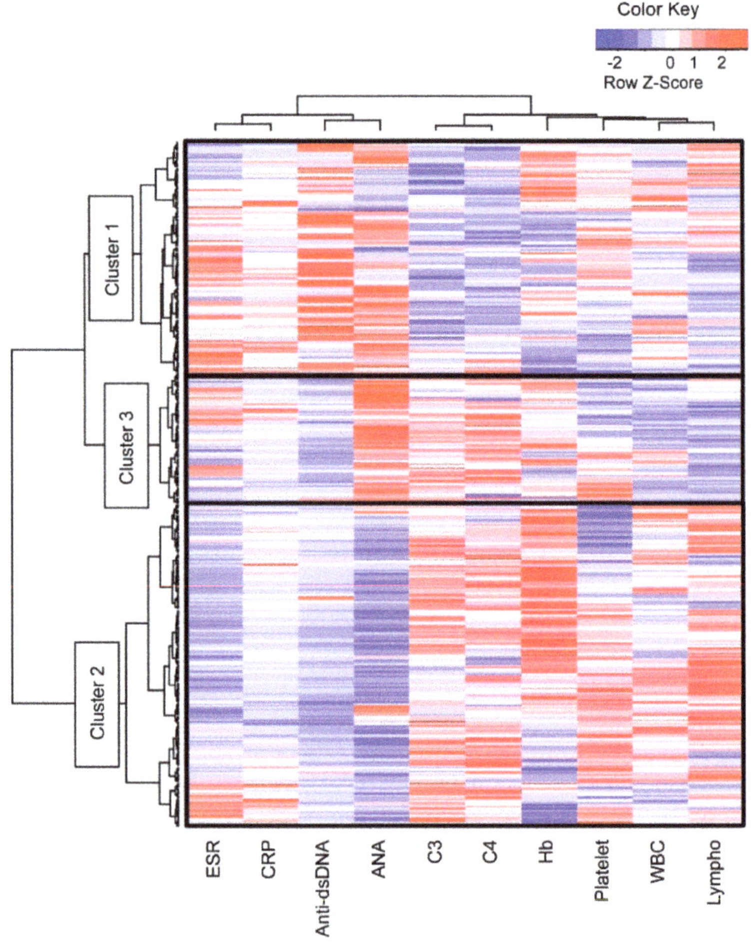

**Figure 1.** Three subgroups of patients with SLE divided via hierarchically cluster analysis. The patients with SLE ($n$ = 389) were divided into three clusters based on the laboratory values at the time of SLE diagnosis. ANA, anti-nuclear antibody; C3, complement 3; C4, complement 4; CRP, C-reactive protein; dsDNA, double-stranded DNA; ESR, erythrocyte sedimentation rate; Hb, hemoglobin, lympho, lymphocyte; SLE, systemic lupus erythematosus; WBC, white blood cell.

**Table 1.** Clinical characteristics of the clusters according to the laboratory findings at the time of SLE diagnosis.

| | Cluster 1 | Cluster 2 | Cluster 3 | $p$-Value [a] | | | |
|---|---|---|---|---|---|---|---|
| | ($n$ = 131) | ($n$ = 183) | ($n$ = 75) | Overall | 1 vs. 2 | 1 vs. 3 | 2 vs. 3 |
| Diagnostic age, years | 31.2 ± 13.2 | 35.6 ± 12.6 | 36.8 ± 12.3 | <0.001 | 0.044 | <0.001 | 0.752 |
| Male: female, $n$ (%) | 10 (7.6):121 (92.4) | 15 (8.2):168 (91.8) | 4 (5.3):71 (94.7) | 0.819 | 1.000 | 0.775 | 0.601 |
| Duration, month [b] | 117.8 ± 48.5 | 126.0 ± 40.6 | 138.4 ± 146.8 | 0.169 | 0.608 | 0.144 | 0.455 |
| WBC count, /μL | 5623.9 ± 3157.4 | 6111.9 ± 2304.6 | 4673.1 ± 1877.5 | <0.001 | 0.219 | 0.029 | <0.001 |
| Lymphocyte count, /μL | 1214.3 ± 628.1 | 1765.5 ± 683.0 | 1171.2 ± 395.7 | <0.001 | <0.001 | 0.881 | <0.001 |
| Hemoglobin, /μL | 11.5 ± 1.8 | 12.5 ± 1.6 | 12.3 ± 1.1 | <0.001 | <0.001 | 0.001 | 0.581 |

**Table 1.** *Cont.*

|  | Cluster 1 | Cluster 2 | Cluster 3 | *p*-Value [a] | | | |
|---|---|---|---|---|---|---|---|
|  | (*n* = 131) | (*n* = 183) | (*n* = 75) | Overall | 1 vs. 2 | 1 vs. 3 | 2 vs. 3 |
| Platelet count, $\times 10^3/\mu L$ | 208.1 ± 76.7 | 238.9 ± 88 | 225.7 ± 66.7 | 0.004 | 0.003 | 0.288 | 0.452 |
| ESR, mm/h | 30.4 ± 26.5 | 16.8 ± 17.9 | 26.7 ± 20.1 | <0.001 | <0.001 | 0.454 | 0.003 |
| CRP, mg/dL | 1.1 ± 3 | 0.5 ± 1.8 | 0.7 ± 2 | 0.097 | 0.078 | 0.554 | 0.756 |
| Complement 3, mg/dL | 71.8 ± 29.4 | 112.7 ± 27.3 | 110.5 ± 21.7 | <0.001 | <0.001 | <0.001 | 0.820 |
| Complement 4, mg/dL | 13 ± 7.5 | 25 ± 9.7 | 27.5 ± 8.6 | <0.001 | <0.001 | <0.001 | 0.100 |
| Anti-dsDNA antibody, IU/mL | 47.8 ± 38.5 | 7.2 ± 10.3 | 7.5 ± 6.8 | <0.001 | <0.001 | <0.001 | 0.996 |
| ANA titer | 1715.1 ± 1135.3 | 463.8 ± 641.6 | 2474.7 ± 731.5 | <0.001 | <0.001 | <0.001 | <0.001 |
| Homogenous (AC-1), *n* (%) | 46 (35.4) | 64 (35) | 11 (14.7) | 0.002 | 1.000 | 0.001 | 0.001 |
| Nucleolar (AC-8,9,10), *n* (%) | 0 (0) | 15 (8.2) | 1 (1.3) | <0.001 | <0.001 | 0.366 | 0.045 |
| Speckled (AC-4,5), *n* (%) | 58 (44.6) | 56 (30.6) | 54 (72) | <0.001 | 0.012 | <0.001 | <0.001 |
| Cytoplasmic (AC-15 to AC-23), *n* (%) | 3 (2.3) | 22 (12) | 1 (1.3) | <0.001 | 0.001 | 1.000 | 0.004 |
| Mixed, *n* (%) | 21 (16.2) | 26 (14.2) | 8 (10.7) | 0.575 | 0.634 | 0.306 | 0.545 |

ANA, anti-nuclear antibody; CRP, C-reactive protein; dsDNA, double-stranded DNA; ESR, erythrocyte sedimentation rate; NA, not available; SLE, systemic lupus erythematosus; WBC, white blood cell. [a] *p*-Values were calculated using ANOVA with Tukey's and Fisher's exact tests. [b] Duration was defined as the period between the initial test and cumulative data collection. Continuous variables are presented as means ± standard deviations.

### 3.2. Three Clusters Were Separated with Significant Statistical Power

An LD analysis (LDA) classification model was developed to identify the patients with SLE and assign them to one of the three clusters. The clusters were classified significantly according to the variables and LD1, LD2, and LD3, and the values were the coefficients of each parameter (Table 2).

**Table 2.** Fisher's linear discriminant functions for clustering.

| Parameters | LD1 | LD2 | LD3 |
|---|---|---|---|
| White blood cell count | 0.001 | 0.001 | 0.000 |
| Erythrocyte sedimentation rate | 0.013 | −0.040 | −0.014 |
| Complement 3 | 0.085 | 0.138 | 0.115 |
| Complement 4 | 0.000 | 0.086 | 0.174 |
| Anti-dsDNA Ab | 0.103 | 0.041 | 0.028 |
| Anti-nuclear antibody level | 0.002 | 0.000 | 0.003 |
| Constant | −10.259 | −11.298 | −14.753 |

LD, linear discriminant.

Each patient's value was identified in the constructed model of canonical discriminant functions (Figure 2). The canonical plot shows that the clusters were separated with an accuracy of 84.5% (72.5% in Cluster 1, 85.3% in Cluster 2, and 92.9% in Cluster 3). Clusters 1 and 3 showed distinctly different positions, and Cluster 2 was found in the intermediate region. This finding indicates that each cluster was characterized by the initial laboratory values that were sufficiently distinctive to allow the construction of discriminator segregating subgroups.

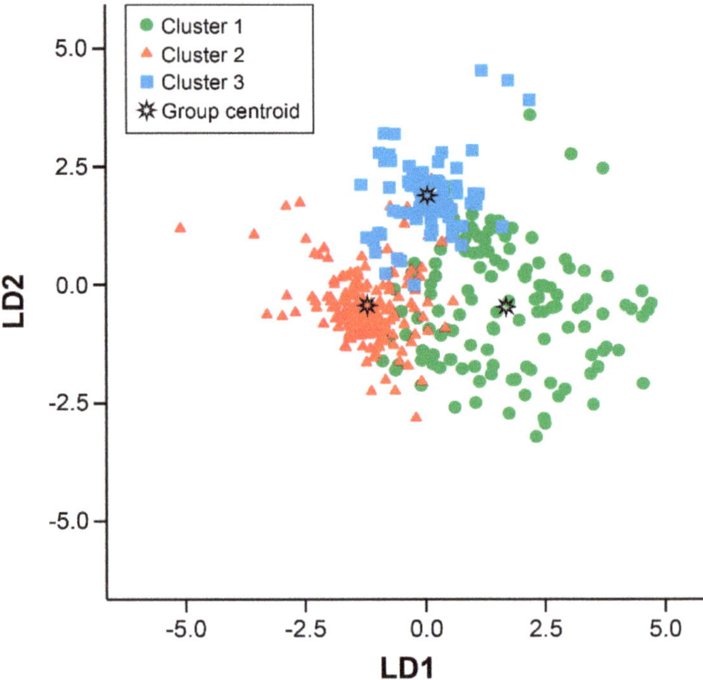



**Figure 2.** Three subgroups of patients with SLE identified at an accuracy of 84.5%. Canonical discriminant function shows that the LD was 72.5% in Cluster 1, 85.3% in Cluster 2, and 92.9% in Cluster 3. LD, linear discriminant; SLE, systemic lupus erythematosus.

### 3.3. Each Cluster Showed Different Manifestations during Follow-Up

The manifestations in each cluster from the time of classification to the time of collection of the medical records were compared (Table 3). Cluster 1 had a higher number of clinical manifestations than Clusters 2 and 3 ($p = 0.002$ and $p < 0.001$, respectively). Further-more, Cluster 1 showed the highest prevalence of malar rash, alopecia, renal disease, azathioprine and cyclophosphamide use, and glucocorticoid use ($p = 0.006$, $p = 0.001$, $p < 0.001$, $p = 0.004$, and $p < 0.001$, respectively); however, the prevalence of oral ulcers was significantly higher in Cluster 2 than in Cluster 1 ($p = 0.042$). The prevalence of arthritis and serositis did not significantly differ between the SLE clusters; however, arthritis had an overall high prevalence in all clusters.

The SLEDAI score, which was calculated at the time of enrollment, was higher in Cluster 1 ($7.2 \pm 4.9$) than in Clusters 2 ($3.0 \pm 3.2$, $p < 0.001$) and 3 ($2.4 \pm 2.7$, $p < 0.001$), while the SLICC/ACR damage index, which was also collected and calculated at the time of enrollment, did not differ. The proportion of patients taking hydroxychloroquine (HCQ) was lower in Cluster 1 (61.8%) than in Clusters 2 (75.4%, $p = 0.013$) and 3 (76.0%, $p = 0.045$). Although the proportion of patients currently taking glucocorticoids was similar in the three clusters, the total and mean doses of glucocorticoids were significantly higher in Cluster 1 than in Cluster 2 ($p = 0.008$ and $p = 0.001$, respectively). The patients in Cluster 1 took azathioprine more frequently than did those in Cluster 3 ($p = 0.001$) and cyclophosphamide more frequently than did those in Clusters 2 and 3 ($p < 0.001$ and $p = 0.019$, respectively).

**Table 3.** Cumulative manifestations and treatment patterns of the SLE clusters.

| | Cluster 1 | Cluster 2 | Cluster 3 | *p*-Value [a] | | | |
|---|---|---|---|---|---|---|---|
| | (*n* = 131) | (*n* = 183) | (*n* = 75) | Overall | 1 vs. 2 | 1 vs. 3 | 2 vs. 3 |
| Number of CMs, *n* (%) | 1.4 ± 1.3 | 0.9 ± 0.9 | 0.8 ± 1 | <0.001 | <0.001 | 0.002 | 0.901 |
| Number of CMs of ≥ 2, *n* (%) | 49 (37.7) | 41 (22.4) | 16 (21.3) | 0.006 | 0.004 | 0.019 | 1.000 |
| Oral ulcer, *n* (%) | 18 (13.7) | 43 (23.5) | 15 (20) | 0.095 | 0.042 | 0.244 | 0.624 |
| Malar rash, *n* (%) | 35 (26.7) | 27 (14.8) | 8 (10.7) | 0.006 | 0.010 | 0.007 | 0.430 |
| Alopecia, *n* (%) | 35 (26.9) | 19 (10.4) | 12 (16) | 0.001 | <0.001 | 0.085 | 0.212 |
| Arthritis, *n* (%) | 42 (32.1) | 52 (28.4) | 16 (21.3) | 0.263 | 0.533 | 0.110 | 0.278 |
| Renal disease, *n* (%) | 46 (35.1) | 24 (13.1) | 12 (16) | <0.001 | <0.001 | 0.004 | 0.556 |
| Serositis, *n* (%) | 2 (1.5) | 0 (0.0) | 0 (0) | 0.150 | 0.173 | 0.535 | NA |
| SLEDAI score * | 7.2 ± 4.9 | 3.0 ± 3.2 | 2.4 ± 2.7 | <0.001 | <0.001 | <0.001 | 0.548 |
| SLICC/ACR damage index | 0.4 ± 0.9 | 0.4 ± 1.0 | 0.4 ± 0.9 | 0.993 | 0.992 | 0.999 | 0.998 |
| Hydroxychloroquine use, *n* (%) | 81 (61.8) | 138 (75.4) | 57 (76) | 0.021 | 0.013 | 0.045 | 1.000 |
| Current glucocorticoid use, *n* (%) | 120 (91.6) | 159 (86.9) | 65 (86.7) | 0.385 | 0.208 | 0.339 | 1.000 |
| Total glucocorticoid dose, mg | 8465.9 ± 10,962 | 5306.0 ± 8645.4 | 5611.1 ± 6466.9 | 0.008 | 0.008 | 0.080 | 0.968 |
| Mean glucocorticoid dose, mg | 67.1 ± 76 | 37.8 ± 54.1 | 51.5 ± 82.3 | 0.001 | 0.001 | 0.254 | 0.305 |
| Azathioprine use, *n* (%) | 35 (26.7) | 22 (12) | 14 (18.7) | 0.004 | 0.001 | 0.235 | 0.170 |
| Cyclophosphamide use, *n* (%) | 19 (14.5) | 4 (2.2) | 3 (4) | <0.001 | <0.001 | 0.019 | 0.418 |
| MMF use, *n* (%) | 17 (13) | 12 (6.6) | 5 (6.7) | 0.127 | 0.074 | 0.240 | 1.000 |
| Methotrexate use, *n* (%) | 14 (10.7) | 31 (16.9) | 13 (17.3) | 0.243 | 0.142 | 0.200 | 1.000 |

CM, cumulative manifestation; MMF, mycophenolate mofetil; SLE, systemic lupus erythematosus; SLEDAI, Systemic Lupus Erythematosus Disease Activity Index; SLICC/ACR, Systemic Lupus International Collaborating Clinics/American College of Rheumatology. [a] *p*-Values were calculated using ANOVA with Tukey's and Fisher's exact tests. Continuous variables are presented as means ± standard deviations. * Most recent clinical visit.

### 3.4. Mortality and Renal Damage within 171 Months Were Not Different between the Clusters

Fifteen patients (3.9%) died, while five patients (1.3%) had progressed to end-stage renal disease (ESRD). There was no difference in the mortality rate during the average follow-up period of 171 months between the clusters (data not shown).

## 4. Discussion

The cluster analysis using laboratory findings that were obtained at the time of SLE diagnosis identified three clusters among patients with SLE. Each cluster shared similar clinical characteristics, including laboratory findings and manifestations. These data allowed the determination of one of the subtypes using LDA based on the initial WBC count, ESR, C3 and C4 levels, anti-dsDNA antibody level, and ANA titers. The LDA classification model demonstrated a high level of spectral discrimination (84.5%), which reflected the utility of the serum levels of inflammatory or autoimmune markers and the homogeneity of patients in each cluster. It is also evident that the three clusters can be considered separately in patients with SLE, and this model can be helpful in understanding the clinical features of patients with SLE in clinical practice.

Several studies have attempted to classify patients with SLE using clinical or immunological features. One study showed two subsets of patients: Most patients in the active disease subgroup were of Black African descent and were diagnosed when younger, while the patients in the other sub-group had different features [26]. In a recent study, four discrete clusters were identified on the basis of patients' symptoms, while disease characteristics, patient-reported outcomes, and treatment received in each cluster were significantly different [17]. A K-means cluster analysis based on patterns of clinical symptoms and mortality identified three clusters, and the cluster with frequent renal and hematologic symptoms showed a higher mortality in Chinese patients with SLE [27]. A cohort of Spanish patients with SLE showed that cardiovascular and musculoskeletal damage among several damage patterns was correlated with mortality [28]. An identification of three clusters using the damage index scale concluded that the cluster with prevalent renal and ocular damage had the highest damage score [19].

In this study, laboratory findings obtained at the time of SLE diagnosis were used to identify the three clusters. The levels of complement and anti-dsDNA antibodies in Cluster 1 were significantly different from those in Clusters 2 and 3. As both abnormal levels were included in the SLEDAI score, the score was also elevated in Cluster 1. Cluster 1 represented patients with SLE with a high disease activity who received a higher dose of glucocorticoids. High levels of anti-dsDNA antibody or low levels of complements indicate an active disease and predict a poor prognosis in patients with SLE [29,30]. Both markers are known to correlate with activity of lupus nephritis (LN) [31].

To summarize the results of clinical manifestations of clusters briefly, cluster 1 had higher numbers of clinical manifestations and SLEDAI scores and malar rash and renal disease more frequently and showed less frequent use of hydroxychloroquine and more frequent use of cyclophosphamide than Cluster 2 or 3. In addition, Cluster 1 had oral ulcer and alopecia more frequently, showed more frequent use of azathioprine, and took higher doses of glucocorticoids than Cluster 2. Clusters 2 and 3 had no difference in clinical manifestations.

In general, most patients with SLE are diagnosed in their 30s, and patients who develop SLE at a younger age have an active disease [26,32]. The diagnostic age was the lowest in Cluster 1, which represents the active disease group. However, the WBC count was not lower in Cluster 1, and the lymphocyte count was lower in Cluster 3, while the hemoglobin level was lower in Cluster 1 than in the other clusters. Cytopenia is the main hematologic manifestation in most SLE classification criteria and is derived from destruction to autoimmune response in patients with SLE [33]. SLE patients with lymphopenia have reduced surface expression of complement regulatory proteins and endogenous production of type 1 interferon [34]. An analysis of autoimmune cytopenia before or at childhood-onset SLE showed that patients with autoimmune cytopenia had a lower incidence of arthritis and a lower 2-year incidence of LN than those without autoimmune cytopenia [35]. Herein, cytopenia is not a typical finding of a particular cluster and might occur through the different etiologies of arthritis and renal involvement in SLE.

The presence of ANA is a typical feature of SLE, and its types and titers vary and have distinct roles, including non-pathological and pathogenic roles [36,37]. As a diagnostic marker for SLE, the ANA titer is known to have a sensitivity of 93% and a specificity of 57%. The titer of ANA and disease status of SLE are not regarded as relevant [38]. A high ANA titer does not indicate active inflammation, while a low ANA has not been ignored. Cluster 3 (mild disease) had the highest ANA titers and the lowest WBC and lymphocyte counts, while the ANA titer was not higher in Cluster 1 (active disease).

More than 10 types of ANA patterns have been reported and categorized into nuclear, cytoplasmic, and cell-cycle-related types. A recent study revealed that 36.5% of checked ANAs were homogeneous (AC-1), 19.9% speckled (AC-4,5), and 17.0% nucleolar (AC-8,9,10) among 9268 patients with positive ANAs [39]. The speckled type is known to be more specific for the diagnosis of SLE; however, its association with disease activity has not been identified. Among the several types of ANA, the homogeneous type was more frequently observed in Clusters 1 and 2 than in Cluster 3 and the speckled type in Clusters 1 and 3 than in Cluster 2 in this study. The speckled type might not be typical in patients with active SLE but was associated with the characteristics of Cluster 3.

Interestingly, the proportion of patients taking HCQ and the doses of glucocorticoids in Cluster 1 differed from those in Clusters 2 and 3. These results suggest that a lower proportion of patients taking HCQ might have a higher disease activity and might receive higher doses of glucocorticoids in Cluster 1. HCQ has been known to prevent disease flare-up or severe manifestations, including LN, and its maintenance has been associated with better prognosis in patients with SLE [40,41]. Some patients could not maintain HCQ owing to adverse effects, including retinopathy, or refused to take the medicine. Our data confirmed that the discontinuation of HCQ could be associated with disease activation in patients with SLE.

A limitation of this study is that there may be biases that arise from research methods that use data collected retrospectively. The data were dependent on medical records, and the follow-up time differed among the patients. In addition, biologic drugs, including belimumab and rituximab, were not included because they were not available for patients with SLE in Korea during the data collection period (2006–2015). The disease status of SLE is changing, and the SLEDAI score was collected only once. The number of cumulative manifestations and SLICC/ACR damage index were included to compensate for such weak points. However, our study suggests that patients with SLE can be classified into three subgroups based on the initial laboratory findings at the time of SLE diagnosis. Since each subgroup herein had different clinical characteristics, clinicians need to consider which subgroup of patients should be included for further management, and clinical trials should be designed to categorize which subgroups the study should enroll in.

### 5. Conclusions

In conclusion, the cluster analysis using the initial laboratory findings divided the patients with SLE into three clusters showing a clear differences in clinical symptoms and drug history. Cluster 1, which had the highest disease activity markers at SLE diagnosis, had a higher number of clinical manifestations within approximately 10 years than Clusters 2 and 3. However, prognosis, including mortality or ESRD, did not differ between the clusters.

**Author Contributions:** Conceptualization, H.-Y.L. and C.-H.S.; methodology, H.-Y.L.; software, H.-Y.L., E.L. and D.Y.; validation, E.L. and D.Y.; formal analysis, H.-Y.L.; investigation, H.-A.K. and C.-H.S.; resources, J.-Y.J., E.L., H.-A.K., D.Y., and C.-H.S.; data curation, E.L. and D.Y.; writing—original draft preparation, J.-Y.J.; writing—review and editing, J.-Y.J. and C.-H.S.; visualization, H.-Y.L.; supervision, C.-H.S.; project administration, C.-H.S.; funding acquisition, C.-H.S. All authors have read and agreed to the published version of the manuscript.

**Funding:** This research was supported by a grant of the Korea Health Technology R&D Project through the Korea Health Industry Development Institute (KHIDI), funded by the Ministry of Health & Welfare, Republic of Korea (grant number: HR16C0001).

**Institutional Review Board Statement:** The study was conducted in accordance with the Declaration of Helsinki and approved by the Institutional Review Board of Ajou University Hospital (AJIRB-MED-MDB-17-147).

**Informed Consent Statement:** Patient informed consent was waived because of the retrospective nature of the study.

**Data Availability Statement:** The data presented in this study are available on request from the corresponding author.

**Conflicts of Interest:** The authors declare no conflict of interest.

## References

1. Lisnevskaia, L.; Murphy, G.; Isenberg, D. Systemic lupus erythematosus. *Lancet* **2014**, *384*, 1878–1888. [CrossRef]
2. Kamen, D.L.; Zollars, E.S. Corticosteroids in lupus nephritis and central nervous system lupus. *Rheum. Dis. Clin. N. Am.* **2016**, *42*, 63–73. [CrossRef] [PubMed]
3. Tarr, T.; Papp, G.; Nagy, N.; Cserép, E.; Zeher, M. Chronic high-dose glucocorticoid therapy triggers the development of chronic organ damage and worsens disease outcome in systemic lupus erythematosus. *Clin. Rheumatol.* **2017**, *36*, 327–333. [CrossRef] [PubMed]
4. Moghaddam, B.; Marozoff, S.; Li, L.; Sayre, E.C.; Zubieta, J.A. All-cause and Cause-specific Mortality in Systemic Lupus Erythematosus: A Population-based Study. *Rheumatology* **2021**, *61*, 367–376. [CrossRef]
5. Tselios, K.; Gladman, D.D.; Sheane, B.J.; Su, J.; Urowitz, M. All-cause, cause-specific and age-specific standardised mortality ratios of patients with systemic lupus erythematosus in Ontario, Canada over 43 years (1971–2013). *Ann. Rheum. Dis.* **2019**, *78*, 802–806. [CrossRef] [PubMed]

6.  Legge, A.; Doucette, S.; Hanly, J.G. Predictors of organ damage progression and effect on health-related quality of life in systemic lupus erythematosus. *J. Rheumatol.* **2016**, *43*, 1050–1056. [CrossRef]

7.  Alarcón, G.S.; McGwin, G., Jr.; Bastian, H.M.; Roseman, J.; Lisse, J.; Fessler, B.J.; Friedman, A.W.; Reveille, J.D. Systemic lupus erythematosus in three ethnic groups. VII [correction of VIII]. Predictors of early mortality in the LUMINA cohort. LUMINA Study Group. *Arthritis Rheum.* **2001**, *45*, 191–202. [CrossRef]

8.  Pimentel-Quiroz, V.R.; Ugarte-Gil, M.F.; Harvey, G.; Wojdyla, D.; Pons-Estel, G.; Quintana, R.; Esposto, A.; García, M.A.; Catoggio, L.J.; Cardiel, M.H.; et al. Factors predictive of serious infections over time in systemic lupus erythematosus patients: Data from a multi-ethnic, multi-national, Latin American lupus cohort. *Lupus* **2019**, *28*, 1101–1110. [CrossRef] [PubMed]

9.  McMahon, M.; Skaggs, B. Pathogenesis and treatment of atherosclerosis in lupus. *Rheum. Dis. Clin. N. Am.* **2014**, *40*, 475–495. [CrossRef]

10. Thanou, A.; Jupe, E.; Purushothaman, M.; Niewold, T.B.; Munroe, M.E. Clinical disease activity and flare in SLE: Current concepts and novel biomarkers. *J. Autoimmun.* **2021**, *119*, 102615. [CrossRef] [PubMed]

11. Ceccarelli, F.; Perricone, C.; Massaro, L.; Cipriano, E.; Alessandri, C.; Spinelli, F.R.; Valesini, G.; Conti, F. Assessment of disease activity in systemic lupus erythematosus: Lights and shadows. *Autoimmun. Rev.* **2015**, *14*, 601–608. [CrossRef]

12. Castrejón, I.; Tani, C.; Jolly, M.; Huang, A.; Mosca, M. Indices to assess patients with systemic lupus erythematosus in clinical trials, long-term observational studies, and clinical care. *Clin. Exp. Rheumatol.* **2014**, *32*, S85–S95.

13. Romero-Diaz, J.; Isenberg, D.; Ramsey-Goldman, R. Measures of adult systemic lupus erythematosus: Updated version of British Isles Lupus Assessment Group (BILAG 2004), European Consensus Lupus Activity Measurements (ECLAM), Systemic Lupus Activity Measure, Revised (SLAM-R), Systemic Lupus Activity Questionnaire for Population Studies (SLAQ), Systemic Lupus Erythematosus Disease Activity Index 2000 (SLEDAI-2K), and Systemic Lupus International Collaborating Clinics/American College of Rheumatology Damage Index (SDI). *Arthritis Care Res.* **2011**, *63* (Suppl. 11), S37–S46.

14. Yu, J.; Zeng, T.; Wu, Y.; Tian, Y.; Tan, L.; Duan, X.; Wu, Q.; Li, H.; Yu, L. Neutrophil-to-C3 ratio and neutrophil-to-lymphocyte ratio were associated with disease activity in patients with systemic lupus erythematosus. *J. Clin. Lab. Anal.* **2019**, *33*, e22633. [CrossRef] [PubMed]

15. Linnik, M.D.; Hu, J.Z.; Heilbrunn, K.R.; Strand, V.; Hurley, F.L.; Joh, T. Relationship between anti-double-stranded DNA antibodies and exacerbation of renal disease in patients with systemic lupus erythematosus. *Arthritis Rheum.* **2005**, *52*, 1129–1137. [CrossRef] [PubMed]

16. Gladman, D.D.; Goldsmith, C.H.; Urowitz, M.B.; Bacon, P.; Fortin, P.; Ginzler, E.; Gordon, C.; Hanly, J.G.; Isenberg, D.; Petri, M.; et al. The Systemic Lupus International Collaborating Clinics/American College of Rheumatology (SLICC/ACR) Damage Index for Systemic Lupus Erythematosus International Comparison. *J. Rheumatol.* **2000**, *27*, 373–376.

17. Touma, Z.; Hoskin, B.; Atkinson, C.; Bell, D.; Massey, O.; Lofland, J.H.; Berry, P.; Karyekar, C.S.; Costenbader, K.H. Systemic lupus erythematosus symptom clusters and their association with patient-reported outcomes and treatment: Analysis of real-world data. *Arthritis Care Res.* 2020. [CrossRef]

18. Lu, Z.; Li, W.; Tang, Y.; Da, Z.; Li, X. Lymphocyte subset clustering analysis in treatment-naive patients with systemic lupus erythematosus. *Clin. Rheumatol.* **2021**, *40*, 1835–1842. [CrossRef] [PubMed]

19. Ahn, G.Y.; Lee, J.; Won, S.; Ha, E.; Kim, H.; Nam, B.; Kim, J.S.; Kang, J.; Kim, J.; Song, G.G.; et al. Identifying damage clusters in patients with systemic lupus erythematosus. *Int. J. Rheum. Dis.* **2020**, *23*, 84–91. [CrossRef] [PubMed]

20. Petri, M.; Orbai, A.-M.; Alarcón, G.S.; Gordon, C.; Merrill, J.T.; Fortin, P.R.; Bruce, I.N.; Isenberg, D.; Wallace, D.J.; Nived, O.; et al. Derivation and validation of the Systemic Lupus International Collaborating Clinics classification criteria for systemic lupus erythematosus. *Arthritis Rheum.* **2012**, *64*, 2677–2686. [CrossRef] [PubMed]

21. Hochberg, M.C. Updating the American College of Rheumatology revised criteria for the classification of systemic lupus erythematosus. *Arthritis Rheum.* **1997**, *40*, 1725. [CrossRef]

22. Chan, E.K.L.; von Mühlen, C.A.; Fritzler, M.J.; Damoiseaux, J.; Infantino, M.; Klotz, W.; Satoh, M.; Musset, L.; García-De La Torre, I.; Carballo, O.G.; et al. The International Consensus on ANA Patterns (ICAP) in 2021-The 6th Workshop and Current Perspectives. *J. Appl. Lab. Med.* **2022**, *7*, 322–330. [CrossRef] [PubMed]

23. Uribe, A.G.; Vila, L.M.; McGwin, G., Jr.; Sanchez, M.L.; Reveille, J.D.; Alarcon, G.S. The Systemic Lupus Activity Measure-revised, the Mexican Systemic Lupus Erythematosus Disease Activity Index (SLEDAI), and a modified SLEDAI-2K are adequate instruments to measure disease activity in systemic lupus erythematosus. *J. Rheumatol.* **2004**, *31*, 1934–1940.

24. Cheadle, C.; Vawter, M.P.; Freed, W.J.; Becker, K.G. Analysis of microarray data using Z score transformation. *J. Mol. Diagn.* **2003**, *5*, 73–81. [CrossRef]

25. Murtagh, F.; Legendre, P. Ward's hierarchical agglomerative clustering method: Which algorithms implement ward's criterion? *J. Classific.* **2014**, *31*, 274–295. [CrossRef]

26. Cervera, R.; Doria, A.; Amoura, Z.; Khamashta, M.; Schneider, M.; Guillemin, F.; Maurel, F.; Garofano, A.; Roset, M.; Perna, A.; et al. Patterns of systemic lupus erythematosus expression in Europe. *Autoimmun. Rev.* **2014**, *13*, 621–629. [CrossRef]

27. To, C.H.; Mok, C.C.; Tang, S.S.K.; Ying, S.K.Y.; Wong, R.W.K.; Lau, C.S. Prognostically distinct clinical patterns of systemic lupus erythematosus identified by cluster analysis. *Lupus* **2009**, *18*, 1267–1275. [CrossRef] [PubMed]

28. Pego-Reigosa, J.M.; Lois, A.; Rúa-Figueroa, Í.; Galindo, M.; Calvo-Alén, J.; de Uña-Álvarez, J.; Barreiro, V.B.; Ruan, J.I.; Olivé, A.; Rodríguez-Gómez, M.; et al. Relationship between damage clustering and mortality in systemic lupus erythematosus in early and late stages of the disease: Cluster analyses in a large cohort from the Spanish Society of Rheumatology Lupus Registry. *Rheumatology* **2016**, *55*, 1243–1250. [CrossRef]

29. Enocsson, H.; Sjöwall, C.; Wirestam, L.; Dahle, C.; Kastbom, A.; Rönnelid, J.; Wetterö, J.; Skogh, T. Four anti-dsDNA antibody assays in relation to systemic lupus erythematosus disease specificity and activity. *J. Rheumatol.* **2015**, *42*, 817–825. [CrossRef]

30. Biesen, R.; Rose, T.; Hoyer, B.F.; Alexander, T.; Hiepe, F. Autoantibodies, complement and type I interferon as biomarkers for personalized medicine in SLE. *Lupus* **2016**, *25*, 823–829. [CrossRef] [PubMed]

31. Soliman, S.; Mohan, C. Lupus nephritis biomarkers. *Clin. Immunol.* **2017**, *185*, 10–20. [CrossRef] [PubMed]

32. Cervera, R. Systemic lupus erythematosus in Europe at the change of the millennium: Lessons from the "Euro-Lupus Project". *Autoimmun. Rev.* **2006**, *5*, 180–186. [CrossRef] [PubMed]

33. Hepburn, A.L.; Narat, S.; Mason, J.C. The management of peripheral blood cytopenias in systemic lupus erythematosus. *Rheumatology* **2010**, *49*, 2243–2254. [CrossRef]

34. García-Valladares, I.; Atisha-Fregoso, Y.; Richaud-Patin, Y.; Jakez-Ocampo, J.; Soto-Vega, E.; Elías-López, D.; Carrillo-Maravilla, E.; Cabiedes, J.; Ruiz-Argüelles, A.; Llorente, L. Diminished expression of complement regulatory proteins (CD55 and CD59) in lymphocytes from systemic lupus erythematosus patients with lymphopenia. *Lupus* **2006**, *15*, 600–605. [CrossRef] [PubMed]

35. Ogbu, E.A.; Chandrakasan, S.; Rouster-Stevens, K.; Greenbaum, L.A.; Sanz, I.; Gillespie, S.E.; Marion, C.; Okeson, K.; Prahalad, S. Impact of autoimmune cytopenias on severity of childhood-onset systemic lupus erythematosus: A single-center retrospective cohort study. *Lupus* **2021**, *30*, 109–117. [CrossRef]

36. Kurien, B.T.; Scofield, R.H. Autoantibody determination in the diagnosis of systemic lupus erythematosus. *Scand. J. Immunol.* **2006**, *64*, 227–235. [CrossRef]

37. Pisetsky, D.S. Antinuclear antibodies in rheumatic disease: A proposal for a function-based classification. *Scand. J. Immunol.* **2012**, *76*, 223–228. [CrossRef]

38. Solomon, D.H.; Kavanaugh, A.J.; Schur, P.H. Evidence-based guidelines for the use of immunologic tests: Antinuclear antibody testing. *Arthritis Rheum.* **2002**, *47*, 434–444. [CrossRef] [PubMed]

39. Eriksson, C.; Kokkonen, H.; Johansson, M.; Hallmans, G.; Wadell, G.; Rantapää-Dahlqvist, S. Autoantibodies predate the onset of systemic lupus erythematosus in northern Sweden. *Arthritis Res. Ther.* **2011**, *13*, R30. [CrossRef]

40. Almeida-Brasil, C.C.; Hanly, J.G.; Urowitz, M.; Clarke, A.E.; Ruiz-Irastorza, G.; Gordon, C.; Ramsey-Goldman, R.; Petri, M.; Ginzler, E.M.; Wallace, D.J.; et al. Flares after hydroxychloroquine reduction or discontinuation: Results from the Systemic Lupus International Collaborating Clinics (SLICC) inception cohort. *Ann. Rheum. Dis.* **2022**, *81*, 370–378. [CrossRef] [PubMed]

41. Pakchotanon, R.; Gladman, D.D.; Su, J.; Urowitz, M.B. More Consistent Antimalarial Intake in First 5 Years of Disease Is Associated with Better Prognosis in Patients with Systemic Lupus Erythematosus. *J. Rheumatol.* **2018**, *45*, 90–94. [CrossRef] [PubMed]

Journal of
*Clinical Medicine*

*Review*

# Significance of Autoantibodies to Ki/SL as Biomarkers for Systemic Lupus Erythematosus and Sicca Syndrome

Michael Mahler [1,*], Chelsea Bentow [1], Mary-Ann Aure [1], Marvin J. Fritzler [2] and Minoru Satoh [3]

1 Werfen Autoimmunity, San Diego, CA 92131, USA; cbentow@werfen.com (C.B.);
maure@werfen.com (M.-A.A.)
2 Department of Medicine, Cumming School of Medicine, University of Calgary, Calgary, AB T2N 1N4, Canada;
fritzler@ucalgary.ca
3 Department of Clinical Nursing, School of Health Sciences, University of Occupational and Environmental
Health, Kitakyushu 807-8555, Japan; satohm@health.uoeh-u.ac.jp
* Correspondence: mmahler@werfen.com; Tel.: +1-858-586-9900

**Abstract:** Anti-Ki/SL antibodies were first described in 1981 and have been associated with systemic lupus erythematosus (SLE) and Sicca syndrome. Despite the long history, very little is known about this autoantibody system, and significant confusion persists. Anti-Ki/SL antibodies target a 32 kDa protein (also known as PSME3, HEL-S-283, PA28γ, REGγ, proteasome activator subunit 3), which is part of the proteasome complex. Depending on the assay used and the cohort studied, the antibodies have been reported in approximately 20% of SLE patients with high disease specificity as compared to non-connective tissue disease controls. The aim of this review is to summarize the history and key publications, and to explore future direction of anti-Ki/SL antibodies.

**Keywords:** Ki/SL; proteasome; autoantibodies; lupus; SLE; Sjögren syndrome

**Citation:** Mahler, M.; Bentow, C.;
Aure, M.-A.; Fritzler, M.J.; Satoh, M.
Significance of Autoantibodies to
Ki/SL as Biomarkers for Systemic
Lupus Erythematosus and Sicca
Syndrome. *J. Clin. Med.* **2022**, *11*,
3529. https://doi.org/10.3390/
jcm11123529

Academic Editors: Christopher
Sjöwall, Ioannis Parodis and Erkan
Demirkaya

Received: 31 March 2022
Accepted: 6 June 2022
Published: 20 June 2022

**Publisher's Note:** MDPI stays neutral
with regard to jurisdictional claims in
published maps and institutional affiliations.

## 1. Introduction

Although known for more than four decades (Figure 1), very few details are known about anti-Ki/SL antibodies, and confusion persists. Historically, the nomenclature of the Ki/SL target antigen included SL (Sicca Lupus), PL-2 and Ki [1,2]. In addition, several other names can be found, including PSME3, HEL-S-283, PA28γ, REGγ, proteasome activator subunit 3. Eventually, it was concluded that this was indeed a single autoantibody system, now named Ki/SL. When anti-Ki antibodies were first described by Tojo et al. [3], and almost in parallel by Harmon et al. [4], as was the convention at the time, Tojo et al. named the novel autoantibody after the index patient Kikuta (Ki) [3], and Harmon et al. [4] choose to link it to the clinical association Sicca/lupus (SL). Early evidence using double immunodiffusion showed that they were identified in approximately 10% of SLE sera and were often associated with anti-Sm autoantibodies.

Initially, some sources confused the Ki with Ku/DNA-PKcs (DNA-dependent phos-phokinase catalytic subunit) [5], but it was clearly demonstrated that anti-Ki/SL autoan-tibodies recognize a 32 kDa protein, a soluble subunit of the nuclear PA-28 (proteasome activator) protein family, which is unrelated to the Ku/DNA-PKcs antigens [2]. The con-fusion from the study by Francoeur et al. [5] arose because the serum that Tojo sent to Francoeur contained both anti-Ku and Ki/SL antibodies. Due to the strong presence of Ku-specific bands in immunoprecipitation (IP), the 32 kDa protein band was overlooked, and it was concluded that anti-Ki and anti-Ku were identical. Unlike other systemic lupus erythematosus (SLE)-related autoantigens, such as Sm and U1RNP, Ki/SL was not asso-ciated with detectable RNA species [2]. Some studies focused on another autoantibody system in SLE termed Ki-67, which added to the confusion [6].

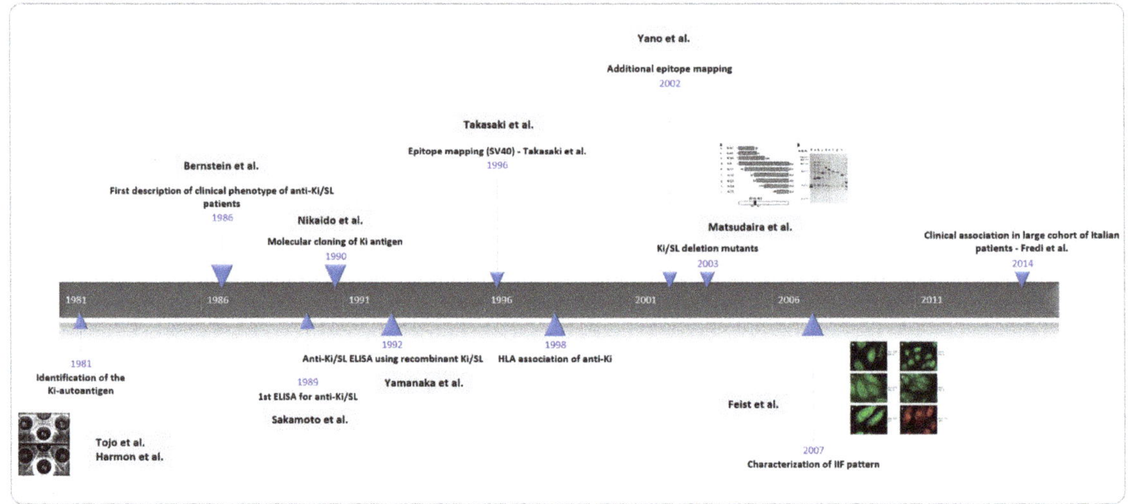

**Figure 1.** Four historical decades of anti-Ki/SL antibodies. The history of anti-Ki/SL antibodies started with the discovery by Tojo and Harmon et al. in 1981, followed by several clinical association and epitope mapping studies. ELISA = enzyme linked immunoassay; HLA = Human Leukocyte Antigen; SL=sicca lupus.

## 2. Materials and Methods

Due to the limited number of studies and the heterogenicity of methods and observations, our aim was to summarize the current knowledge in a narrative review using the search terms (Ki+ autoantibodies; SL+ autoantibodies, Ki/SL+ antibodies) instead of a systematic literature review.

## 3. Clinical and Demographic Association of Anti-Ki/SL Antibodies

Although there are no meta-data available as of today, mostly due to the limited number of studies and the heterogenicity of the methods used to detect anti-Ki/SL antibodies, we concluded that anti-Ki/SL antibodies are mostly found in SLE patients followed by patients with Sjögren syndrome (SjS) or Sicca syndrome [7,8], depending on the clinical definition. Especially in SLE, autoantibodies to a wide range of antigens have been reported, and anti-Ki/SL is part of the ever-expanding list [9]. High prevalence of anti-Ki/SL antibodies was also observed in patients with the overlap syndrome [3] and systemic sclerosis (SSc) [8,10]; however, in these studies, the number of patients was relatively small. In one of the earliest and largest clinical and serological studies of 516 connective tissue disease (CTD) patients, anti-Ki/SL autoantibodies were found in 12% of SLE patients, 14% of patients with mixed connective tissue disease (MCTD), 18% of patients with vasculopathies and 3% of patients with SjS [11]. Early clinical correlation studies focused on SLE patients indicated that anti-Ki/SL autoantibodies were associated with malar rash and multiple ANA specificities [7]. Another report of clinical, serological and HLA data from 119 SLE patients found no clear clinical associations with anti-Ki/SL antibodies, except for a higher frequency of non-infective fever [12], Sicca syndrome and skin involvement [13]. Fredi et al. [14] focused on anti-Ki-SL antibodies in SLE patients and reported, based on multivariate analysis, that anti-Ki/SL was significantly associated with male sex ($p$ = 0.017), an observation, which is in line with the early work by Riboldi et al. [11], Cavazzana et al. [7] and Fredi et al. [14]. Although no systematic study has been conducted until today, it appears that anti-Ki/SL antibodies can be found in patients with a wide range of ethnicities [15].

When more sensitive ELISA methods, using purified native Ki/SL antigens, were used to analyze the clinical and serologic features of SLE, a higher prevalence of central nervous system involvement was noted [10]. Outside SLE and other CTD, anti-proteasome antibodies have been studied in psoriasis patients [16].

## 4. Case Reports and Longitudinal Analysis of anti-Ki/SL Antibodies

Several case reports have been published on patients exhibiting anti-Ki/SL antibodies [13,17–20], including a patient with fatal CTD overlap syndrome [13], a patient with SSc/dermatomyositis (DM) overlap syndrome, an individual with anti-centromere positive pulmonary-renal syndrome [18], a case with SLE with epileptic seizures and chorea during prednisolone treatment [16], an individual with SSc with interstitial pneumonia and various autoantibodies (improvement by intravenous cyclophosphamide therapy) [20] (Table 1). In addition to the studies measuring anti-Ki/SL antibodies during a single timepoint (mostly at diagnosis), one case report also provided longitudinal analysis. In this case of a female SLE patient, the titer of anti-Ki/SL antibody rose before the onset of pericarditis and pleuritis, suggesting that anti-Ki/SL titers might reflect disease activity [8]. Although case reports and case series do not allow us to draw strong conclusions about clinical utility, they provide valuable reference points for future studies.

**Table 1.** Overview of case studies including the measurement of anti-Ki/SL antibodies.

| Case Study | Diagnosis | Comments | Ref |
| --- | --- | --- | --- |
| Ishiyama 1996 | SSc/ILD | - | [20] |
| Wakasugi 1996 | SLE/epileptic seizure/chorea | - | [19] |
| Oide 2001 | Pulmonary-renal syndrome | - | [18] |
| Miyachi 2002 | SSc/DM overlap | anti-Ku and anti-Ki/SL | [17] |

DM = dermatomyositis; ILD = interstitial lung disease; SLE = systemic lupus erythematosus; SSc = systemic sclerosis.

## 5. Epitope Distribution on the Proteasome Complex and on Ki/SL

Ki/SL is part of the human proteasome macromolecular complex, which is a known target of several autoantibodies [2,21–25]. Studies aimed to identify the reactive epitope of autoantibodies on the Ki/SL antigen [26–28]. Using different methods, including recombinant protein fragments and synthetic peptides, multiple epitopes were mapped to different regions of the protein (see Figure 2) that were associated with distinctive immune responses and certain clinical subtypes [5,15]. Interestingly, a short peptide sequence (named KILT) was identified [26,28], which bound antibodies in 18/49 (36.7%) anti-Ki/SL positive serum samples. A preliminary analysis indicates that KILT exhibited different clinical associations when compared to the full-length protein, a finding that needs to be validated in larger cohorts. Similarly, patients with antibodies that react with both N- and C-terminal areas are reported to have higher prevalence of the Sicca syndrome [27].

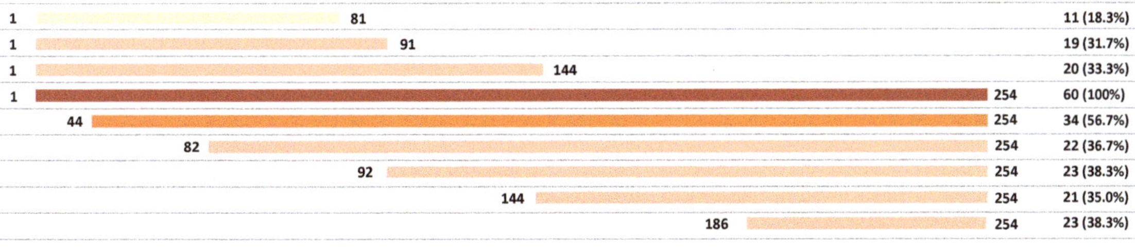


(a)

**Figure 2.** *Cont.*

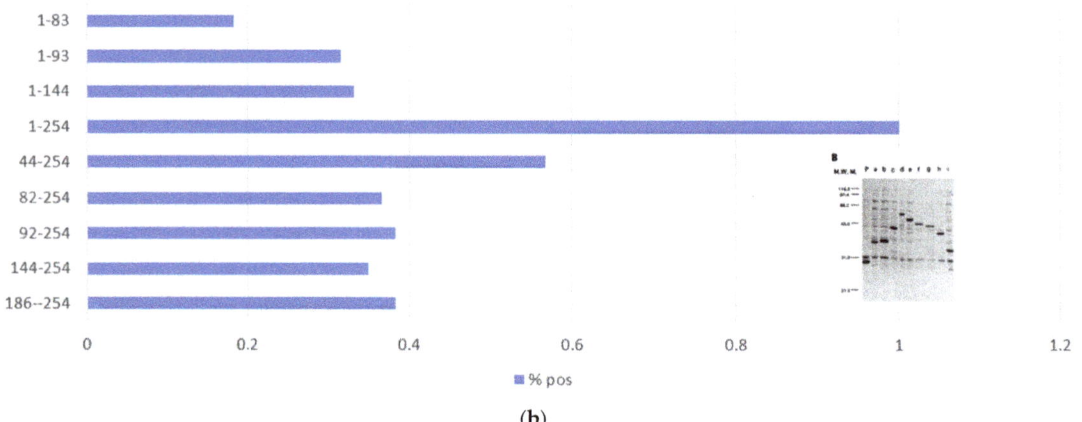


**(b)**

**Figure 2.** Epitope distribution on the Ki/SL antigen. (a) shows a visual representation of the recombinant truncated fragments (and full-length) of the Ki/SL antigen and the corresponding reactivity study by Matsudaira et al. [27] (Panel (**b**)) Shows the fraction of patients reacting with the recombinant fragments.

## 6. Detection Methods for Anti-Ki/SL Antibodies

### 6.1. Indirect Immunofluorescence Pattern of Anti-Ki/SL Antibodies

The characteristic indirect immunofluorescent (IIF) staining pattern of anti-Ki/SL antibodies was reported to be diffuse speckled nuclear on HEp-2 cells, although some substrates showed nucleolar staining as well [29] (Figure 3). Interestingly, antibodies to PA28a showed cytoplasmic staining, which is consistent with the reported localization of the protein and also with the moderate (~40%) homology between Ki/SL and PA28a, as the cognate antibodies are apparently not cross-reactive [30]. More specifically, although 13/27 (48%) of anti-Ki/SL also reacted with PA28a, it is unlikely that this represents cross-reactivity. Until the present, only one study that investigated the reactivity of anti-PA28a and anti-Ki/SL in the same cohort of patients [30] found that the prevalence of the two autoantibodies was comparable. Anti-Ki/SL antibodies have not been addressed by the International Consensus of ANA Patterns (ICAP) [31]; however, the described pattern is similar to AC-04 and/or AC-05. Along those lines, it is of relevance that more and more sub-patterns are being added to the consensus list [32]. Interestingly, anti-Ki/SL antibodies frequently occur at high titers, both using IIF as well as solid-phase assays, such as ELISA (unpublished data).

### 6.2. Other Detection Methods for Anti-Ki/SL Antibodies

Historically, anti-Ki/SL antibodies were initially detected by double immunodiffusion (DID) and IP [5]. The first ELISA was based on a native Ki/SL antigen purified from rabbit thymus by ammonium sulfate precipitation and affinity chromatography, followed by high-pressure liquid chromatography gel filtration [10]. In total, 30 out of 140 (21.4%) patients with SLE had anti-Ki/SL antibody by ELISA, whereas 11 (7.9%) were positive by DID. In the early 1990s, when an ELISA system utilizing a recombinant human protein was used to test samples from 220 patients with various CTDs, anti-Ki/SL antibodies were detected in 18.9% of SLE sera [8]. Consequently, the method rather than the source of antigen (recombinant *vs.* native) affects the prevalence of the antibodies in disease cohorts.

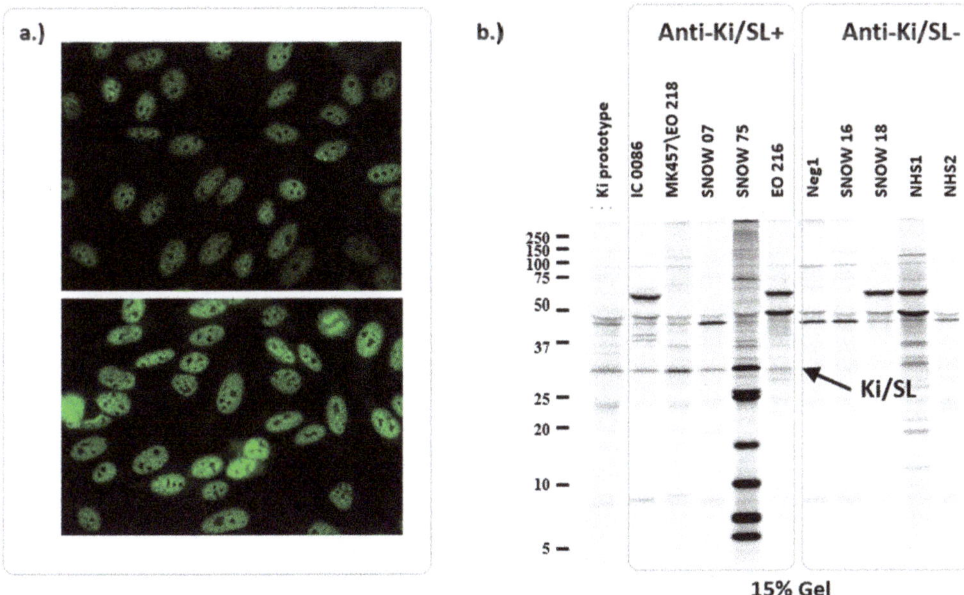


**Figure 3.** Detection methods for anti-Ki/SL antibodies. (**a**). Indirect immunofluorescence patterns on HEp-2 slides showing a nuclear speckled pattern. (**b**). Immunoprecipitation pattern shows the immunoprecipitation (IP) bands associated with the presence of anti-Ki/SL antibodies.

## 7. Co-Expression of Anti-Ki/SL and Other Autoantibodies

Anti-Ki/SL antibodies have been associated with several other autoantibodies, including anti-Sm [5], anti-Ro [2], anti-Ku, as well as anti-proliferating cell nuclear antigen (PCNA) [2,11] (Table 2). However, no clear consensus has been established, as some studies resulted in conflicting findings. As an example, a study by Fredi et al. [14] identified anti-Ki/SL antibodies in 31 patients, of which about one-half had no accompanying antibodies.

**Table 2.** Prevalence of anti-Ki/SL antibodies in different diseases.

| Disease | Tojo et al. 1981 | Bernstein et al. 1986 | Riboldi et al. 1987 | Boey et al. 1988 | Sakamoto et al. 1989 | Yamanaka et al. 1992 | Fredi et al. 2014 |
|---|---|---|---|---|---|---|---|
| Method | DID | CIE | CIE | DID | ELISA | ELISA | CIE |
| SLE | 30/255 (11.8%) | 20/300 (6.7%) | 27/217 (12.4%) | 8/94 (8.5%) | 30/140 (21.4%) | 21/111 (18.9%) | 31/540 (5.8%) |
| SjS | | | 1/38 (2.6%) | | | 2/25 (8.0%) | |
| SS | | 2/60 (3.3%) | | | | | |
| SSc | 0/90 (0.0%) | | 0/119 (0.0%) | | 3/25 (12.0%) | 2/30 (6.7%) | |
| PM/DM | 0/29 (0.0%) | | 0/14 (0.0%) | | (0.0%) | 1/30 (3.3%) | |
| RA | 0/33 (0.0%) | 2/70 (2.9%) | 0/37 (0.0%) | | (1.4%) | 2/50 (4.0%) | |
| OS | 7/36 (19.4%) | | | | | | |
| PN | 0/6 (0.0%) | | | | | | |
| MCTD | | 1/50 (2.0%) | 3/21 (14.3%) | | 1/12 (8.3%) | | |
| HI | | | 0/28 (0.0%) | | (0.0%) | | |
| PBC | | 1/135 (0.7%) | | | | | |
| ITP | | 1/110 (0.9%) | | | | | |

**Table 2.** *Cont.*

| Disease | Tojo et al. 1981 | Bernstein et al. 1986 | Riboldi et al. 1987 | Boey et al. 1988 | Sakamoto et al. 1989 | Yamanaka et al. 1992 | Fredi et al. 2014 |
|---|---|---|---|---|---|---|---|
| Method | DID | CIE | CIE | DID | ELISA | ELISA | CIE |
| VAS | | | 2/11 (18.2%) | | | | |
| pRP | | | 0/59 (0.0%) | | | | |
| **Demographics** | | | | | | | |
| Male sex | | | yes | | | | yes |
| Other associations | Arthritis/pericarditis, Sm | White SLE, Ro(SS-A), PCNA | PCNA | | CNS, Sm | | |

Abbreviations: DM, dermatomyositis; HI, healthy individuals; ITP, idiopathic thrombocytopenic purpura; MCTD, mixed connective tissue disease; OS, overlap syndrome; PBC, primary biliary cholangitis; PM, polymyositis; PN, periarteritis nodosa; SS, Sicca syndrome; RA, rheumatoid arthritis; SjS, Sjögren's syndrome; SLE, systemic lupus erythematosus; SSc, systemic sclerosis.

## 8. Future Directions

Future studies should re-evaluate the serological and clinical associations of anti-Ki/SL antibodies and also include experiments to shed more light on the potential associations with disease activity and treatment response in SLE patients. Along those lines, it is noteworthy that protease inhibitors have shown promise in treatment of refractory SLE [33,34]. Whether this is related to the proteasome levels or activity in serum or with the presence of anti-Ki/SL antibodies is a matter of future studies. Ideally, such investigations of the clinical phenotypes should be performed on inception cohorts of SLE patients, such as the SLICC cohort [35]. Lastly, with the intent to identify pre-clinical autoimmune conditions (e.g., early SLE), studies of cohorts, such as the US military, might provide valuable insights [36].

**Author Contributions:** Conceptualization, M.M., M.S., C.B., M.-A.A. and M.J.F.; methodology, M.M., M.S., C.B., M.-A.A. and M.J.F.; software, M.M., M.S., C.B., M.-A.A. and M.J.F.; validation, M.M., M.S., C.B., M.-A.A. and M.J.F., M.M., M.S., C.B., M.-A.A. and M.J.F. and M.M., M.S., C.B., M.-A.A. and M.J.F.; formal analysis, M.M., M.S., C.B., M.-A.A. and M.J.F.; investigation, M.M., M.S., C.B., M.-A.A. and M.J.F.; resources, M.M., M.S., C.B., M.-A.A. and M.J.F.; data curation, M.M., M.S., C.B., M.-A.A. and M.J.F.; writing—original draft preparation, M.M., M.S., C.B., M.-A.A. and M.J.F.; writing—review and editing, M.M., M.S., C.B., M.-A.A. and M.J.F.; visualization, M.M., M.S., C.B., M.-A.A. and M.J.F.; supervision, M.M., M.S., C.B., M.-A.A. and M.J.F.; project administration, M.M., M.S., C.B., M.-A.A. and M.J.F.; funding acquisition, M.M., M.S., C.B., M.-A.A. and M.J.F. All authors have read and agreed to the published version of the manuscript.

**Funding:** This research received no external funding.

**Acknowledgments:** We thank Carmen Andalucia (Werfen) for support with literature search.

**Conflicts of Interest:** Michael Mahler, Chelsea Bentow, are employees of Werfen, a diagnostic company. No stocks or shares. Marvin Fritzler is a consultant to Werfen and is Medical Director of Mitogen Diagnostics Corporation.

## References

1. Bernstein, R.M.; Bunn, C.C.; Hughes, G.R.; Francoeur, A.M.; Mathews, M.B. Cellular protein and RNA antigens in autoimmune disease. *Mol. Biol. Med.* **1984**, *2*, 105–120. [PubMed]
2. Bernstein, R.M.; Morgan, S.H.; Bunn, C.C.; Gainey, R.C.; Hughes, G.R.; Mathews, M.B. The SL autoantibody-antigen system: Clinical and biochemical studies. *Ann. Rheum. Dis.* **1986**, *45*, 353–358. [CrossRef] [PubMed]
3. Tojo, T.; Kaburaki, J.; Hayakawa, M.; Okamoto, T.; Tomii, M.; Homma, M. Precipitating antibody to a soluble nuclear antigen "Ki" with specificity for systemic lupus erythematosus. *Ryumachi* **1981**, *21*, 129–140. [PubMed]
4. Harmon, C.; Peebles, C.; Tan, E.M. SL-a new precipitating system. *Arthritis Rheumatol.* **1981**, *24*, S122.
5. Francoeur, A.M.; Peebles, C.L.; Gompper, P.T.; Tan, E.M. Identification of Ki (Ku, p70/p80) autoantigens and analysis of anti-Ki autoantibody reactivity. *J. Immunol.* **1986**, *136*, 1648–1653.
6. Muro, Y.; Kano, T.; Sugiura, K.; Hagiwara, M. Low frequency of autoantibodies against Ki-67 antigen in Japanese patients with systemic autoimmune diseases. *J. Autoimmun.* **1997**, *10*, 499–503. [CrossRef]

7.    Cavazzana, I.; Franceschini, F.; Vassalini, C.; Danieli, E.; Quinzanini, M.; Airo, P.; Cattaneo, R. Clinical and serological features of 35 patients with anti-Ki autoantibodies. *Lupus* **2005**, *14*, 837–841. [CrossRef]
8.    Yamanaka, K.; Takasaki, Y.; Nishida, Y.; Shimada, K.; Shibata, M.; Hashimoto, H. Detection and quantification of anti-Ki antibodies by enzyme-linked immunosorbent assay using recombinant Ki antigen. *Arthritis Rheum.* **1992**, *35*, 667–671. [CrossRef]
9.    Sherer, Y.; Gorstein, A.; Fritzler, M.J.; Shoenfeld, Y. Autoantibody explosion in systemic lupus erythematosus: More than 100 different antibodies found in SLE patients. *Semin. Arthritis Rheum.* **2004**, *34*, 501–537. [CrossRef]
10.   Sakamoto, M.; Takasaki, Y.; Yamanaka, K.; Kodama, A.; Hashimoto, H.; Hirose, S. Purification and characterization of Ki antigen and detection of anti-Ki antibody by enzyme-linked immunosorbent assay in patients with systemic lupus erythematosus. *Arthritis Rheum.* **1989**, *32*, 1554–1562. [CrossRef]
11.   Riboldi, P.; Asero, R.; Origgi, L.; Crespi, S. The SL/Ki system in connective tissue diseases: Incidence and clinical associations. *Clin. Exp. Rheumatol.* **1987**, *5*, 29–33. [PubMed]
12.   Matsunaga, K.; Yawata, M.; Tsuji, T.; Tani, K. HLA class II antigens associated with anti-Ki autoantibody positive connective tissue disease in Japanese. *J. Rheumatol.* **1998**, *25*, 1446–1447. [PubMed]
13.   Parodi, A.; Nigro, A.; Rebora, A. Anti-SL-Ki antibody in a patient with fatal connective tissue overlap disease. *Br. J. Dermatol.* **1989**, *121*, 243–246. [CrossRef] [PubMed]
14.   Fredi, M.; Cavazzana, I.; Quinzanini, M.; Taraborelli, M.; Cartella, S.; Tincani, A.; Franceschini, F. Rare autoantibodies to cellular antigens in systemic lupus erythematosus. *Lupus* **2014**, *23*, 672–677. [CrossRef] [PubMed]
15.   Boey, M.L.; Peebles, C.L.; Tsay, G.; Feng, P.H.; Tan, E.M. Clinical and autoantibody correlations in Orientals with systemic lupus erythematosus. *Ann. Rheum. Dis.* **1988**, *47*, 918–923. [CrossRef] [PubMed]
16.   Colmegna, I.; Sainz, B., Jr.; Citera, G.; Maldonado-Cocco, J.A.; Garry, R.F.; Espinoza, L.R. Anti-20S proteasome antibodies in psoriatic arthritis. *J. Rheumatol.* **2008**, *35*, 674–676.
17.   Miyachi, K.; Hankins, R.W.; Mimori, T.; Okano, Y.; Akizuki, M. Prospective study of a systemic sclerosis/dermatomyositis overlap patient presenting with anti-Ku and anti-Ki antibodies. *Mod. Rheumatol.* **2002**, *12*, 253–255. [CrossRef]
18.   Oide, T.; Iwamura, A.; Yamamoto, H.; Aizawa, K.; Inoue, K.; Itoh, N.; Ikeda, S. Elderly-onset anticentromere antibody-positive pulmonary-renal syndrome: Report of an autopsy case. *Nihon Kokyuki Gakkai Zasshi* **2001**, *39*, 498–503.
19.   Wakasugi, M.; Sato, T.; Maruyama, Y.; Ueno, M.; Arakawa, M. A case of systemic lupus erythematosus diagnosed 7 years after epileptic seizure and developed chorea during prednisolone treatment. *Ryumachi* **1996**, *36*, 545–550.
20.   Ishiyama, K.; Suwa, A.; Ohta, S.; Moriguchi, M.; Suzuki, T.; Miyachi, K.; Hara, M.; Kashiwazaki, S. A case of systemic sclerosis associated with interstitial pneumonia with various autoantibodies: Improvement by intravenous cyclophosphamide therapy. *Nihon Rinsho Meneki Gakkai Kaishi* **1996**, *19*, 512–518. [CrossRef]
21.   Feist, E.; Dorner, T.; Kuckelkorn, U.; Scheffler, S.; Burmester, G.; Kloetzel, P. Diagnostic importance of anti-proteasome antibodies. *Int. Arch. Allergy Immunol.* **2000**, *123*, 92–97. [CrossRef] [PubMed]
22.   Kordonouri, O.; Meyer, K.; Egerer, K.; Hartmann, R.; Scheffler, S.; Burmester, G.R.; Kuckelkorn, U.; Danne, T.; Feist, E. Prevalence of 20S proteasome, anti-nuclear and thyroid antibodies in young patients at onset of type 1 diabetes mellitus and the risk of autoimmune thyroiditis. *J. Pediatric Endocrinol. Metab.* **2004**, *17*, 975–981. [CrossRef] [PubMed]
23.   Brychcy, M.; Kuckelkorn, U.; Hausdorf, G.; Egerer, K.; Kloetzel, P.M.; Burmester, G.R.; Feist, E. Anti-20S proteasome autoantibodies inhibit proteasome stimulation by proteasome activator PA28. *Arthritis Rheum.* **2006**, *54*, 2175–2183. [CrossRef] [PubMed]
24.   Feist, E.; Brychcy, M.; Hausdorf, G.; Hoyer, B.; Egerer, K.; Dorner, T.; Kuckelkorn, U.; Burmester, G.R. Anti-proteasome autoantibodies contribute to anti-nuclear antibody patterns on human larynx carcinoma cells. *Ann. Rheum. Dis.* **2007**, *66*, 5–11. [CrossRef] [PubMed]
25.   Feist, E.; Burmester, G.R.; Kruger, E. The proteasome—victim or culprit in autoimmunity. *Clin. Immunol.* **2016**, *172*, 83–89. [CrossRef]
26.   Takasaki, Y.; Yano, T.; Hirokawa, K.; Takeuchi, K.; Ando, S.; Takahashi, T.; Shimada, K.; Hashimoto, H. An epitope on Ki antigen recognized by autoantibodies from lupus patients shows homology with the SV40 large T antigen nuclear localization signal. *Arthritis Rheum.* **1996**, *39*, 855–862. [CrossRef]
27.   Matsudaira, R.; Takeuchi, K.; Takasaki, Y.; Yano, T.; Matsushita, M.; Hashimoto, H. Relationships between autoantibody responses to deletion mutants of Ki antigen and clinical manifestations of lupus. *J. Rheumatol.* **2003**, *30*, 1208–1214.
28.   Yano, T.; Takasaki, Y.; Takeuchi, K.; Hirokawa, K.; Yamanaka, K.; Hashimoto, H. Anti-Ki antibodies recognize an epitope homologous with SV40 nuclear localization signal: Clinical significance and reactivities in various immunoassays. *Mod. Rheumatol.* **2002**, *12*, 50–55. [CrossRef]
29.   Matsushita, M.; Matsudaira, R.; Ikeda, K.; Nawata, M.; Tamura, N.; Takasaki, Y. Anti-proteasome activator 28alpha is a novel anti-cytoplasmic antibody in patients with systemic lupus erythematosus and Sjogren's syndrome. *Mod. Rheumatol.* **2009**, *19*, 622–628. [CrossRef]
30.   Matsushita, M.; Takasaki, Y.; Takeuchi, K.; Yamada, H.; Matsudaira, R.; Hashimoto, H. Autoimmune response to proteasome activator 28alpha in patients with connective tissue diseases. *J. Rheumatol.* **2004**, *31*, 252–259.
31.   Damoiseaux, J.; von Muhlen, C.A.; Garcia-De La Torre, I.; Carballo, O.G.; de Melo, C.W.; Francescantonio, P.L.; Fritzler, M.J.; Herold, M.; Mimori, T.; Satoh, M.; et al. International consensus on ANA patterns (ICAP): The bumpy road towards a consensus on reporting ANA results. *Autoimmun. Highlights* **2016**, *7*, 1. [CrossRef] [PubMed]

32. Röber, N.; Dellavance, A.; Ingénito, F.; Reimer, M.L.; Carballo, O.G.; Conrad, K.; Chan, E.K.L.; Andrade, L.E.C. Strong Association of the Myriad Discrete Speckled Nuclear Pattern With Anti-SS-A/Ro60 Antibodies: Consensus Experience of Four International Expert Centers. *Front. Immunol.* **2021**, *12*, 730102. [CrossRef] [PubMed]
33. Walhelm, T.; Gunnarsson, I.; Heijke, R.; Leonard, D.; Trysberg, E.; Eriksson, P.; Sjöwall, C. Clinical Experience of Proteasome Inhibitor Bortezomib Regarding Efficacy and Safety in Severe Systemic Lupus Erythematosus: A Nationwide Study. *Front. Immunol.* **2021**, *12*, 756941. [CrossRef] [PubMed]
34. Segarra, A.; Arredondo, K.V.; Jaramillo, J.; Jatem, E.; Salcedo, M.T.; Agraz, I.; Ramos, N.; Carnicer, C.; Valtierra, N.; Ostos, E. Efficacy and safety of bortezomib in refractory lupus nephritis: A single-center experience. *Lupus* **2020**, *29*, 118–125. [CrossRef]
35. Petri, M.; Orbai, A.M.; Alarcon, G.S.; Gordon, C.; Merrill, J.T.; Fortin, P.R.; Bruce, I.N.; Isenberg, D.; Wallace, D.J.; Nived, O.; et al. Derivation and validation of the Systemic Lupus International Collaborating Clinics classification criteria for systemic lupus erythematosus. *Arthritis Rheum.* **2012**, *64*, 2677–2686. [CrossRef]
36. Arbuckle, M.R.; McClain, M.T.; Rubertone, M.V.; Scofield, R.H.; Dennis, G.J.; James, J.A.; Harley, J.B. Development of autoantibodies before the clinical onset of systemic lupus erythematosus. *N. Engl. J. Med.* **2003**, *349*, 1526–1533. [CrossRef] [PubMed]

Journal of
*Clinical Medicine*

*Article*

# Trends in Hospital Admissions and Death Causes in Patients with Systemic Lupus Erythematosus: Spanish National Registry

Víctor Moreno-Torres [1,*], Carlos Tarín [2], Guillermo Ruiz-Irastorza [3], Raquel Castejón [1], Ángela Gutiérrez-Rojas [1], Ana Royuela [4], Pedro Durán-del Campo [1], Susana Mellor-Pita [1], Pablo Tutor [1], Silvia Rosado [1], Enrique Sánchez [1], María Martínez-Urbistondo [1], Carmen de Mendoza [1], Miguel Yebra [1] and Juan-Antonio Vargas [1]

[1]  Internal Medicine Department, Health Research Institute Puerta de Hierro-Segovia de Arana (IDIPHIM), Hospital Universitario Puerta de Hierro Majadahonda, 28222 Madrid, Spain; rcastejon.hpth@salud.madrid.org (R.C.); Angelagutierrezrojas@gmail.com (Á.G.-R.); Pedrodurandc@hotmail.com (P.D.-d.C.); Susanamellor@hotmail.com (S.M.-P.); pablo.tutor@hotmail.com (P.T.); Silvia.rosado@salud.madrid.org (S.R.); sanchezchica@gmail.com (E.S.); mmurbistondo@gmail.com (M.M.-U.); cmendoza.cdm@gmail.com (C.d.M.); myebrab@hotmail.com (M.Y.); juanantonio.vargas@salud.madrid.org (J.-A.V.)
[2]  Basic Medical Sciences, Faculty of Medicine, Universidad CEU San Pablo, 28003 Madrid, Spain; carlos.tarincerezo@ceu.es
[3]  Autoimmune Diseases Research Unit, BioCruces Bizkaia Health Research Institute, UPV/EHU, 48903 Bizkaia, Spain; r.irastorza@outlook.es
[4]  Clinical Biostatistics Unit, Health Research Institute Puerta de Hierro-Segovia de Arana, CIBERESP, 28222 Madrid, Spain; aroyuela@idiphim.org
*   Correspondence: victor.moreno.torres.1988@gmail.com

**Citation:** Moreno-Torres, V.; Tarín, C.; Ruiz-Irastorza, G.; Castejón, R.; Gutiérrez-Rojas, Á.; Royuela, A.; Durán-del-Campo, P.; Mellor-Pita, S.; Tutor, P.; Rosado, S.; et al. Trends in Hospital Admissions and Death Causes in Patients with Systemic Lupus Erythematosus: Spanish National Registry. *J. Clin. Med.* **2021**, *10*, 5749. https://doi.org/10.3390/jcm10245749

Academic Editor: Roberta Gualtierotti

Received: 22 November 2021
Accepted: 6 December 2021
Published: 8 December 2021

**Publisher's Note:** MDPI stays neutral with regard to jurisdictional claims in published maps and institutional affiliations.



**Abstract:** Background: the admission and death causes of SLE patients might have changed over the last years. Methods: Analysis of the Spanish National Hospital Discharge database. All individuals admitted with SLE, according to ICD-9, were selected. The following five admission categories were considered: SLE, cardiovascular disease (CVD), neoplasm, infection, and venous-thromboembolic disease (VTED), along four periods of time (1997–2000, 2001–2005, 2006–2010, and 2011–2015). Results: The admissions (99,859) from 43.432 patients with SLE were included. The absolute number of admissions increased from 15,807 in 1997–2000 to 31,977 in 2011–2015. SLE decreased as a cause of admission (from 47.1% to 20.8%, $p < 0.001$), while other categories increased over the time, as follows: 5% to 8.6% for CVD, 8.2% to 13% for infection, and 1.4% to 5.5% for neoplasm ($p < 0.001$ for all). The admission mortality rate rose from 2.22% to 3.06% ($p < 0.001$) and the causes of death evolved in parallel with the admission categories. A significant trend to older age was observed over time in the overall population and deceased patients ($p < 0.001$). Conclusions: Better control of SLE over the past two decades has led to a decrease in early admissions, and disease chronification. As a counterpart, CVD, infections, and neoplasm have become the main causes of admissions and mortality.

**Keywords:** systemic lupus erythematosus; cardiovascular disease; infections; neoplasm; mortality; hospital admissions

## 1. Introduction

Systemic lupus erythematosus (SLE) is an extraordinarily complex disease with a wide variety of clinical features and phenotypes [1]. SLE is a chronic disease mainly affecting young women who suffer from chronic inflammation, disease flares, cumulative drug toxicity, and frequent hospital admissions [2–5]. Since the last century, huge advances have been made in the diagnosis and management of lupus [6]. In parallel with the development of more efficient and less harmful treatments, great concern has emerged related to the long-term complications of SLE, such as the high prevalence of cardiovascular disease, the risk of infections, and the impact on quality of life [7]. Therefore, important changes in the relative weights of the different causes of hospital admissions of lupus patients are likely to have taken place over the last years.

In light of the foregoing, our objective was to analyze the trends in the hospital admissions and mortality of Spanish SLE patients over the last two decades.

## 2. Materials and Methods

We performed a registry study with a case series design, where the hospital admissions and death causes in SLE patients were the main outcomes. We analyzed the data extracted from the Spanish Hospital Discharge Database (SNHDD), a national registry belonging to the Spanish Ministry of Health. The SNHDD includes demographic and epidemiological data and up to 20 discharge diagnoses carried out during the admission of patients, as coded by the International Classification of Diseases (ICD-9), between 1 January 1997 and 31 December 2015. The study was approved by local research ethics committees (PI_80-21) in accordance with the Declaration of Helsinki. Before making it available to researchers, the database was anonymized and all potential patient identifiers were eliminated.

We selected hospital admissions of patients with a diagnosis within the CIE-9-ES code 710.0 (systemic lupus erythematosus), regardless of its position within the diagnoses coding list. According to the database, the main diagnosis during admission or at discharge was the cause of the admission and/or death. Thus, all the main diagnoses were decoded and classified into the following 5 main groups: active SLE; cardiovascular disease (CVD, comprising coronary disease, cerebrovascular disease, heart failure, arterial thromboembolism, hypertensive kidney disease, arteriosclerosis or peripheral arterial disease); neoplasm (including solid organ, hematological, benign or unknown origin neoplasm); infection (classified according to the foci); venous thromboembolic disease (VTED). Only these admission causes were evaluated. In order to analyze the epidemiological trends over time, we grouped admissions within the following four periods: 1997–2000, 2001–2005, 2006–2010 and 2011–2015.

### Statistical Analysis

Quantitative variables were expressed as mean and standard deviation or as median plus interquartile range (IQR), as appropriate; qualitative variables were expressed as percentages. Numerical variables were compared using the t-test or Mann–Whitney's U test, and these tests were also used for the analysis of the average age and stay of the patients in the different periods. Normality (Shapiro's) and homoscedasticity (Levene's) tests were performed to characterize the populations and therefore Kruskal–Wallis test with post hoc FDR correction was carried out. Categorical variables were compared using the chi-square test. For all the analyses, a significance level of 0.05 was set. Analysis was performed using R and R Studio 1.3.1093.

## 3. Results

### 3.1. Population Characteristics

From a total of 66,462,136 nationwide hospital admissions recorded during the study period, 99,859 involved 43,432 patients with a diagnosis of SLE, according to ICD-9, who were, thus, the object of the analysis. The main descriptive variables are shown in Table 1. The mean age was 46.5 years, 83.3% of the patients were female, the average stay lasted 9.1 days, and the readmission rate was 17.6%. Overall, 2786 individuals (6.41%) died, with an overall admission mortality of 2.79%. During the study period, the mean age upon admission rose from 41.1 years in 1997–2000 to 51 years in 2011–2015, as did the mortality rate (from 4.22% to 5.67%) and the mortality during admission (from 2.22% to 3.06%) ($p < 0.001$ for all). By contrast, a decrease was observed in the average length of stay and in the readmission rates, from 9.4 to 8.5 days and from 18.3% to 16.4%, respectively ($p < 0.001$ for both).

**Table 1.** Main demographic characteristics in SLE hospitalized patients.

|  | Overall | 1997–2000 | 2001–2005 | 2006–2010 | 2011–2015 |
|---|---|---|---|---|---|
| Patients (*n*) | 43,432 | 8304 | 12,348 | 15,051 | 17,257 |
| Gender female (%, CI) | 83.3 (83.1–83.6) | 82.8 (82.2–83.4) | 83 (82.5–83.5) | 83.4 (83–83.9) | 83 (82.6–83.4) |
| Age (years) (Mean, SD) | 46.5 (18.7) | 41.1 (18.1) | 44 (18.3) * | 46.6 (18.5) * | 51 (18.3) * T |
| Admissions (*n*) | 99,859 | 15,807 | 24,204 | 27,781 | 31,977 |
| Average stay (days) (Mean, SD) | 9.1 (13.1) | 9.4 (12.7) | 9.5 (14.2) | 9.3 (12.8) | 8.5 (12.6) * T |
| Readmission rate (%, CI) | 17.6 (17.4–17.8) | 18.3 (17.7–18.9) | 18.6 (18.1–19.1) | 17.7 (17.3–18.2) * | 16.4 (16–16.8) * T |
| Deaths (*n*, %) | 2786 (6.41) | 351 (4.22) | 659 (5.34) | 798 (5.30) * | 978 (5.67) * T |
| Mortality rate per admission (%, CI) | 2.79 (2.7–2.9) | 2.22 (2.0–2.46) | 2.72 (2.52–2.94) * | 2.86 (2.67–3.07) | 3.06 (2.87–3.25) T |

SD: standard deviation, CI: confidence interval. *: $p < 0.05$ when compared with the previous period. T: $p < 0.001$ when compared with the first period.

### 3.2. Hospital Admissions

Admissions caused by active SLE, CVD, infections, neoplasm or VTED are shown in Table 2. During the whole study period, active SLE was the main cause of admission (31.6% of cases), while infections accounted for 10.9%, CVD for 7.1%, neoplasms for 4.2%, and VTED for 1.1%. Admissions due to SLE decreased over the years, being the cause of 47.1% of the admissions in 1997–2000, 38.7% in 2001–2005, 29% in 2006–2010, and 20.8% in 2011–2015 ($p < 0.0001$ when comparing each period with the previous). On the contrary, hospitalizations due to CVD (from 5% in the first period to 8.6% in the last), infection (from 8.2% to 13%), and neoplasm (from 2.4% to 5.5%) increased over the time ($p < 0.001$ in all cases). Admissions due to VTED did not suffer significant variations. Overall, patients whose admissions were attributable to SLE were younger (mean 37.6 years) when compared to the other causes ($p < 0.001$ for all), and their mean age rose from the first to the fourth period in all the subgroups ($p < 0.001$).

**Table 2.** Rate of SLE hospitalized individuals by study period and clinical conditions.

|  | Overall | 1997–2000 | 2001–2005 | 2006–2010 | 2011–2015 |
|---|---|---|---|---|---|
| **Active SLE** | | | | | |
| Admissions *n* (%) | 31,539 (31.6) | 7440 (47.1) | 9354 (38.7) * | 8088 (29) * | 6657 (20.8) * T |
| Age (years) (Mean, SD) | 37.6 (16.5) ** | 35.9 (16.3) | 37.1 (16.3) | 37.9 (16.5) | 40 (16.8) T |
| **Cardiovascular disease** | | | | | |
| Admissions *n* (%) | 7065 (7.1) | 789 (5) | 1448 (6) * | 2096 (7.5) * | 2736 (8.6) * T) |
| Age (years) (Mean, SD) | 59.5 (16.9) | 55.5 (17.6) | 57.1 (16.6) | 59.6 (16.7) | 61.6 (16.6) T |
| **Infection** | | | | | |
| Admissions *n* (%) | 10,865 (10.9) | 1292 (8.2) | 2295 (9.5) * | 3137 (11.3) * | 4141 (13) * T |
| Age (years) (Mean, SD) | 52.1 (19) | 47.9 (18.8) | 49.7 (18.9) | 50.8 (19.1) | 55.6 (18.4) T |
| **Neoplasm** | | | | | |
| Admissions *n* (%) | 4182 (4.2) | 386 (2.4) | 785 (3.2) * | 1238 (4.4) * | 1773 (5.5) * T |
| Age (years) (Mean, SD) | 56.5 (14.3) | 54.8 (14.4) | 55.8 (14.9) | 55.3 (14.2) | 57.9 (13.9) T |
| **Venous thrombo-embolic disease** | | | | | |
| Admissions *n* (%) | 1069 (1.1) | 174 (1.1) | 277 (1.1) | 268 (1) | 350 (1.1) |
| Age (years) (Mean, SD) | 50.1 (19.0) | 45.1 (18) | 47.9 (18.4) | 51 (18.9) | 53.6 (19.3) T |

SLE: systemic lupus erythematosus. SD: standard deviation. *: $p < 0.05$ when compared with the previous period. T: $p < 0.05$ when compared with the first period. **: When SLE was compared to the other causes.

### 3.3. Deaths

During the study period, 2786 patients died (6.41% of all the patients). Altogether, the mortality rate per admission was 2.79%. As shown in Table 1, the mortality rates, both overall and per admission, increased significantly after the year 2000.

SLE accounted for 13% of all deaths; CVD caused 18.5% of the deaths; infections 18.7%; neoplasm 11.7% and VTED 1.44% (Table 3). SLE decreased as a cause of death, from 24.2% in 1997–2000, 18.1% in 2001–2005, 12% in 2006–2010 to 6.4% in 2011–2015 ($p < 0.001$ when comparing each period with the previous). On the other hand, CVD (from 15.4% to 20.4%, $p = 0.04$), infection (from 14.3% to 21.1%, $p = 0.005$), and neoplasm (from 7.4% to 13.8%, $p = 0.002$) increased as causes of death from the first to the fourth period of study. Again, VTED remained unchanged throughout the study period.

**Table 3.** Causes of mortality during admission and mean age of SLE patients.

| | Overall | 1997–2000 | 2001–2005 | 2006–2010 | 2011–2015 |
|---|---|---|---|---|---|
| | | | Active SLE | | |
| Deaths (*n*, % of all deaths) | 363 (13) | 85 (24.2) | 119 (18.1) * | 96 (12) * | 63 (6.4) * T |
| Mortality rate (%) | 1.2 | 1.1 | 1.3 | 1.2 | 0.9 |
| Age (years) (Mean, SD) | 54.5 (19.7) ** | 53.5 (19.9) | 52.3 (19.6) | 55.7 (18.9) | 58.5(20.5) |
| | | | Cardiovascular disease | | |
| Deaths (*n*, % of all deaths) | 515 (18.5) | 54 (15.4) | 110 (16.7) | 146 (18.3) | 199 (20.4) T |
| Mortality rate (%) | 7.3 | 6.8 | 7.6 | 7 | 7.3 |
| Age (years) (Mean, SD) | 67 (16.6) | 60 (19.5) | 63.5 (16.9) | 66.9 (16) | 71 (14.8) T |
| | | | Infection | | |
| Deaths (*n*, % of all deaths) | 522 (18.7) | 50 (14.3) | 116 (17.6) | 150 (18.8) | 206 (21.1) T |
| Mortality rate (%) | 4.8 | 3.9 | 5.1 | 4.8 | 5 |
| Age (years) (Mean, SD) | 64.5 (17.4) | 62.5 (18.6) | 62.5 (18.3) | 61.9 (18.2) | 68 (15.4) T |
| | | | Neoplasm | | |
| Deaths (*n*, % of all deaths) | 327 (11.7) | 26 (7.4) | 66 (10) | 100 (12.5) | 135 (13.8) T |
| Mortality rate (%) | 7.8 | 6.7 | 8.4 | 8.1 | 7.6 |
| Age (years) (Mean, SD) | 63.6 (13.6) | 63 (11.7) | 67.1 (12.6) | 61.6 (14.6) | 63.6 (13.4) |
| | | | Venous thrombo-embolic disease | | |
| Deaths (*n*, % of all deaths) | 40 (1.4) | 4 (1.1) | 13 (2) | 7 (0.9) | 16 (1.6) |
| Mortality rate (%) | 3.7 | 2.3 | 4.7 | 2.6 | 4.6 |
| Age (years) (Mean, SD) | 62.3 (16.1) | 54.8 (5.8) | 59.5 (17.8) | 68 (16.5) | 68.9 (15.3) |

SLE: systemic lupus erythematosus. SD: standard deviation. *: $p < 0.05$ when compared with the previous period. T: $p < 0.05$ when compared with the first period. **: $p < 0.01$ when SLE was compared to the other causes.

Comparisons of the mortality rates, for each cause of admission, are shown in Table 3. The overall mortality rate was 1.2% for admissions attributable to SLE, significantly lower than in admissions due to CVD (7.3%), infections (4.8%), neoplasm (7.8%), and VTED (3.7%) ($p < 0.001$). There were no significant variations in the mortality rates when the different periods were compared. Patients who died because of SLE were significantly younger (mean age of 54.5 years) than those who died due to other causes ($p < 0.001$ for all). Indeed, the mean age of deceased patients significantly rose during the study period (from 58.5 years to 67.2, $p < 0.001$), reflecting the older age of patients dying from causes other than lupus activity (Table 3).

*3.4. Subgroup Analysis*

The subgroups comprised in the main categories were studied (Table S1). Overall, coronary artery disease, cerebrovascular disease, and heart failure were the main determinants of CVD admissions and deaths. As shown in Figure 1A, all the CVD subgroups, except for arterial thromboembolism, increased as causes of admission overtime ($p < 0.001$ when the first and the fourth period were compared). On the other hand, no significant variations were found when the different cardiovascular causes of death were compared between different periods (Figure 1B). Respiratory infections were by far the main cause of admission and death (4.1% and 7% of all the admissions and deaths, respectively) in the infection group. All the infectious causes of admission increased when the first and the last period were compared ($p < 0.02$), except for tuberculosis, which actually decreased over time (from 1.5 % to 0.2%, $p < 0.001$) (Figure 2A). It is noteworthy that only sepsis significantly increased as a cause of death (from 0.3% to 1.2%, $p < 0.001$) (Figure 2B). The increasing admissions due to solid organ neoplasm (from 1.1% to 3.8%, $p < 0.001$) and cancer-related deaths (from 6% to 10.5%, $p < 0.01$) were the main determinants of the highest burden in morbidity and mortality of the neoplasms in the subsequent periods (Figure 3). Hematological neoplasms were, however, a less important cause of admissions and death.

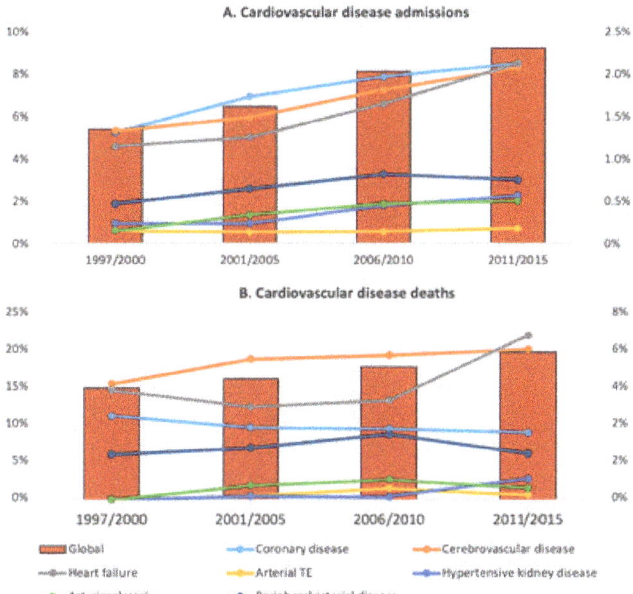


**Figure 1.** Cardiovascular disease admissions and deaths. The figure shows the trends in admissions (**A**) and deaths (**B**) for cardiovascular disease. Data are expressed as a percentage of total admissions and deaths for cardiovascular disease (left ordered axis) and for the subgroups that comprise cardiovascular disease (right ordered axis).

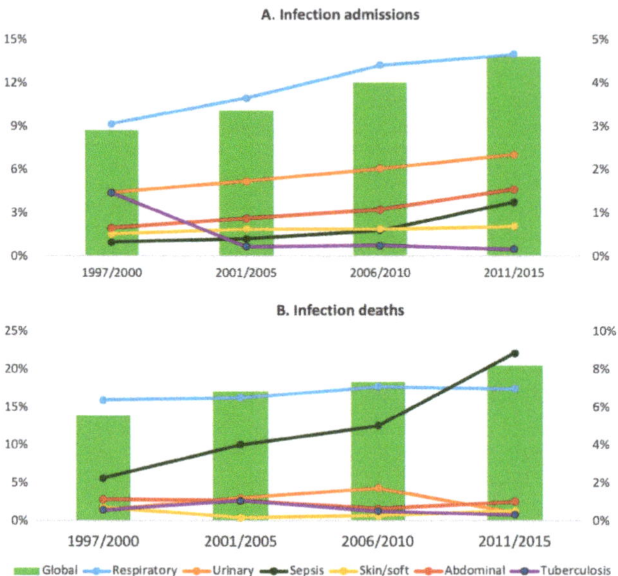



**Figure 2.** Infectious disease admissions and deaths. The figure shows the trends in admissions (**A**) and deaths (**B**) because of infection. Data are expressed as a percentage of total admissions and deaths for the overall infections (left ordered axis) and for the subgroups that comprise infectious disease (right ordered axis).

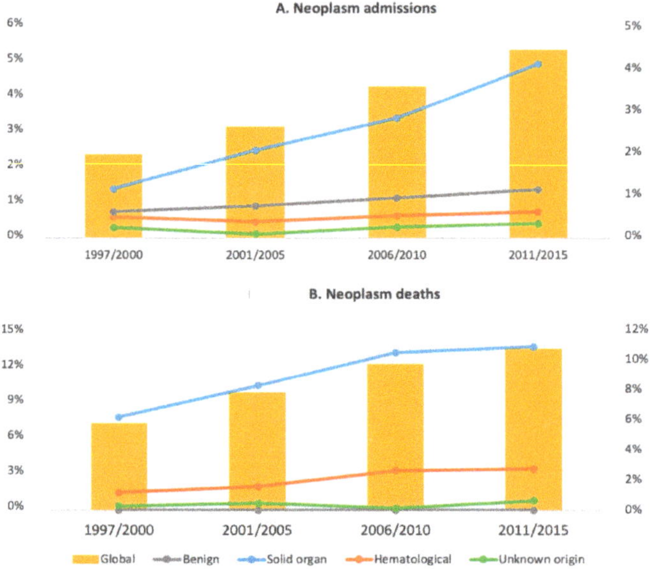



**Figure 3.** Neoplasm admissions and deaths. The figure shows the trends in admissions (**A**) and deaths (**B**) because of neoplasm. Data are expressed as a percentage of total admissions and deaths for the overall neoplasms (left ordered axis) and for the benign, solid organ, hematological and unknown origin neoplasm (ordered axis).

## 4. Discussion

Robust evidence supports that mortality is two- to five-fold higher in SLE patients than in the general population [4,8–11]. In fact, SLE has been identified as the 10th highest cause of death among 15- to 24-year-old women [12]. However, mortality rates and specific causes of death have varied throughout the last decades [4,13–16]. Other groups have previously analyzed the causes of admission in SLE patients worldwide, with different results limited in many cases by sample size, monocentric design, and short follow-up periods [17–25]. Our study analyzes data from 99.859 admissions of SLE patients between 1997 and 2015, thus offering quite a wide view of the whole picture in Spain.

According to our analysis, hospital admissions and deaths due to active SLE decreased dramatically from the late 1990's to the 2010–2015 period. Several studies have identified SLE itself as the main cause of hospital admission, with a wide range between 8.1 and 86.3% [17–26]. On the other hand, other studies have shown a significant downward trend in SLE-related admissions [19,27–29]. Recently, Anastisou et al. reported a decreasing risk of inpatient death in US lupus patients from 2006 to 2016 [29].

In our study, the reduction in SLE-related admissions and deaths was accompanied by a parallel increase in admissions and deaths due to CVD, infections, and neoplasm. In particular, CVD admissions, mainly determined by coronary and cerebrovascular disease and heart failure, rose from 5% to 8.6% of the total, and more than 20% of the deaths in the last period were of cardiovascular cause. Such an increase in the cardiovascular burden of SLE has been previously reported [12,15,16], and Piga et al. have also confirmed an increasing trend in admissions due to stroke and acute coronary syndrome between 2001 and 2012 [28].

According to our data, infections were the other main cause of death among Spanish SLE patients between 2011 and 2015. Previous studies have identified infections as an important cause of admission [19,23,25,30] and death [4,15,16,31–34]. Selvananda et al. reported that infections were concurrent with SLE flares in 41.1% of admissions, reflecting that both situations tend to coincide, and emphasizing the importance of the rational use of immunosuppressants [24]. Similarly to our findings, a recent study in the US showed that hospitalization rates due to infection significantly increased in the period 2015–2016 compared to 1998–2000, with sepsis overtaking pneumonia as the most common infection [35]. Both infections and CVD have been a major concern in longstanding SLE, with a significant impact on morbidity and mortality [1,2,36,37]. Infections are more common within the early phases of the disease, and are strongly determined by immunosuppressive treatment, whilst CVD tends to happen later and is mostly related to chronic inflammation, irreversible organ damage, and cumulative drug toxicity [3,23,38,39].

Despite the increased risk of malignancy in SLE patients, neoplasms have not yet been uniformly identified as a main cause of admission or death [40,41]. We found, in this study, that neoplasms increased as a cause of admission from 2.4% in 1997–2000 to 5.5% in 2010–2015 and became the third cause of death (from 7.4% to 13.8%). These changes were mainly due to solid organ tumors and might also be attributable to the ageing of SLE patients previously exposed to immunosuppression, a well-known risk factor for cancer development [42]. In addition, some of these drugs, such as azathioprine, cyclophosphamide, and mycophenolate, have been related to solid and hematological neoplasms themselves [43]. Similar evolution has been found from series published in the early 2000s when 2.3% of patients developed malignancy and 5.9% died because of cancer [3], and more recent studies, in which neoplasms caused over 13% of deaths in lupus patients [9,30], even being the first cause of death in some of them [44]. However, other authors have found neoplasms to be a stable, or even decreasing, cause of admission [19,28]. Obviously, these striking findings should be assessed deeply, and they merit further epidemiological studies to clarify the impact of neoplasm in SLE patients.

No significant variations in VTED-related admissions and deaths were found over the study period, similar to what has already been described [10,17].

Overall, the mortality rate in our population was 6.41%, with a 2.79% mortality rate per admission, which is in agreement with the literature [5,6,19,26,30]. However, it is noteworthy that in our registry, mortality rates rose after the year 2000, despite the improved survival observed in SLE patients over the last decades by other studies [6,13–15]. This apparent paradox could be explained by the increasing age of the admitted patients over the study periods, with the resultant decrease in SLE-related deaths relative to other causes. In other words, patients live longer due to better control of the disease, leading to late mortality related to CVD and malignancy.

Our study presents some limitations. Firstly, all the data come from a nationwide database, so the diagnoses could not be verified by the authors. Secondly, this analysis was performed considering hospital admissions, with a resultant limitation in power and potential selection bias. However, we mainly evaluated categorical variables, such as admission due to SLE, cardiovascular disease, infection, neoplasms, and deaths, which are difficult to misclassify. In parallel, SLE prevalence could not be properly assessed in the databases. Therefore, the rate of the events could not be elucidated and only admissions or deaths could have been compared. Secondly, we were unable to retrieve additional variables, such as CV risk factors, treatments, and the presence of antiphospholipid syndrome or specific organic SLE involvement. In this regard, data regarding hydroxychloroquine treatment, steroid-sparing immunotherapies, and prior steroid exposure would have yielded interesting data that could have supported our conclusions. Finally, we were only able to study the period between 1997 and 2015, due to the change in the registry diagnostic codification to ICD-10 after that year. On the other hand, our study offers a nationwide analysis with a large sample size and a long study period, with consistent results confirming the different trends shown in recent, smaller studies.

## 5. Conclusions

In conclusion, our large-scale study confirms the chronification and ageing of SLE patients. As a direct consequence, CVD, infections, and neoplasms have risen as admission causes, and have even surpassed lupus itself as the main cause of mortality in the last two decades. It is, thus, important to implement actions directed to mitigate the impact of such groups of diseases, including the extensive use of antimalarials and of lower doses of oral glucocorticoids, which can decrease infections and cardiovascular damage [45,46].

**Supplementary Materials:** The following are available online at https://www.mdpi.com/article/10.3390/jcm10245749/s1, Table S1: admission and death causes by categories.

**Author Contributions:** Conceptualization, V.M.-T., C.T., G.R.-I., R.C., Á.G.-R., A.R., P.D.-d.C., S.M.-P., P.T., S.R., E.S., M.M.-U., C.d.M., M.Y. and J.-A.V.; Data curation, V.M.-T., C.T., G.R.-I. and A.R.; Formal analysis, V.M.-T., C.T. and A.R.; Funding acquisition, V.M.-T., R.C. and J.-A.V.; Investigation, V.M.-T., G.R.-I., R.C. and M.Y.; Methodology, V.M.-T., C.T., G.R.-I., R.C., A.R. and M.M.-U.; Project administration, V.M.-T., R.C. and J.-A.V.; Resources, V.M.-T., R.C., C.d.M. and J.-A.V.; Software, V.M.-T., C.T. and A.R.; Supervision, V.M.-T., G.R.-I., R.C., Á.G.-R., P.D.-d.C., P.T., S.R., E.S., M.M.-U., C.d.M., M.Y. and J.-A.V.; Validation, V.M.-T., C.T., G.R.-I., R.C., S.M.-P., C.d.M., M.Y. and J.-A.V.; Visualization, V.M.-T., G.R.-I., R.C. and M.Y.; Writing—original draft, V.M.-T., G.R.-I. and M.Y.; Writing—review & editing, V.M.-T., G.R.-I., R.C., Á.G.-R., P.D.-d.C., S.M.-P., P.T., S.R., E.S., M.M.-U., C.d.M., M.Y. and J.-A.V. All authors have read and agreed to the published version of the manuscript.

**Funding:** This work has been supported by a grant from Instituto de Salud Carlos III (Expedient number CM19/00223).

**Institutional Review Board Statement:** The study was approved by the Hospital Puerta de Hierro Research Ethics Commission (PI_80-21) in accordance with the Declaration of Helsinki.

**Informed Consent Statement:** Patient consent was waived according to the Spanish law and ethics committee.

**Data Availability Statement:** The data presented in this study are available on request from the corresponding author. The data are not publicly available since the database analysis has to be approved by the Spanish Ministry of Health.

**Conflicts of Interest:** The authors declare no conflict of interest.

# References


1. Lisnevskaia, L.; Murphy, G.; Isenberg, D. Systemic lupus erythematosus. *Lancet* **2014**, *384*, 1878–1888. [CrossRef]
2. Kaul, A.; Gordon, C.; Crow, M.K.; Touma, Z.; Urowitz, M.B.; van Vollenhoven, R.; Ruiz-Irastorza, G.; Hughes, G. Systemic lupus erythematosus. *Nat. Rev. Dis. Primers* **2016**, *2*, 16039. [CrossRef]
3. Cervera, R.; Khamashta, M.A.; Font, J.; Sebastiani, G.D.; Gil, A.; Lavilla, P.; Mejía, J.C.; Aydintug, A.O.; Chwalinska-Sadowska, H.; de Ramón, E.; et al. European Working Party on Systemic Lupus Erythematosus. Morbidity and mortality in systemic lupus erythematosus during a 10-year period: A comparison of early and late manifestations in a cohort of 1000 patients. *Medicine* **2003**, *82*, 299–308. [CrossRef]
4. Bernatsky, S.; Boivin, J.F.; Joseph, L.; Manzi, S.; Ginzler, E.; Gladman, D.D.; Urowitz, M.; Fortin, P.R.; Petri, M.; Barr, S.; et al. Mortality in systemic lupus erythematosus. *Arthritis Rheumatol.* **2006**, *54*, 2550–2557. [CrossRef] [PubMed]
5. Yen, E.Y.; Singh, R.R. Brief Report: Lupus-An Unrecognized Leading Cause of Death in Young Females: A Population-Based Study Using Nationwide Death Certificates, 2000–2015. *Arthritis Rheumatol.* **2018**, *70*, 1251–1255. [CrossRef] [PubMed]
6. Fanouriakis, A.; Kostopoulou, M.; Alunno, A.; Aringer, M.; Bajema, I.; Boletis, J.N.; Cervera, R.; Doria, A.; Gordon, C.; Govoni, M.; et al. 2019 update of the EULAR recommendations for the management of systemic lupus erythematosus. *Ann. Rheum. Dis.* **2019**, *78*, 736–745. [CrossRef] [PubMed]
7. Arnaud, L.; Tektonidou, M.G. Long-term outcomes in systemic lupus erythematosus: Trends over time and major contributors. *Rheumatology* **2020**, *59*, v29–v38. [CrossRef] [PubMed]
8. Fors Nieves, C.E.; Izmirly, P.M. Mortality in Systemic Lupus Erythematosus: An Updated Review. *Curr. Rheumatol. Rep.* **2016**, *18*, 21. [CrossRef] [PubMed]
9. Ingvarsson, R.F.; Landgren, A.J.; Bengtsson, A.A.; Jönsen, A. Good survival rates in systemic lupus erythematosus in southern Sweden, while the mortality rate remains increased compared with the population. *Lupus* **2019**, *28*, 1488–1494. [CrossRef] [PubMed]
10. Yurkovich, M.; Vostretsova, K.; Chen, W.; Aviña-Zubieta, J.A. Overall and cause-specific mortality in patients with systemic lupus erythematosus: A meta-analysis of observational studies. *Arthritis Care Res.* **2014**, *66*, 608–616. [CrossRef]
11. Tselios, K.; Gladman, D.D.; Sheane, B.J.; Su, J.; Urowitz, M. All-cause, cause-specific and age-specific standardised mortality ratios of patients with systemic lupus erythematosus in Ontario, Canada over 43 years (1971–2013). *Ann. Rheum. Dis.* **2019**, *7*, 802–806. [CrossRef]
12. Singh, R.R.; Yen, E.Y. SLE mortality remains disproportionately high, despite improvements over the last decade. *Lupus* **2018**, *27*, 1577–1581. [CrossRef] [PubMed]
13. Yen, E.Y.; Shaheen, M.; Woo, J.M.P.; Mercer, N.; Li, N.; McCurdy, D.K.; Karlamangla, A.; Singh, R.R. 46-Year Trends in Systemic Lupus Erythematosus Mortality in the United States, 1968 to 2013: A Nationwide Population-Based Study. *Ann. Intern. Med.* **2017**, *167*, 777–785. [CrossRef] [PubMed]
14. Goss, L.B.; Ortiz, J.R.; Okamura, D.M.; Hayward, K.; Goss, C.H. Significant Reductions in Mortality in Hospitalized Patients with Systemic Lupus Erythematosus in Washington State from 2003 to 2011. *PLoS ONE* **2015**, *10*, e0128920. [CrossRef]
15. Urowitz, M.B.; Gladman, D.D.; Tom, B.D.; Ibañez, D.; Farewell, V.T. Changing patterns in mortality and disease outcomes for patients with systemic lupus erythematosus. *J. Rheumatol.* **2008**, *35*, 2152–2158. [CrossRef] [PubMed]
16. Lee, Y.H.; Choi, S.J.; Ji, J.D.; Song, G.G. Overall and cause-specific mortality in systemic lupus erythematosus: An updated meta-analysis. *Lupus* **2016**, *25*, 727–734. [CrossRef] [PubMed]
17. Lee, J.; Dhillon, N.; Pope, J. All-cause hospitalizations in systemic lupus erythematosus from a large Canadian referral centre. *Rheumatology* **2013**, *52*, 905–909. [CrossRef] [PubMed]
18. Rosa, G.P.D.; Ortega, M.F.; Teixeira, A.; Espinosa, G.; Cervera, R. Causes and factors related to hospitalizations in patients with systemic lupus erythematosus: Analysis of a 20-year period (1995–2015) from a single referral centre in Catalonia. *Lupus* **2019**, *28*, 1158–1166. [CrossRef]
19. Anastasiou, C.; Dulai, O.; Baskaran, A.; Proudfoot, J.; Verhaegen, S.; Kalunian, K. Immunosuppressant use and hospitalisations in adult patients with systemic lupus erythematosus admitted to a tertiary academic medical centre. *Lupus Sci. Med.* **2018**, *5*, e000249. [CrossRef] [PubMed]
20. Levy, O.; Markov ADrob, Y.; Maslakov, I.; Tishler, M.; Amit-Vazina, M. All-cause hospitalizations in systemic lupus erythematosus from a single medical center in Israel. *Rheumatol. Int.* **2018**, *38*, 1841–1846. [CrossRef]
21. Liang, H.; Pan, H.F.; Tao, J.H.; Ye, D.Q. Causes and Factors Associated with Frequent Hospitalization in Chinese Patients with Systemic Lupus Erythematosus: An Ambispective Cohort Study. *Med. Sci. Monit.* **2019**, *25*, 8061–8068. [CrossRef] [PubMed]
22. Dzifa, D.; Boima, V.; Yorke, E.; Yawson, A.; Ganu, V.; Mate-Kole, C. Predictors and outcome of systemic lupus erythematosus (SLE) admission rates in a large teaching hospital in sub-Saharan Africa. *Lupus* **2018**, *27*, 336–342. [CrossRef] [PubMed]
23. Busch, R.W.; Kay, S.D.; Voss, A. Hospitalizations among Danish SLE patients: A prospective study on incidence, causes of admission and risk factors in a population-based cohort. *Lupus* **2018**, *27*, 165–171. [CrossRef]

24. Selvananda, S.; Chong, Y.Y.; Thundyil, R.J. Disease activity and damage in hospitalized lupus patients: A Sabah perspective. *Lupus* **2020**, *29*, 344–350. [CrossRef] [PubMed]
25. Gu, K.; Gladman, D.D.; Su, J.; Urowitz, M.B. Hospitalizations in Patients with Systemic Lupus Erythematosus in an Academic Health Science Center. *J. Rheumatol.* **2017**, *44*, 1173–1178. [CrossRef] [PubMed]
26. Krishnan, E. Hospitalization and mortality of patients with systemic lupus erythematosus. *J. Rheumatol.* **2006**, *33*, 1770–1774.
27. Rees, F.; Doherty, M.; Grainge, M.; Davenport, G.; Lanyon, P.; Zhang, W. The incidence and prevalence of systemic lupus erythematosus in the UK, 1999–2012. *Ann. Rheum. Dis.* **2016**, *75*, 136–141. [CrossRef] [PubMed]
28. Piga, M.; Casula, L.; Perra, D.; Sanna, S.; Floris, A.; Antonelli, A.; Cauli, A.; Mathieu, A. Population-based analysis of hospitalizations in a West-European region revealed major changes in hospital utilization for patients with systemic lupus erythematosus over the period 2001–2012. *Lupus* **2016**, *25*, 28–37. [CrossRef] [PubMed]
29. Anastasiou, C.; Trupin, L.; Glidden, D.V.; Li, J.; Gianfrancesco, M.; Shiboski, S.; Schmajuk, G.; Yazdany, J. Mortality among Hospitalized Individuals with Systemic Lupus Erythematosus in the United States between 2006 and 2016. *Arthritis Care Res.* **2020**. [CrossRef]
30. Anver, H.; Dubey, S.; Fox, J. Changing trends in mortality in systemic lupus erythematosus? An analysis of SLE inpatient mortality at University Hospital Coventry and Warwickshire NHS Trust from 2007 to 2016. *Rheumatol. Int.* **2019**, *39*, 2069–2075. [CrossRef] [PubMed]
31. Jakes, R.W.; Bae, S.C.; Louthrenoo, W.; Mok, C.C.; Navarra, S.V.; Kwon, N.W. Systematic review of the epidemiology of systemic lupus erythematosus in the Asia-Pacific region: Prevalence, incidence, clinical features, and mortality. *Arthritis Care Res.* **2012**, *64*, 159–168. [CrossRef] [PubMed]
32. Fei, Y.; Shi, X.; Gan, F.; Li, X.; Zhang, W.; Li, M.; Hou, Y.; Zhang, X.; Zhao, Y.; Zeng, X.; et al. Death causes and pathogens analysis of systemic lupus erythematosus during the past 26 years. *Clin. Rheumatol.* **2014**, *33*, 57–63. [CrossRef] [PubMed]
33. Zhen, J.; Ling-Yun, S.; Yao-Hong, Z.; Xiang-Dang, W.; Jie-Ping, P.; Miao-Jia, Z.; Juan, T.; Yu, Z.; Kui-Lin, T.; Jing, L.; et al. Death-related factors of systemic lupus erythematosus patients associated with the course of disease in Chinese populations: Multicenter and retrospective study of 1958 inpatients. *Rheumatol. Int.* **2013**, *33*, 1541–1546. [CrossRef] [PubMed]
34. Lorenzo-Vizcaya, A.; Isenberg, D. Analysis of trends and causes of death in SLE patients over a 40-years period in a cohort of patients in the United Kingdom. *Lupus* **2021**, *30*, 702–706. [CrossRef] [PubMed]
35. Singh, J.A.; Cleveland, J.D. Hospitalized Infections in Lupus: A Nationwide Study of Types of Infections, Time Trends, Health Care Utilization, and In-Hospital Mortality. *Arthritis Rheumatol.* **2021**, *73*, 617–630. [CrossRef]
36. Symmons, D.P.; Gabriel, S.E. Epidemiology of CVD in rheumatic disease, with a focus on RA and SLE. *Nat. Rev. Rheumatol.* **2011**, *7*, 399–408. [CrossRef] [PubMed]
37. Borchers, A.T.; Keen, C.L.; Shoenfeld, Y.; Gershwin, M.E. Surviving the butterfly and the wolf: Mortality trends in systemic lupus erythematosus. *Autoimmun. Rev.* **2004**, *3*, 423–453. [CrossRef] [PubMed]
38. Urowitz, M.B.; Ibañez, D.; Gladman, D.D. Atherosclerotic vascular events in a single large lupus cohort: Prevalence and risk factors. *J. Rheumatol.* **2007**, *34*, 70–75. [PubMed]
39. Danza, A.; Ruiz-Irastorza, G. Infection risk in systemic lupus erythematosus patients: Susceptibility factors and preventive strategies. *Lupus* **2013**, *22*, 1286–1294. [CrossRef]
40. Ladouceur, A.; Clarke, A.E.; Ramsey-Goldman, R.; Bernatsky, S. Malignancies in systemic lupus erythematosus: An update. *Curr. Opin. Rheumatol.* **2019**, *31*, 678–681. [CrossRef] [PubMed]
41. Bernatsky, S.; Ramsey-Goldman, R.; Labrecque, J.; Joseph, L.; Boivin, J.F.; Petri, M.; Zoma, A.; Manzi, S.; Urowitz, M.B.; Gladman, D.; et al. Cancer risk in systemic lupus: An updated international multi-centre cohort study. *J. Autoimmun.* **2013**, *42*, 130–135. [CrossRef] [PubMed]
42. Schreiber, R.D.; Old, L.J.; Smyth, M.J. Cancer immunoediting: Integrating immunity's roles in cancer suppression and promotion. *Science* **2011**, *331*, 1565–1570. [CrossRef]
43. Vial, T.; Descotes, J. Immunosuppressive drugs and cancer. *Toxicology* **2003**, *185*, 229–240. [CrossRef]
44. Alonso, M.D.; Llorca, J.; Martinez-Vazquez, F.; Miranda-Filloy, J.A.; Diaz de Teran, T.; Dierssen, T.; Rodriguez, T.R.V.; Gomez-Acebo, I.; Blanco, R.; Gonzalez-Gay, M.A. Systemic lupus erythematosus in northwestern Spain: A 20-year epidemiologic study. *Medicine* **2011**, *90*, 350–358. [CrossRef] [PubMed]
45. Ugarte, A.; Danza, A.; Ruiz-Irastorza, G. Glucocorticoids and antimalarials in systemic lupus erythematosus: An update and future directions. *Curr. Opin. Rheumatol.* **2018**, *30*, 482–489. [CrossRef]
46. Ruiz-Arruza, I.; Lozano, J.; Cabezas-Rodriguez, I.; Medina, J.-A.; Ugarte, A.; Erdozain, J.-G.; Ruiz-Irastorza, G. Restrictive Use of Oral Glucocorticoids in Systemic Lupus Erythematosus and Prevention of Damage Without Worsening Long-Term Disease Control: An Observational Study. *Arthritis Care Res.* **2018**, *70*, 582–591. [CrossRef] [PubMed]

Journal of
*Clinical Medicine*

*Article*

# Incidence of COVID-19 Hospitalisation in Patients with Systemic Lupus Erythematosus: A Nationwide Cohort Study from Denmark

René Cordtz [1,2,*,†], Salome Kristensen [1,3,†], Louise Plank Holm Dalgaard [1], Rasmus Westermann [1], Kirsten Duch [1,4], Jesper Lindhardsen [5], Christian Torp-Pedersen [6,7] and Lene Dreyer [1,2,3]

1   Department of Rheumatology, Aalborg University Hospital, 9000 Aalborg, Denmark; sakr@rn.dk (S.K.); louise.dalgaard@rn.dk (L.P.H.D.); raw@rn.dk (R.W.); k.duch@rn.dk (K.D.); l.dreyer@rn.dk (L.D.)
2   Department of Rheumatology, Center for Rheumatology and Spine Diseases, Gentofte Hospital, 2900 Hellerup, Denmark
3   Department of Clinical Medicine, Aalborg University, 9000 Aalborg, Denmark
4   Unit of Clinical Biostatistics, Aalborg University Hospital, 9000 Aalborg, Denmark
5   Lupus and Vasculitis Clinic, Center for Rheumatology and Spine Diseases, Rigshospitalet, 2100 Copenhagen, Denmark; jesper.lindhardsen@regionh.dk
6   Department of Cardiology, Nordsjællands Hospital, 3400 Hillerod, Denmark; christian.torp-pedersen.01@regionh.dk
7   Department of Public Health, University of Copenhagen, 1014 Copenhagen, Denmark
*   Correspondence: r.cordtz@rn.dk
†   Shared first authorship/contributed equally.

**Citation:** Cordtz, R.; Kristensen, S.; Dalgaard, L.P.H.; Westermann, R.; Duch, K.; Lindhardsen, J.; Torp-Pedersen, C.; Dreyer, L. Incidence of COVID-19 Hospitalisation in Patients with Systemic Lupus Erythematosus: A Nationwide Cohort Study from Denmark. *J. Clin. Med.* **2021**, *10*, 3842. https://doi.org/10.3390/jcm10173842

Academic Editors: Christopher Sjöwall and Ioannis Parodis

Received: 30 July 2021
Accepted: 23 August 2021
Published: 27 August 2021



**Abstract:** Background: Patients with systemic lupus erythematosus (SLE) have an increased risk of infections due to impaired immune functions, disease activity, and treatment. This study investigated the impact of having SLE on the incidence of hospitalisation with COVID-19 infection. Methods: This was a nationwide cohort study from Denmark between 1 March 2020 to 2 February 2021, based on the linkage of several nationwide registers. The adjusted incidence of COVID-19 hospitalisation was estimated for patients with SLE compared with the general population in Cox-regression models. Among SLE patients, the hazard ratio (HR) for hospitalisation was analysed as nested case-control study. Results: Sixteen of the 2533 SLE patients were hospitalised with COVID-19 infection. The age-sex adjusted rate per 1000 person years was 6.16 (95% CI 3.76–10.08) in SLE patients, and the corresponding hazard ratio was 2.54 (95% CI 1.55–4.16) compared with the matched general population group after adjustment for comorbidities. Among SLE patients, hydroxychloroquine treatment was associated with a HR for hospitalisation of 0.61 (95% CI 0.19–1.88), and 1.06 (95% CI 0.3–3.72) for glucocorticoid treatment. Conclusion: Patients with SLE were at increased risk of hospitalisation with COVID-19.

**Keywords:** systemic lupus erythematosus; COVID-19; hydroxychloroquine; glucocorticoids

## 1. Introduction

Patients with systemic lupus erythematosus (SLE) are considered at higher risk of infections compared with the general population owing to SLE-related innate immune perturbations and the use of immunosuppressive drugs [1]. This raised the question about whether patients with SLE might be at increased risk for contracting SARS-CoV-2 and a more severe clinical course once infected. In a cross-sectional survey study of 165 SLE patients from the Lombardy and Emilia-Romagna regions of Italy, 2% had confirmed coronavirus disease 2019 (COVID-19) [2]. The corresponding proportions in the background population were lower (0.76 and 0.47% in Lombardy and Emilia-Romagna, respectively), but as the authors pointed out, a potential bias could be a higher test frequency among SLE patients although no information was available to disprove or confirm such a tendency. In a prior study describing the incidence and severity of COVID-19 hospitalisation in patients with inflammatory rheumatic diseases in Denmark during the first wave of the epidemic, we found a statistically non-significantly 40% increased risk of hospital admission in patients with connective tissue disease including SLE patients [3]. Only a few studies have investigated the risk of COVID hospitalisation specifically in patients with SLE. In a survey study by Ramirez et al., 417 patients with SLE responded, and the frequency of COVID-19 hospitalisation was 0.24%, compared to 0.43% in the general population of Lombardy [4]. However, the information provided by the responders could not be verified and the reported proportions were not age- and sex-standardised.

In the first period of the COVID-19 pandemic, treatment with hydroxychloroquine (HCQ) was suggested as an inhibitor of SARS-CoV-2 in vitro, but recent studies involving SLE patients have demonstrated that the doses used in treatment of SLE are not protective against severe COVID-19 infection [5–10]. On the other hand glucocorticoids might increase the risk of hospitalisation and a subsequent severe outcome [11,12]. Furthermore, studies suggest that pausing the SLE treatment during COVID-19 infection leads to a flare up in the SLE [13–15].

Using the nationwide registers in Denmark, this study aimed to investigate the impact of having SLE on the incidence of hospitalisation with COVID-19 infection compared with the general population, and secondarily aimed at investigating the potential association between treatment with HCQ or glucocorticoids and the risk of being hospitalised with COVID-19 among SLE patients.

## 2. Materials and Methods

### 2.1. Study Design

This was a population-based observational cohort study investigating the incidence of COVID-19 hospitalisation in patients with SLE from 1 March 2020 to 2 February 2021, based on the linkage of several Danish nationwide registers.

### 2.2. Data Sources, Study Population and Exposures

The Danish Civil Registration System contains information on all residents of Denmark including the unique civil registration number, which allows for linkage between registers. The Civil Registration System was used to identify the primary cohort consisting of all individuals aged 18 years or older alive on 1 March 2020 of the entire Danish population [16], and to obtain information on age and sex, used for the matching procedures, and vital status during follow-up.

SLE patients were identified from the Danish National Patient Register (DNPR) using the International Classification of Diseases 10th Edition (ICD-10) code M32, except M32.0, according to the algorithm suggested by Hermansen et al., 2016 [17,18]. This case definition requires that a first registration of SLE in the DNPR should be followed by (a) 1 year of out-patient follow up or (b) consecutive inpatient admissions coded with an SLE diagnosis with over 3-month intervals during the first year of follow up. Patients who fulfilled either (a), (b), or both prior to 1 March 2020 constituted the SLE group. Each SLE patient was matched with up to 1000 individuals of the same age and sex from the general population corresponding to the onset of the COVID-19 epidemic in Denmark, 1 March 2020, while requiring that the matched controls had no history of inflammatory rheumatic diseases.

For descriptive purposes, information on redeemed prescriptions of HCQ, azathioprine, methotrexate, glucocorticoid, warfarin, clopidogrel, and acetylsalicylic acid was obtained from the Danish National Database of Reimbursed Prescriptions (DNDRP) using Anatomical Therapeutic Chemical Classification System (ATC)-codes [19] within 1 year prior to 1 March 2020. In the nested case control study, information on redeemed prescriptions of HCQ and glucocorticoid in the 6 months leading up to date of hospitalisation for cases, and matching dates for corresponding controls, was obtained using the DNDRP. Information on treatment with rituximab, belimumab, cyclophosphamide, and mycophenolate mofetil in a hospital setting was obtained from DNPR within 1 year prior to 1 March 2020.

### 2.3. Outcome Information

COVID-19 hospitalisation was obtained through DNPR using ICD-10 codes created by the Danish Ministry of Health specifically for the pandemic in accordance with the definition established by the World Health Organization (ICD-10 codes B34.2A, B97.2 and B97.2A). These codes have recently been validated in a Danish setting [20]. Hospitalisation was defined as a registration with the abovementioned ICD-10 codes and with a further requirement that the hospital-stay lasted at least 24 h.

Secondarily, the number of patients experiencing a severe outcome of COVID-19 hospitalisation was obtained for the SLE and the general population comparator group. A severe outcome was defined as the composite of either acute respiratory distress syndrome, admission to an intensive care unit, and/or death.

### 2.4. Other Covariates

Chronic lung disease, cardiovascular disease (ischemic heart disease, heart failure, hypertension, and stroke), diabetes mellitus type I and II (DM), hospital registered diagnosis of obesity, and cancer were chosen as comorbidities of interest. These diagnoses were identified using ICD-10 codes in the DNPR and/or redeemed prescriptions registered in the Danish Prescription Register of relevant drugs for each comorbidity (see Supplementary Table S1) [21].

In the nested case-control analysis, lupus nephritis was used as a matching parameter to account for SLE disease severity among cases and controls. Lupus nephritis was defined according to the register-based definition suggested by Hermansen et al., which had a positive predictive value of 90% for lupus nephritis. Thus, we identified patients registered with a nephritis diagnosis (ICD-10 codes N00-06, N08.2, N08.5, N16.2, N16.4, N16.8, N18, N19, N26, M32.1B) in the DNPR between the date of SLE diagnosis and date of case/control matching.

### 2.5. Statistical Analysis

The incidence of hospitalisation in SLE patients compared with the general population was found by following the cohort from 1 March 2020 to 2 February 2021, the date of a COVID-19 hospitalisation or date of death, whichever occurred first. For each group, the age- and sex standardised incidence rate of hospitalisation per 1000 person years was estimated. The incidence of hospitalisation with COVID-19 in SLE patients was compared with the matched general population group in a Cox-regression model with age as underlying time scale, and stratified by sex to estimate a hazard ratio (HR) with 95% confidence interval (95%CI). An additional Cox model was adjusted for the following comorbidities: chronic lung disease, DM, cardiovascular disease, obesity, and cancer.

Furthermore, a Cox model with covariates age (restricted cubic spline with four degrees of freedom), sex, and group (SLE vs. general population) was used to estimate the predicted probability of being hospitalised with COVID-19 stratified according to age (40-, 60-, and 80-year-olds), sex, and SLE or non-SLE status. The predicted probability was plotted with cumulative absolute risk of COVID-19 hospitalisation in % on the $y$-axis and months since start of the pandemic in Denmark (1 March 2020) on the $x$-axis.

Lastly, in a nested case-control design of the SLE patients, each SLE patient admitted with COVID-19 (cases) was matched with up to five SLE patients who were not hospitalised (controls) on age (3-year intervals), sex, time at risk during the pandemic, and history of nephritis corresponding to the date of COVID-19 hospitalisation for cases and controls. Baseline characteristics were presented with count and percentage for discrete variables unless fewer than three patients were observed. Continuous variables were presented with median and interquartile range (IQR). HR was calculated using conditional logistic regression to account for the dependence of matching and presented with 95% CI. Both separate crude analyses for HCQ and glucocorticoid, and subsequently an adjusted model including both HCQ and glucocorticoid were performed.

## 3. Results

From 1 March 2020, 2533 individuals with SLE were followed up; 88.5% were women with a median age of 55.4 years (Table 1). SLE patients had higher prevalence of all comorbidities compared with the matched general population group.

**Table 1.** Demographics, comorbidities, and medication in SLE and the general population at the start of follow-up.

| Group | Systemic Lupus Erythematosus | General Population |
| --- | --- | --- |
| *n* | 2533 | 2,532,914 |
| Age in years, median (interquartile range) | 55.4 (44.1 to 66.5) | 55.5 (44.1 to 66.6) |
| Women, *n* (%) | 2242 (88.5%) | 2,241,914 (88.5%) |
| Disease duration in years, median (interquartile range) | 12.6 (6.1 to 21.7) | - |
| Lupus nephritis, *n* (%) | 205 (8.1%) | - |
| Cardiovascular disease, *n* (%) | 1070 (42.2%) | 447,760 (17.7%) |
| Lung disease, *n* (%) | 595 (23.5%) | 330,635 (13.1%) |
| Diabetes mellitus, *n* (%) | 202 (8.0%) | 165,528 (6.5%) |
| Cancer, *n* (%) | 221 (8.7%) | 201,063 (7.9%) |
| Diagnosed with obesity, *n* (%) | 300 (11.8%) | 251,883 (9.9%) |
| Treated with, *n* (%) | | |
|   Hydroxychloroquine | 1170 (46.2%) | 1830 (0.1%) |
|   Azathioprine | 202 (8.0%) | 3418 (0.1%) |
|   Methotrexate | 118 (4.7%) | 7282 (0.3%) |
|   Glucocorticoids | 685 (27.0%) [1] | 67,738 (2.7%) |
|   Cyclophosphamide | 8 (0.3%) | 439 (0%) |
|   Mycophenolate mofetil | 54 (2.1%) | 336 (0%) |
|   Rituximab | 42 (1.7%) | 1494 (0.1%) |
|   Belimumab | 30 (1.2%) | 0 (0%) |
|   Warfarin | 306 (12.1%) | 21,613 (0.9%) |
|   Clopidogrel | 142 (5.6%) | 68,210 (2.7%) |
|   Acetylsalicylic acid | 440 (17.4%) | 136,380 (5.4%) |

[1] Based on redeemed glucocorticoid-prescriptions 12 months prior to index, the average daily dosage redeemed was estimated to: 0–5 mg 45.5%, 5–10 mg 41.3% and $\geq$10 mg 13.2%. SLE: systemic lupus erythematosus.

Of the 2533 patients with SLE, 46.2% were treated with HCQ and 28.9% received glucocorticoids with the corresponding proportions in the general population group being 0.1 and 3.6%, respectively.

Sixteen of the 2533 SLE patients were hospitalised with COVID-19 infection during follow-up, while the corresponding number of patients in the general population group was 5069 (0.63% vs. 0.20%), Table 2. The age- and sex-standardised incidence rate was threefold higher in SLE patients than in the general population. The age- and sex-adjusted HR was 3.23 (95% CI 1.98 to 5.28). The estimate was slightly attenuated following adjustment for comorbidities: 2.54 (95% CI 1.55 to 4.16).

The predicted cumulative incidence increased with age and was higher among men than women. Also, the incidence was higher for SLE patients compared with the general population across all age and sex specific strata (Figure 1).

**Table 2.** Numbers, incidence rates and hazard ratios for hospitalisation with COVID-19 infection among SLE patients and the general population.

| Analysis | Systemic Lupus Erythematosus | General Population |
|---|---|---|
| *N* hospitalised with COVID-19 | 16 | 5069 |
| Person years of observation | 2616.7 | 2,634,850.9 |
| Age- and sex-adjusted rates per 1000 person years (95% CI) | 6.16 (3.76 to 10.08) | 1.91 (1.86 to 1.96) |
| HR (95% CI) for hospitalisation with COVID-19 adjusted for sex with age as underlying time scale | 3.20 (1.96 to 5.24) | 1 (Reference) |
| HR (95% CI) for hospitalisation with COVID-19 adjusted for sex and comorbidities with age as underlying time scale | 2.62 (1.55 to 4.16) | 1 (Reference) |

*N*: Numbers, 95% CI: 95% Confidence Interval; HR, hazard ratio.

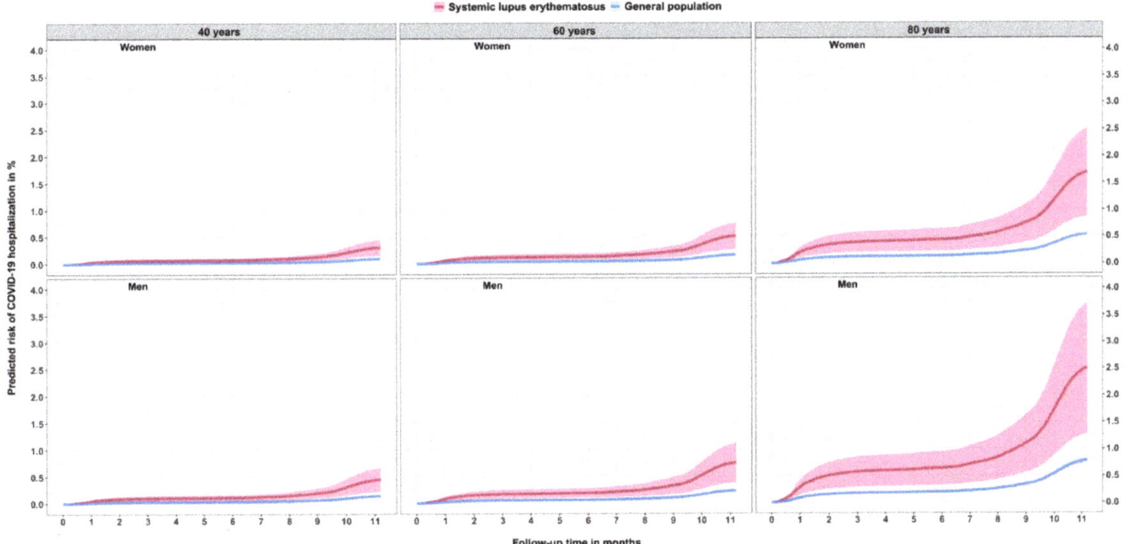


**Figure 1.** Predicted risk of COVID-19 hospitalisation in % follow-up in months for patients with systemic lupus erythematosus and the general population stratified by sex and age.

The proportion of patients experiencing a severe outcome during their COVID-19 hospitalisation was similar between the SLE and the general population groups, but due to GPDR regulations, the absolute number of events was too low to present here.

In the nested case-control analyses, the adjusted HR for COVID-19 hospitalisation among SLE patients treated with HCQ was 0.61 (95% CI 0.19 to 1.88) compared with non-HCQ treated patients, whereas the corresponding HR for glucocorticoid treated compared with non-glucocorticoid treated SLE patients was 1.06 (95% CI 0.30 to 3.72) (see Table 3).

**Table 3.** Numbers, incidence rates and hazard ratios for hospitalisation with COVID-19 infection among hospitalised patients with SLE matched with controls from the SLE population.

| Group | SLE Cases Hospitalised with COVID-19 | Matched Controls from SLE Population not Hospitalised with COVID-19 |
| --- | --- | --- |
| *N* | 16 | 79 |
| Age in years, median (interquartile range) | 69.1 (55.5–78.8) | 67.3 (52.6 to 78.9) |
| Women, *n* (%) | 11 (68.8%) | 55 (69.9%) |
| Disease duration in years, median (interquartile range) | 12.6 (9.9–22.3) | 14.9 (5.6–23.9) |
| Lupus nephritis, *n* (%) | ≤3 | 15 (19%) |
| Cardiovascular disease, *n* (%) | 7 (43.8%) | 22 (27.8%) |
| Lung disease, *n* (%) | 5 (31.2%) | 15 (19%) |
| Diabetes Mellitus, *n* (%) | 4 (25%) | 13 (16.5%) |
| Cancer, *n* | ≤3 | ≤3 |
| Diagnosed with obesity, *n* | ≤3 | ≤3 |
| Hydroxychloroquine, *n* (%) | 5 (31.2%) | 34 (43%) |
| Glucocorticoids, *n* (%) | 4 (25%) | 19 (24.1%) |
| Crude HR (95% CI) for COVID-19 hospitalisation in hydroxychloroquine treated compared with non-hydroxychloroquine treated | 0.61 (0.19 to 1.88) | 1 (Reference) |
| Crude HR (95% CI) for COVID-19 hospitalisation in glucocorticoid treated compared with non-glucocorticoid treated | 1.06 (0.30 to 3.72) | 1 (Reference) |
| Adjusted * HR (95% CI) for COVID-19 hospitalisation in hydroxychloroquine treated compared with non-hydroxychloroquine treated | 0.60 (0.19 to 1.87) | 1 (Reference) |
| Adjusted * HR (95% CI) for COVID-19 hospitalisation in glucocorticoid treated compared with non-glucocorticoid treated | 1.12 (0.32 to 3.96) | 1 (Reference) |

*N*: numbers, OR: odds ratio, CI: confidence interval. * Matched on age (3-year intervals), sex, time at risk, and history of lupus nephritis/yes/no), and model with both glucocorticoid (yes/no) and hydroxychloroquine treatment (yes/no).

## 4. Discussion

This nationwide study showed an increased risk of hospitalisation with COVID-19 for the 2533 SLE patients compared with the age- and sex-matched group from the background population. Among the SLE patients, there was no association between HCQ nor glucocorticoid treatment and the risk of being hospitalised with COVID-19.

Several case reports and small case series on SLE and COVID-19 have been published, but to date, only few cohort- or registry-based studies focusing on the incidence of hospitalisation of patients with SLE infected with COVID-19 exist.

To ensure sufficient data on COVID-19 in rheumatic patients, the Global Rheumatology Alliance established a COVID-19 register with support from the ACR and EULAR, allowing clinicians to register information on patients with rheumatic disease and COVID-19 [22]. In a report of the first 600 patients, there were 85 SLE patients registered. In unadjusted chi-squared analysis, differences in hospitalisation status by disease revealed that a higher percentage of the cohort who were hospitalised had SLE (17%) versus those who were not hospitalised (11%). However, given the mechanism of collection of the case information and the mix of rheumatic diseases in the cohort, it is impossible to conclude if the SLE patients in that study were more likely to be admitted with COVID-19 than non-SLE individuals.

In a French study based on hospitalisation data of all inpatients during the first 6 months of the pandemic 1411 patients with SLE were hospitalised with COVID-19 [23]. Among these 17% needed treatment in intensive care unit and 9.5% died. Furthermore, severe infection in patients with SLE was associated with age, male gender, and comorbidities such as hypertension and chronic kidney disease. However, the population was not compared with age- and sex-matched individuals from the general population. In this study only few patients suffered from severe COVID-19 infection and cannot be concluded on.

In the beginning of the COVID-19 pandemic it was suggested that treatment with HCQ could have a prophylactic effect on infection with SARS-CoV2 based on in vitro studies [24]. However, large in vivo studies have since shown that routine treatment with HCQ in patients with rheumatic disease does not protect against infection nor hospitalisation with SARS-CoV2 [25,26] *per se*. Similarly, we found no association between routine treatment with HCQ and a lower likelihood of being admitted with COVID-19.

In a previous study, of 17 HCQ-treated patients with SLE, 71% were taking glucocorticoids, most of them below 10 mg prednisone equivalent, and 41% received other immunosuppressants [12]. Almost half of these patients were admitted with COVID-19 to intensive care and 2 of the 17 patients died. These findings along with others [6] have raised the question whether long-term glucocorticoid treatment might have played a role.

In the present study, having redeemed a prescription for glucocorticoid in the 6 months leading up to the index date for cases and controls was not associated with increased risk of hospital admission in SLE patients. We recognise that the absolute number of patients treated with HCQ and glucocorticoids in the present study is too low for any clear association to be identified, but overall, there was no indication of either protective or harmful effects of HCQ and glucocorticoid treatment in the SLE group.

Strengths: The main strengths of the present study are the complete follow-up of a nationwide cohort of patients with SLE using a validated identification algorithm for SLE in combination with the use of registers with high degree of completeness. Additionally, the overall high positive predictive value of 99% (95% CI 99–100) for COVID-19 hospital diagnosis, which was consistently high among all subgroups of sex, age groups, and calendar period, is another important strength [20]. Lastly, the ability to create an age- and sex-matched comparator group is a strength of the present study, as one of the most consistent problems in the existing literature on SLE and COVID-19 has been the lack of non-rheumatic comparator groups or insufficient age- and sex-standardised rates and proportions when comparing with the general population.

Limitations: It is possible that patients with SLE are admitted with a lower threshold than people from the general population due to their immunosuppressed state, although we believe this to be less likely due to the outcome definition that excluded patients who stayed <24 h in-hospital. Rather, we may have underestimated the true risk of hospitalisation related to COVID-19 infection in patients with SLE, as these patients potentially exhibit more behavioural social distancing precautions compared with the general population and therefore are less likely to contract COVID-19. Indeed, a Danish study reported that patients with inflammatory rheumatic diseases self-isolated more than others of the same age, and it is likely to be the same for patients with SLE [27]. Using ICD-10 codes for case definition could potentially be a limitation; however, as mentioned, the ICD-10 codes for SLE and COVID-19 in the DNPR have been validated with high PPVs [17,20]. Also, information regarding serologic parameters and disease activity could have been of interest to study predictors of hospitalisation and severe outcome, yet such information could not be collected.

Another limitation is that our dataset was rather small in terms of outcomes (hospitalisations) among SLE patients, especially when it came to specific treatment types for SLE. Furthermore, no information on kidney and lung function was available, and nor could we collect specific information regarding the prothrombotic state that patients with SLE experience due to antiphospholipid syndromes and other potential SLE-intrinsic prothrombotic risk factors [28]. It cannot be ruled out that the increased incidence of hospitalisation in SLE patients could to some extent be explained by either of these factors.

## 5. Conclusions

In this unselected nationwide cohort of Danish SLE patients, there was an approximately threefold increased incidence of hospitalisation with COVID-19 compared with age- and sex-matched controls from the general population after adjustment for several confounders. There was no obvious impact on the risk of hospitalisation associated with glucocorticoid nor HCQ treatment in this cohort, but the number of hospitalisations was too low to draw any definite conclusion, and we encourage further investigations into whether SLE in itself, as well as specific SLE treatment modalities, pose any modulation of risk for both hospitalisation and/or developing poor subsequent outcomes of COVID-19 infection.

**Supplementary Materials:** The following are available online at https://www.mdpi.com/article/10.3390/jcm10173842/s1, Table S1: Definition of comorbidities.

**Author Contributions:** Conceptualization, R.C., S.K., K.D., J.L. and L.D.; data curation, R.C., J.L. and C.T.-P.; formal analysis, R.C. and K.D.; funding acquisition, C.T.-P. and L.D.; investigation, R.C., S.K., L.P.H.D., R.W. and L.D.; methodology, R.C., S.K., R.W., K.D., J.L., C.T.-P. and L.D.; project administration, C.T.-P. and L.D.; resources, C.T.-P.; software, K.D. and J.L.; supervision, C.T.-P. and L.D.; validation, L.P.H.D.; visualization, R.C., S.K. and J.L.; writing—original draft, R.C., S.K. and L.P.H.D.; writing—review & editing, R.C., S.K., R.W., K.D., J.L., C.T.-P. and L.D. All authors have read and agreed to the published version of the manuscript.

**Funding:** The study was sponsored by Aalborg University Hospital, Denmark. The funder had no role in the design and conduct of the study; collection, management, analysis, and interpretation of the data; preparation, review, or approval of the manuscript; and decision to submit the manuscript for publication.

**Institutional Review Board Statement:** According to Danish legislation, no ethics approval is needed for register-based studies. The study has been approved by the Data Protection Committee of Northern Jutland, Denmark (2020-032).

**Informed Consent Statement:** Not applicable.

**Data Availability Statement:** According to Danish legislation, none of the original data can be shared.

**Conflicts of Interest:** R.C., S.K., L.P.H.D., K.D., R.W. and J.L. have no disclosures or conflicts of interests. C.T.-P. has received grants for studies from Bayer and Novo Nordisk not relevant to the current study. L.D. has received research grant/research support from BMS, and speakers' bureau from Eli Lilly and Galderma. The funders had no role in the design of the study; in the collection, analyses, or interpretation of data; in the writing of the manuscript, or in the decision to publish the results.

## References

1. Danza, A.; Ruiz-Irastorza, G. Infection risk in systemic lupus erythematosus patients: Susceptibility factors and preventive strategies. *Lupus* **2013**, *22*, 1286–1294. [CrossRef] [PubMed]
2. Cassione, E.B.; Zanframundo, G.; Biglia, A.; Codullo, V.; Montecucco, C.; Cavagna, L. COVID-19 infection in a northern-Italian cohort of systemic lupus erythematosus assessed by telemedicine. *Ann. Rheum. Dis.* **2020**, *79*, 1382–1383. [CrossRef]
3. Cordtz, R.; Lindhardsen, J.; Soussi, B.G.; Vela, J.; Uhrenholt, L.; Westermann, R.; Kristensen, S.; Nielsen, H.; Torp-Pedersen, C.; Dreyer, L. Incidence and severeness of COVID-19 hospitalization in patients with inflammatory rheumatic disease: A nationwide cohort study from Denmark. *Rheumatology* **2020**, keaa897. [CrossRef] [PubMed]

4.   Ramirez, G.A.; Gerosa, M.; Beretta, L.; Bellocchi, C.; Argolini, L.M.; Moroni, L.; Della Torre, E.; Artusi, C.; Nicolosi, S.; Caporali, R.; et al. COVID-19 in systemic lupus erythematosus: Data from a survey on 417 patients. *Semin. Arthritis Rheum.* **2020**, *50*, 1150–1157. [CrossRef]

5.   Ning, R.; Meng, S.; Tang, F.; Yu, C.; Xu, D.; Luo, X.; Sun, H. A case of SLE with COVID-19 and multiple infections. *Open Med.* **2020**, *15*, 1054–1060. [CrossRef] [PubMed]

6.   Horisberger, A.; Moi, L.; Ribi, C.; Comte, D. Impact of COVID-19 pandemic on SLE: Beyond the risk of infection. *Lupus Sci. Med.* **2020**, *7*, e000408. [CrossRef] [PubMed]

7.   Sookaromdee, P.; Wiwanitkit, V. Hydroxychloroquine, TTP, COVID-19, and SLE. *Turk. J. Hematol.* **2021**, *38*, 99–100. [CrossRef]

8.   Goyal, M. SLE patients are not immune to covid-19: Importance of sending the right message across. *Ann. Rheum. Dis.* **2021**, *80*, e23. [CrossRef] [PubMed]

9.   Zurita, M.F.; Arreaga, A.I.; Chavez, A.A.L.; Zurita, L. SARS-CoV-2 Infection and COVID-19 in 5 Patients in Ecuador after Prior Treatment with Hydroxychloroquine for Systemic Lupus Erythematosus. *Am. J. Case Rep.* **2020**, *21*, 1–5. [CrossRef]

10.  Konig, M.F.; Kim, A.H.; Scheetz, M.H.; Graef, E.R.; Liew, J.W.; Simard, J.; Machado, P.M.; Gianfrancesco, M.; Yazdany, J.; Langguth, D.; et al. Baseline use of hydroxychloroquine in systemic lupus erythematosus does not preclude SARS-CoV-2 infection and severe COVID-19. *Ann. Rheum. Dis.* **2020**, *79*, 1386–1388. [CrossRef]

11.  Wallace, B.; Washer, L.; Marder, W.; Kahlenberg, J.M. Patients with lupus with COVID-19: University of Michigan experience. *Ann. Rheum. Dis.* **2021**, *80*, e35. [CrossRef]

12.  Mathian, A.; Mahevas, M.; Rohmer, J.; Roumier, M.; Cohen-Aubart, F.; Amador-Borrero, B.; Barrelet, A.; Chauvet, C.; Chazal, T.; Delahousse, M.; et al. Clinical course of coronavirus disease 2019 (COVID-19) in a series of 17 patients with systemic lupus erythematosus under long-term treatment with hydroxychloroquine. *Ann. Rheum. Dis.* **2020**, *79*, 837–839. [CrossRef] [PubMed]

13.  Chuah, S.L.; Teh, C.L.; Akbar, S.A.W.M.; Cheong, Y.K.; Singh, B.S.M. Impact of COVID-19 pandemic on hospitalisation of patients with systemic lupus erythematosus (SLE): Report from a tertiary hospital during the peak of the pandemic. *Ann. Rheum. Dis.* **2020**, annrheumdis-2020-218475. [CrossRef]

14.  Smeele, H.T.; Perez-Garcia, L.F.; Grimminck, K.; Schoenmakers, S.; Mulders, A.G.; Dolhain, R.J. Systemic lupus erythematosus and COVID-19 during pregnancy. *Lupus* **2021**, *30*, 1188–1191. [CrossRef] [PubMed]

15.  Thanou, A.; Sawalha, A.H. SARS-CoV-2 and Systemic Lupus Erythematosus. *Curr. Rheumatol. Rep.* **2021**, *23*, 1–8. [CrossRef] [PubMed]

16.  Schmidt, M.; Pedersen, L.; Sørensen, H.T. The Danish Civil Registration System as a tool in epidemiology. *Eur. J. Epidemiol.* **2014**, *29*, 541–549. [CrossRef] [PubMed]

17.  Hermansen, M.-L.F.; Lindhardsen, J.; Torp-Pedersen, C.; Faurschou, M.; Jacobsen, S. Incidence of Systemic Lupus Erythematosus and Lupus Nephritis in Denmark: A Nationwide Cohort Study. *J. Rheumatol.* **2016**, *43*, 1335–1339. [CrossRef] [PubMed]

18.  Schmidt, M.; Schmidt, S.A.J.; Sandegaard, J.L.; Ehrenstein, V.; Pedersen, L.; Sørensen, H.T. The Danish National Patient Registry: A review of content, data quality, and research potential. *Clin. Epidemiol.* **2015**, *7*, 449–490. [CrossRef] [PubMed]

19.  Johannesdottir, S.A.; Horváth-Puhó, E.; Ehrenstein, V.; Schmidt, M.; Pedersen, L.; Sørensen, H. Existing data sources for clinical epidemiology: The Danish National Database of Reimbursed Prescriptions. *Clin. Epidemiol.* **2012**, *4*, 303–313. [CrossRef]

20.  Bodilsen, J.; Leth, S.; Nielsen, S.L.; Holler, J.G.; Benfield, T.; Omland, L.H. Positive Predictive Value of ICD-10 Diagnosis Codes for COVID-19. *Clin. Epidemiol.* **2021**, *13*, 367–372. [CrossRef]

21.  Fosbøl, E.L.; Butt, J.H.; Østergaard, L.; Andersson, C.; Selmer, C.; Kragholm, K.; Schou, M.; Phelps, M.D.; Gislason, G.H.; Gerds, T.A.; et al. Association of Angiotensin-Converting Enzyme Inhibitor or Angiotensin Receptor Blocker Use With COVID-19 Diagnosis and Mortality. *JAMA-J. Am. Med. Assoc.* **2020**, *324*, 168–177. [CrossRef]

22.  Gianfrancesco, M.; Hyrich, K.L.; Al-Adely, S.; Carmona, L.; Danila, M.I.; Gossec, L.; Izadi, Z.; Jacobsohn, L.; Katz, P.; Lawson-Tovey, S.; et al. Characteristics associated with hospitalisation for COVID-19 in people with rheumatic disease: Data from the COVID-19 Global Rheumatology Alliance physician-reported registry. *Ann. Rheum. Dis.* **2020**, *79*, 859–866. [CrossRef] [PubMed]

23.  Mageau, A.; Aldebert, G.; Van Gysel, D.; Papo, T.; Timsit, J.-F.; Sacre, K. SARS-CoV-2 infection among inpatients with systemic lupus erythematosus in France: A nationwide epidemiological study. *Ann. Rheum. Dis.* **2021**, *80*. [CrossRef] [PubMed]

24.  Liu, J.; Cao, R.; Xu, M.; Wang, X.; Zhang, H.; Hu, H.; Li, Y.; Hu, Z.; Zhong, W.; Wang, M. Hydroxychloroquine, a less toxic derivative of chloroquine, is effective in inhibiting SARS-CoV-2 infection in vitro. *Cell Discov.* **2020**, *6*, 1–4. [CrossRef] [PubMed]

25.  Geleris, J.; Sun, Y.; Platt, J.; Zucker, J.; Baldwin, M.; Hripcsak, G.; Labella, A.; Manson, D.K.; Kubin, C.; Barr, R.G.; et al. Observational Study of Hydroxychloroquine in Hospitalized Patients with Covid-19. *N. Engl. J. Med.* **2020**, *382*, 2411–2418. [CrossRef]

26.  Rosenberg, E.S.; Dufort, E.M.; Udo, T.; Wilberschied, L.A.; Kumar, J.; Tesoriero, J.; Weinberg, P.; Kirkwood, J.; Muse, A.; DeHovitz, J.; et al. Association of Treatment With Hydroxychloroquine or Azithromycin With In-Hospital Mortality in Patients With COVID-19 in New York State. *JAMA* **2020**, *323*, 2493–2502. [CrossRef]

27. Glintborg, B.; Jensen, D.V.; Engel, S.; Terslev, L.; Jensen, M.P.; Hendricks, O.; Østergaard, M.; Rasmussen, S.H.; Adelsten, T.; Colic, A.; et al. Self-protection strategies and health behaviour in patients with inflammatory rheumatic diseases during the COVID-19 pandemic: Results and predictors in more than 12 000 patients with inflammatory rheumatic diseases followed in the Danish DANBIO registry. *RMD Open* **2021**, *7*, e001505. [CrossRef]
28. Manzano, E.M.; Fernández-Bello, I.; Sanz, R.J.; Marhuenda, Á.R.; López-Longo, F.J.; Acuña, P.; Román, M.T.Á.; Yuste, V.J.; Butta, N.V. Insights into the Procoagulant Profile of Patients with Systemic Lupus Erythematosus without Antiphospholipid Antibodies. *J. Clin. Med.* **2020**, *9*, 3297. [CrossRef]

Journal of
*Clinical Medicine*

*Article*

# Usefulness of Clinical and Laboratory Criteria for Diagnosing Autoimmune Liver Disease among Patients with Systemic Lupus Erythematosus: An Observational Study

Rebecca Heijke [1], Awais Ahmad [2], Martina Frodlund [3], Lina Wirestam [3], Örjan Dahlström [4], Charlotte Dahle [2], Stergios Kechagias [5] and Christopher Sjöwall [3,*]

[1]  Clinic of Internal Medicine, Region Jönköping County, SE-553 05 Jönköping, Sweden; rebecca.heijke@gmail.com
[2]  Department of Biomedical and Clinical Sciences, Division of Inflammation and Infection/Clinical Immunology, Linköping University, SE-581 85 Linköping, Sweden; awais.ahmad@regionostergotland.se (A.A.); charlotte.dahle@regionostergotland.se (C.D.)
[3]  Department of Biomedical and Clinical Sciences, Division of Inflammation and Infection/Rheumatology, Linköping University, SE-581 85 Linköping, Sweden; martina.frodlund@regionostergotland.se (M.F.); lina.wirestam@liu.se (L.W.)
[4]  Department of Behavioural Sciences and Learning, Swedish Institute for Disability Research, Linköping University, SE-581 85 Linköping, Sweden; orjan.dahlstrom@liu.se
[5]  Department of Health, Division of Diagnostics and Specialist Medicine/Gastroenterology and Hepatology, Medicine and Caring Sciences, Linköping University, SE-581 85 Linköping, Sweden; stergios.kechagias@liu.se
*   Correspondence: christopher.sjowall@liu.se; Tel.: +46-10-1032416

**Citation:** Heijke, R.; Ahmad, A.; Frodlund, M.; Wirestam, L.; Dahlström, Ö.; Dahle, C.; Kechagias, S.; Sjöwall, C. Usefulness of Clinical and Laboratory Criteria for Diagnosing Autoimmune Liver Disease among Patients with Systemic Lupus Erythematosus: An Observational Study. *J. Clin. Med.* **2021**, *10*, 3820. https://doi.org/10.3390/jcm10173820

Academic Editor: Angelo Valerio Marzano

Received: 10 June 2021
Accepted: 18 August 2021
Published: 26 August 2021

**Publisher's Note:** MDPI stays neutral with regard to jurisdictional claims in published maps and institutional affiliations.



**Abstract:** Abnormal liver function tests are frequently observed during follow-up of patients with systemic lupus erythematosus (SLE) but data on co-existence with autoimmune liver diseases (AILD) are scarce. This retrospective study aimed to describe the prevalence of autoimmune hepatitis (AIH) and primary biliary cholangitis (PBC) among well-characterized subjects with SLE. We also evaluated whether the presence of autoantibodies to complement protein 1q (C1q) and/or ribosomal P protein (anti-ribP) are, directly or inversely, associated with AIH, as proposed in some reports. The number of screened patients was 287 (86% females), and all cases were included in a regional Swedish cohort. Each subject of the study population met the 1982 American College of Rheumatology classification criteria and/or the Fries' diagnostic principle. By applying the simplified diagnostic AIH criteria combined with persistent transaminasemia, 40 (13.9%) cases reached at least "probable AIH". However, merely 8 of these had been diagnosed with AIH (overall AIH prevalence 2.8%). Neither anti-C1q nor anti-ribP associated significantly with AIH. By applying the recent PBC guidelines, 6 (2.1%) cases were found, but only 3 of them had actually been diagnosed with PBC and one additional subject was not identified by the guidelines (overall PBC prevalence 1.4%). Compared to prevalence data from the general Swedish population, both AIH and PBC were highly overrepresented in our study population. The sensitivity of the diagnostic AIH criteria was impeccable but the specificity was less impressive, mainly due to positive ANA and hypergammaglobulinemia. Based on our findings, among subjects with SLE, the AIH criteria are less useful and liver biopsy combined with detection of other AILD-associated autoantibodies should be performed.

**Keywords:** abnormal liver function tests; autoimmune liver diseases; autoimmune hepatitis; hepatic involvement; liver biopsy; primary biliary cholangitis; systemic lupus erythematosus

## 1. Introduction

Although involvement of joints, skin, mucous membranes, serosa and kidneys is common among patients with systemic lupus erythematosus (SLE) virtually any organ system may be affected. However, hepatic involvement has not been considered a primary organ manifestation in SLE as it is not included in any of the commonly used and recently updated classification criteria [1,2]. Nevertheless, the British Isles Lupus Assessment Group's (BILAG) disease activity index includes 'lupus hepatitis' as a separate item but liver disease is not reflected in the more widely used SLE disease activity score 2000 (SLEDAI-2K) [1,3–6]. Still, abnormal liver function tests (LFTs) at any time-point are common in patients with SLE; reported numbers range from 9–60% depending on study population and limitations applied for abnormal values [7–11]. Potential causes of abnormal LFTs in SLE are numerous and include drug-induced liver injury (DILI), steatosis, viral hepatitis, vascular thrombosis and autoimmune liver disease (AILD). However, clinically significant AILD associated with SLE has been reported to be rare, and hepatic liver involvement appears to have a limited influence on mortality [11–16].

Observations of 'lupus-associated hepatitis' or 'lupus hepatitis' usually refer to an asymptomatic transaminasemia, consistent with SLE disease activity [9,17], which normalizes during glucocorticoid treatment [17]. Lupus hepatitis has been reported in 3–9% of patients [9,18,19]. The histopathological findings of lupus hepatitis are variable and non-specific, but mild portal inflammatory infiltrate, lobular necrosis and fatty infiltration are frequently found [9]. One study reported that intense deposits of complement protein 1q (C1q) were found in the majority of cases with lupus hepatitis, but unfortunately neither circulating C1q levels or anti-C1q antibodies—which often parallel SLE disease activity—were investigated [20–23]. Presence of autoantibodies to ribosomal P protein (anti-ribP) has been reported to associate with lupus hepatitis or autoimmune hepatitis (AIH), but contradictory results have also been published [17,18,24,25].

Whereas lupus hepatitis has been considered as a manifestation of SLE, AIH is regarded as a separate disease. However, the two conditions often share several features, including hypergammaglobulinemia, arthralgia and presence of antinuclear antibodies (ANA) [26]. Similarly to SLE, AIH also has a female predominance. The histopathology is characterized by progressive hepatocellular necrosis and inflammation, which untreated may lead to cirrhosis and end-stage liver disease. Although AIH prevalence data are uncertain, epidemiological studies from Scandinavia, Spain and New Zealand have estimated a prevalence of 12–25 per 100,000 inhabitants [27–30].

Data on co-existence of SLE and AIH are scarce [19]. Differential diagnostics may be challenging since elevated LFTs, hypergammaglobulinemia, a positive ANA test and response to glucocorticoid treatment are characteristic of both conditions. Efe et al. found that up to two thirds of SLE patients with abnormal LFTs fulfilled the simplified AIH criteria whereas only approximately 14% had histopathology compatible with AIH [7]. This illustrates that a liver biopsy is often necessary for a definitive diagnosis of AIH among patients with SLE [31].

Primary biliary cholangitis (PBC) constitutes another AILD with female predominance, which is chronic, often progresses, and can result in end-stage liver disease [32]. The prevalence of PBC in Sweden has been estimated to be approximately 15 per 100,000 inhabitants [33]. PBC is typically associated with the presence of anti-mitochondrial antibodies of M2 type (AMA-M2), which, in addition to persistently elevated serum alkaline phosphatase (ALP) and liver histology consistent with PBC, constitutes one of the three diagnostic criteria [34]. In addition, other subtypes of ANA, such as anti-speckled 100-kDa (Sp100), anti-promyelocytic leukemia protein (PML) and anti-glycoprotein 210-kDa (gp210), are also strongly associated with PBC [35]. Whereas the association of PBC with primary Sjögren's syndrome (SS) is well documented [36,37], co-existence with SLE has been reported as rare [38]. In contrast, however, we recently showed that PBC-associated autoantibodies in SLE are relatively common [39].

The primary aim of the present study was to describe the prevalence of AILD in well-characterized Swedish SLE patients from a tertiary referral center. This was done with support from an experienced hepatologist (S.K.) using different grounds for the diagnoses of AIH and PBC. We further aimed to test whether the presence of anti-C1q and/or anti-ribP antibodies were associated with AIH.

## 2. Materials and Methods

### 2.1. Study Population

The study population consisted of 287 patients (248 women, 39 men) diagnosed with SLE (detailed in Table 1). All subjects had been included in the prospective and observational research program *Clinical Lupus Register in North-Eastern Gothia* (Swedish acronym 'KLURING') at the Rheumatology unit, University Hospital in Linköping [40]. Most (284 of 287 (99%)) patients met the Fries' diagnostic principle and 243 of 287 (84.7%) fulfilled the 1982 American College of Rheumatology (ACR) classification criteria (mean number of fulfilled ACR criteria was 4.8, range 3–9) [5,41]. The study population corresponds to virtually all prevalent and incident adult SLE cases in the catchment area of Region Östergötland (approximately 365,000 adult inhabitants) between September 2008 and May 2020. The medical records of all patients were retrospectively reviewed with a focus on AILD.

**Table 1.** Characteristics of the included patients with SLE. Patients with confirmed AILD are shown in separate columns.

| Background Variables | Total, $n$ = 287 | AIH, $n$ = 8 | PBC, $n$ = 4 |
|---|---|---|---|
| Females, $n$ (%) | 248 (86.4) | 7 (87.5) | 3 (75.0) |
| Age at cohort inclusion, mean years (range) | 49.0 (20–86) | 47.8 (22–70) | 53.2 (28–69) |
| SLE duration at cohort inclusion, mean years (range) | 9.4 (0–45) | 8.0 (0–19) | 11.0 (1–25) |
| Caucasian ethnicity, $n$ (%) | 255 (88.8) | 7 (87.5) | 4 (100.0) |
| Ever smoker (former or current), $n$ (%) | 121 (42.1) | 5 (62.5) | 2 (50.0) |
| Disease variables | | | |
| Secondary Sjögren's syndrome (defined by classification [#]), $n$ (%) | 65 (22.6) | 2 (25.0) | 2 (50.0) |
| Antiphospholipid syndrome (defined by classification [§]), $n$ (%) | 56 (19.5) | 1 (12.5) | 0 (0) |
| Patients meeting ≥4 ACR-82 criteria, $n$ (%) | 243 (84.7) | 6 (75.0) | 3 (75.0) |
| Number of fulfilled ACR-82 criteria, mean (range) | 4.8 (3–9) | 4.8 (3–9) | 3.8 (3–4) |
| Clinical phenotypes (ACR-82 defined), $n$ (%) | | | |
| (1) Malar rash | 114 (39.7) | 4 (50.0) | 0 (0) |
| (2) Discoid rash | 43 (15.0) | 0 (0) | 0 (0) |
| (3) Photosensitivity | 147 (51.2) | 4 (50.0) | 3 (75.0) |
| (4) Oral ulcers | 34 (11.8) | 1 (12.5) | 1 (25.0) |
| (5) Arthritis | 220 (76.7) | 6 (75.0) | 3 (75.0) |
| (6) Serositis | 110 (38.3) | 3 (37.5) | 3 (75.0) |
| Pleuritis | 106 (36.9) | 3 (37.5) | 2 (50.0) |
| Pericarditis | 98 (34.1) | 2 (25.0) | 2 (50.0) |
| (7) Renal disorder | 80 (27.9) | 2 (25.0) | 0 (0) |
| (8) Neurologic disorder | 16 (5.6) | 0 (0) | 0 (0) |
| Seizures | 15 (5.2) | 0 (0) | 0 (0) |
| Psychosis | 3 (1.0) | 0 (0) | 0 (0) |
| (9) Hematologic disorder | 174 (60.6) | 5 (62.5) | 0 (0) |
| Hemolytic anemia | 12 (4.2) | 0 (0) | 0 (0) |
| Leukocytopenia | 86 (30.0) | 2 (25.0) | 0 (0) |
| Lymphopenia | 112 (39.0) | 4 (50.0) | 0 (0) |
| Thrombocytopenia | 31 (10.8) | 0 (0) | 0 (0) |
| (10) Immunological disorder | 152 (53.0) | 4 (50.0) | 1 (25.0) |
| Anti-dsDNA antibody (anti-dsDNA) | 139 (48.4) | 4 (50.0) | 1 (25.0) |
| Anti-Smith antibody (anti-Sm) | 19 (6.6) | 1 (12.5) | 0 (0) |
| (11) Antinuclear antibody (IF-ANA) * | 284 (99.0) | 8 (100) | 4 (100) |

* Positive by immunofluorescence (IF) microscopy. [#] According to Vitali C, et al. [42] [§] According to Miyakis S, et al. [43].

### 2.2. Data Collection

AILD diagnoses, attributed by gastroenterologists among the 287 SLE cases, were retrieved from medical records, reviewed by a hepatologist (S.K.) and considered as 'golden standard'. Furthermore, we applied the simplified diagnostic AIH criteria and the recent PBC guidelines from the American Association for the Study of Liver Diseases and examined their diagnostic performance in our study population [44,45]. In addition, we recorded the presence of secondary SS (diagnosis confirmed by a rheumatologist and defined according to classification criteria) and antiphospholipid syndrome (APS) [42,43]. Data on liver biopsies and liver imaging were also retrieved.

### 2.3. Laboratory Analyses

All patients had undergone continuous monitoring of liver enzyme values as was previously described [39]. Levels of IgG (normal range 6.7–15 g/L) and IgM (0.27–2.1 g/L) in plasma were recorded. IgG-ANA was detected by indirect immunofluorescence (IF) microscopy on fixed HEp-2 cells and anti-ribP antibodies were analyzed by FIDIS™ Connective profile, Solonium software v 1.7.1.0 (Theradiag, Croissy-Beaubourg, France) at the Clinical Immunology laboratory, University Hospital in Linköping [46,47]. Anti-C1q antibodies were analyzed by ELISA [48]. Hepatitis B surface antigen (HBsAg) and antibodies against hepatitis C virus (anti-HCV) were assessed by routine methods at the Clinical Microbiology laboratory, University Hospital in Linköping. Autoantibodies associated with AILD were analyzed as previously described [39].

### 2.4. Statistics

Associations between laboratory variables (categorical) and AILD were examined with Fisher's exact test for significance and using Φ as a measure of association. $p$-values $\leq 0.05$ were considered statistically significant. Statistical analyses were performed using the SPSS software version 26.0.0.0 (SPSS Inc., Chicago, IL, USA). Sensitivity (proportion of subjects correctly identified with AILD), specificity (proportion of subjects correctly identified without AILD), accuracy (proportion of correctly classified subjects), positive predictive value (PPV; proportion of AILD-classified subjects that are true AILD) and negative predictive value (NPV; proportion of non–AILD-classified subjects that are true non-AILD), were calculated, including 95% confidence intervals (CI) using the Wilson score method.

### 2.5. Ethical Approval

Oral and written informed consent were obtained from all participants. The study protocol was approved by the regional ethics review board in Linköping (Decision number M75–08/2008).

### 3. Results

Of the 287 screened patients, 182 (63.4%) had occasional or persistent elevations of aspartate aminotransferase (AST), alanine transaminase (ALT), alkaline phosphatase (ALP), and/or γ-glutamyl transferase (GGT) during follow-up. Less than 10% had elevated LFTs for >3 months. All subjects with confirmed AILD ($n$ = 12) were found among the patients with elevated LFTs. In the subgroup with elevated LFTs ($n$ = 182), 4.4% had a confirmed diagnosis of AIH and 2.2% of PBC.

As illustrated in Figure 1A, we applied the diagnostic AIH criteria to the study population [44]. The prevalence of AIH in the entire study population was 2.8% ($n$ = 8); and establishment of the diagnosis included liver biopsy in 6 of 8 cases. The sensitivity and the NPV of the AIH criteria was high, but the specificity and PPV were lower (Table 2). Among the entire study population, 102 of 226 (45.1%) had hypergammaglobulinemia at least once. HBsAg and anti-HCV were absent in 77 of 102 (75.5%) and IF-ANA was positive (titer > 1:80) in 284 of 287 (99.0%) patients. According to the AIH criteria, 46 (16.0%) reached at least "probable AIH", even without histopathological evaluation, whereof 40 cases had a history of elevated LFTs (AST and/or ALT). Of note, plasma IgG had not been measured in

61 subjects, which clearly limited the possibility of reaching "probable AIH". However, a closer review of those reaching "probable AIH" revealed a different explanation than AIH in a majority of cases with elevated LFTs. Still, the presence of hypergammaglobulinemia ($\geq$16 g/L) at any time-point during follow-up was significantly associated with a confirmed diagnosis of AIH ($\Phi$ = 0.16, $p$ = 0.015).

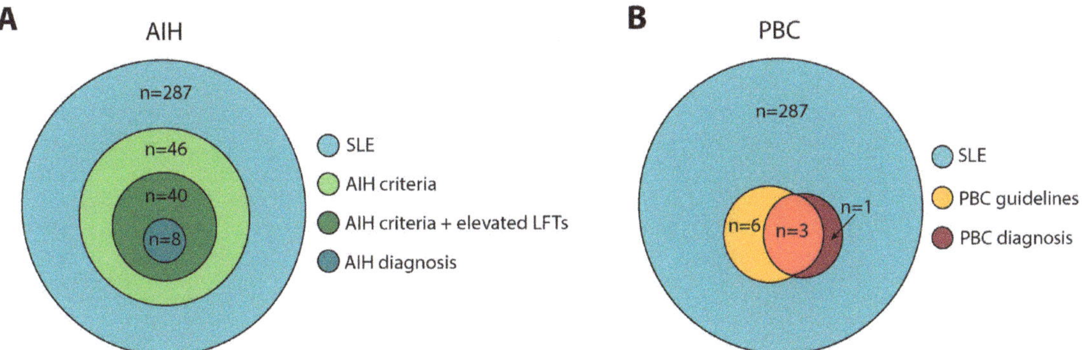


**Figure 1.** Diagrams illustrating the performance of the diagnostic AIH criteria (**A**) and PBC guidelines (**B**) among our study population. AIH = autoimmune hepatitis; SLE = systemic lupus erythematosus; PBC = primary biliary cholangitis; LFTs = liver function tests.

**Table 2.** Performance of the simplified diagnostic AIH criteria and the recent PBC guidelines [44,45] in the study population to identify SLE patients with confirmed AILD. Cut-off for the AIH criteria was set at $\geq$6 points (representing "probable AIH"). 95% confidence intervals are shown in parentheses.

|  | **AIH** | **PBC** |
| --- | --- | --- |
| Sensitivity | 1.00 (0.60–1.00) | 0.75 (0.28–0.97) |
| Specificity | 0.86 (0.82–0.90) | 0.99 (0.97–1.00) |
| Accuracy | 0.87 (0.82–0.90) | 0.99 (0.96–1.00) |
| PPV | 0.17 (0.09–0.31) | 0.50 (0.19–0.81) |
| NPV | 1.00 (0.98–1.00) | 1.00 (0.98–1.00) |

PPV = Positive Predictive Value; NPV = Negative Predictive Value; AIH = autoimmune hepatitis; PBC = primary biliary cholangitis.

Only 4 subjects had a confirmed diagnosis of PBC (1.4%), which included histopathological evaluation in 2 of 4. As shown in Figure 1B, we used the recent diagnostic guidelines for PBC by requiring elevation of ALP in combination with presence of typical PBC-associated antibodies (AMA-M2/Sp100/gp210) [45]. This procedure yielded 6 cases (2.1%). The specificity and the NPV of the guidelines for PBC were impressive, but the sensitivity and PPV among our study population were lower (Table 2).

Anti-C1q antibodies were detected in 60 of 260 subjects (23.1%), but the presence of anti-C1q did not associate significantly with AIH. Anti-ribP antibodies were detected in 17 of 247 patients (6.9%), but in none of those with a confirmed diagnosis of AIH. Altogether, 9 of 17 (52.9%) anti-ribP positive individuals had a history of elevated LFTs, which did not significantly differ from anti-ribP negative subjects. SLE with secondary SS did not associate with either AIH or PBC, although a non-significant trend was observed for the latter ($\Phi$ = 0.09, $p$ = 0.16). APS was found in one patient with AIH (12.5%), but in none with PBC.

## 4. Discussion

Continuous supervision of LFTs is mandatory for patients using disease-modifying anti-rheumatic drugs (DMARDs), such as methotrexate, azathioprine, mycophenolate mofetil, leflunomide or cyclosporine. Abnormal LFTs in the absence of liver-related symptoms are very common during follow-up of patients with SLE. The cause may be obvious but is most often uncertain. Occasionally, the LFTs normalize without specific interventions or with a minor increase of the daily glucocorticoid dose. However, LFTs can also be persistently elevated, which often leads to both further investigations and interruption of a needed and well-functioning DMARD treatment. These cases may be challenging and often require interdisciplinary discussions between rheumatologists, hepatologists and immunologists.

In this study, we aimed to describe the prevalence of AILD among well-characterized SLE patients from a tertiary referral center. We used a previously described cohort in which we had the possibility to longitudinally monitor LFTs, liver imaging, biopsies, concomitant diagnoses, drugs as well as additional laboratory data, including results from autoantibody testing [39,40]. All data were discussed and evaluated by specialists in rheumatology (C.S.), hepatology (S.K.) and immunology (C.D.). The prevalence rates achieved in our study population for established AIH (2.8% among all, and 4.4% among those with elevated LFTs) and PBC (1.4% among all, and 2.2% among those with elevated LFTs) are in line with, or close to, what previously has been published. The recent review by González-Regueiro et al. mentions a prevalence of AIH of approximately 5–10% among SLE patients with abnormal LFTs [31]. For PBC, a lower prevalence (2.5–5%) has been observed [38].

To set these figures into a context, the AILD prevalence rates in our study population can be compared with prevalence data from the general Swedish population. Danielsson-Borssén et al. determined the point prevalence of AIH in 2009 to 17.3/100,000 inhabitants (0.17‰), which strongly contrasts to our finding of 2.8% in SLE [27]. Only older prevalence data from the time span 1973–1982 are available for PBC, when Danielsson et al. reported 15.1/100,000 inhabitants (0.15‰), which is considerably lower than the 1.4% found among patients with SLE herein [33]. Based on these findings, we conclude that both AIH and PBC are over-represented among patients with SLE. Similarly, AIH has been associated with an increased risk of developing systemic autoimmune diseases [49]. This is also in line with the overall empirical knowledge that "one autoimmune disease predisposes to another". Interestingly, the recent years' genetic advances have taught us that identical risk genes are shared by several different autoimmune conditions [50,51].

We also took the opportunity to challenge the simplified diagnostic AIH criteria and the recent guidelines for PBC diagnosis. The AIH criteria have previously been criticized for poor performance in SLE, especially with regard to low specificity [7]. However, to the best of our knowledge, their performance has not systematically been evaluated in a population of well-characterized SLE cases. Basically, our findings confirm the observation by Efe et al. and emphasize that liver biopsy is often needed for a definitive diagnosis of AIH among patients with SLE [31]. The PPV of reaching "probable AIH" herein was only 17%. However, plasma IgG $\geq$ 16 g/L associated significantly with confirmed AIH. This is of relevance as initially high levels of IgG have been associated with poor outcome in patients with combined SLE/AIH [52]. Although the sensitivity and PPV were mediocre, the recent diagnostic guidelines for PBC overall performed slightly better than the AIH criteria in our study population. This was mostly explained by the high diagnostic specificity for the PBC-associated ANA subtypes, compared to the non-specified ANA detected by IF microscopy, which is valid as one of the criteria for AIH.

In SLE, the high prevalence of ANA and hypergammaglobulinemia confers that the other diagnostic markers, i.e., SMA antibodies, antibodies against soluble liver antigen (SLA) or LKM, play a more important role for diagnosing AIH. These antibody specificities as well as PBC-specific ANA, such as Sp100, anti-PML and anti-gp210, in addition to AMA-M2, should therefore be included in the screening algorithm when AILD is suspected. However, in most of our cases reaching the level of "probable AIH", these autoantibody specificities had not been requested since their LFTs were not persistently elevated and the clinical suspicion of AILD had therefore not been raised. Furthermore, it is worth noting that for detection of SMA, a serum dilution that results in a cut-off corresponding to the 95th percentile of a healthy population should be used and must be evaluated by each laboratory. According to the diagnostic AIH criteria a serum dilution of 1:40 is recommended, but is actually based on experience from "the old days" when the quality of the microscopes were much lower than today's modern equipment [44,47].

Levels of the SLE-associated antibodies anti-C1q and anti-ribP are known to fluctuate over time [22,23,47]. An etiological role for anti-ribP in triggering both lupus hepatitis and AIH has been proposed [18,24,25]. Herein, we considered all patients once positive for anti-ribP or anti-C1q as positive and potential seroconversion over time was neglected. Still, we did not confirm the previously reported association between anti-ribP and AIH. However, according to a detailed review by Bessone et al., the association remains both uncertain and controversial [17]. This study has some limitations. It cannot be excluded that the actual prevalence of AILD in SLE is even greater than what we have estimated here [13]. Guided examination of AILD is usually driven by elevated LFTs, usually over a longer period of time, as the risk of progression to cirrhosis is associated with raised LFTs [53,54]. However, in our study population, none of the cases who reached "probable AIH" had persistently elevated ($\geq$6 months) LFTs. Otherwise, the latter would normally have resulted in a liver biopsy. Nevertheless, from an international perspective, it is worthwhile underlining that the Nordic reference limits regarding AST and ALT are unusually high [55]. Thus, based on the liberal LFT reference intervals we apply in Sweden, it is not impossible that subclinical AILD to some extent may pass under the radar. Another limitation was the ethnic composition of the study population—almost 90% of the enrolled patients were of Caucasian origin. Ethnicity is indeed known to affect both disease severity and manifestations of SLE [56]. Thus, extrapolation of our results to other populations should be done with caution. In contrast, the study has several strengths. For instance, the fact that Swedish health care is universally available to all residents significantly reduces the risk of selection bias and ensures a high coverage of cases. The well-characterized cohort of SLE patients, longitudinally followed by a limited number of experienced rheumatologists at a single tertiary referral center, constitutes another advantage [40,47].

## 5. Conclusions

In conclusion, both AIH and PBC are overrepresented in SLE. When the diagnostic AIH criteria were applied, even higher numbers were achieved, but the use of these criteria cannot be recommended in patients with SLE. Instead, liver biopsy and detection of autoantibodies with higher diagnostic specificity for AILD could aid in the search (particularly for AIH) among individuals with SLE.

**Author Contributions:** Conceptualization, R.H., S.K. and C.S.; methodology, R.H., Ö.D., C.D., S.K. and C.S.; R.H., A.A., L.W. and C.S.; software, R.H., Ö.D., L.W., S.K. and C.S.; validation, C.D., S.K. and C.S.; formal analysis, R.H., Ö.D. and C.S.; investigation, R.H., M.F. and C.S.; resources, S.K. and C.S.; data curation, R.H. and C.S.; writing—original draft preparation, C.S.; writing—review and editing, R.H., A.A., M.F., L.W., Ö.D., C.D., S.K. and C.S.; visualization, L.W.; supervision, C.D., S.K. and C.S.; project administration, C.S.; funding acquisition, S.K. and C.S. All authors have read and agreed to the published version of the manuscript.

**Funding:** This work was supported by grants from the Medical Research Council of South-Eastern Sweden (FORSS), the Swedish Rheumatism Association, the Region Östergötland (ALF Grants), the Gustafsson Foundation, the King Gustaf V's 80-year Anniversary foundation and the King Gustaf V and Queen Victoria's Freemasons foundation.

**Institutional Review Board Statement:** The study was conducted according to the guidelines of the Declaration of Helsinki, and approved by the regional ethics review board in Linköping (Decision number M75–08/2008).

**Informed Consent Statement:** Oral and written informed consent were obtained from all subjects involved in the study.

**Data Availability Statement:** Not applicable.

**Conflicts of Interest:** The authors declare no conflict of interest.

## References

1. Petri, M.; Orbai, A.M.; Alarcon, G.S.; Gordon, C.; Merrill, J.T.; Fortin, P.R.; Bruce, I.N.; Isenberg, D.; Wallace, D.J.; Nived, O.; et al. Derivation and validation of the Systemic Lupus International Collaborating Clinics classification criteria for systemic lupus erythematosus. *Arthritis Rheum.* **2012**, *64*, 2677–2686. [CrossRef]
2. Aringer, M.; Costenbader, K.; Daikh, D.; Brinks, R.; Mosca, M.; Ramsey-Goldman, R.; Smolen, J.S.; Wofsy, D.; Boumpas, D.T.; Kamen, D.L.; et al. 2019 European League Against Rheumatism/American College of Rheumatology classification criteria for systemic lupus erythematosus. *Arthritis Rheumatol.* **2019**, *71*, 1400–1412. [CrossRef] [PubMed]
3. Isenberg, D.A.; Rahman, A.; Allen, E.; Farewell, V.; Akil, M.; Bruce, I.N.; D'Cruz, D.; Griffiths, B.; Khamashta, M.; Maddison, P.; et al. BILAG 2004. Development and initial validation of an updated version of the British Isles Lupus Assessment Group's disease activity index for patients with systemic lupus erythematosus. *Rheumatology* **2005**, *44*, 902–906. [CrossRef] [PubMed]
4. Gladman, D.D.; Ibanez, D.; Urowitz, M.B. Systemic lupus erythematosus disease activity index 2000. *J. Rheumatol.* **2002**, *29*, 288–291. [PubMed]
5. Tan, E.M.; Cohen, A.S.; Fries, J.F.; Masi, A.T.; McShane, D.J.; Rothfield, N.F.; Schaller, J.G.; Talal, N.; Winchester, R.J. The 1982 revised criteria for the classification of systemic lupus erythematosus. *Arthritis Rheum.* **1982**, *25*, 1271–1277. [CrossRef] [PubMed]
6. Yee, C.S.; Cresswell, L.; Farewell, V.; Rahman, A.; Teh, L.S.; Griffiths, B.; Bruce, I.N.; Ahmad, Y.; Prabu, A.; Akil, M.; et al. Numerical scoring for the BILAG-2004 index. *Rheumatology* **2010**, *49*, 1665–1669. [CrossRef]
7. Efe, C.; Purnak, T.; Ozaslan, E.; Ozbalkan, Z.; Karaaslan, Y.; Altiparmak, E.; Muratori, P.; Wahlin, S. Autoimmune liver disease in patients with systemic lupus erythematosus: A retrospective analysis of 147 cases. *Scand. J. Gastroenterol.* **2011**, *46*, 732–737. [CrossRef]
8. Huang, D.; Aghdassi, E.; Su, J.; Mosko, J.; Hirschfield, G.M.; Gladman, D.D.; Urowitz, M.B.; Fortin, P.R. Prevalence and risk factors for liver biochemical abnormalities in Canadian patients with systemic lupus erythematosus. *J. Rheumatol.* **2012**, *39*, 254–261. [CrossRef]
9. Piga, M.; Vacca, A.; Porru, G.; Cauli, A.; Mathieu, A. Liver involvement in systemic lupus erythematosus: Incidence, clinical course and outcome of lupus hepatitis. *Clin. Exp. Rheumatol.* **2010**, *28*, 504–510.
10. Takahashi, A.; Abe, K.; Saito, R.; Iwadate, H.; Okai, K.; Katsushima, F.; Monoe, K.; Kanno, Y.; Saito, H.; Kobayashi, H.; et al. Liver dysfunction in patients with systemic lupus erythematosus. *Intern. Med.* **2013**, *52*, 1461–1465. [CrossRef]
11. Runyon, B.A.; LaBrecque, D.R.; Anuras, S. The spectrum of liver disease in systemic lupus erythematosus. Report of 33 histologically proved cases and review of the literature. *Am. J. Med.* **1980**, *69*, 187–194. [CrossRef]
12. Chowdhary, V.R.; Crowson, C.S.; Poterucha, J.J.; Moder, K.G. Liver involvement in systemic lupus erythematosus: Case review of 40 patients. *J. Rheumatol.* **2008**, *35*, 2159–2164. [CrossRef] [PubMed]
13. Gibson, T.; Myers, A.R. Subclinical liver disease in systemic lupus erythematosus. *J. Rheumatol.* **1981**, *8*, 752–759. [PubMed]
14. Grover, S.; Rastogi, A.; Singh, J.; Rajbongshi, A.; Bihari, C. Spectrum of histomorphologic findings in liver in patients with SLE: A review. *Hepat. Res. Treat.* **2014**, *2014*, 562979. [CrossRef] [PubMed]
15. Matsumoto, T.; Yoshimine, T.; Shimouchi, K.; Shiotu, H.; Kuwabara, N.; Fukuda, Y.; Hoshi, T. The liver in systemic lupus erythematosus: Pathologic analysis of 52 cases and review of Japanese Autopsy Registry Data. *Hum. Pathol.* **1992**, *23*, 1151–1158. [CrossRef]
16. Ippolito, A.; Petri, M. An update on mortality in systemic lupus erythematosus. *Clin. Exp. Rheumatol.* **2008**, *26*, S72–S79.
17. Bessone, F.; Poles, N.; Roma, M.G. Challenge of liver disease in systemic lupus erythematosus: Clues for diagnosis and hints for pathogenesis. *World J. Hepatol.* **2014**, *6*, 394–409. [CrossRef] [PubMed]
18. Arnett, F.C.; Reichlin, M. Lupus hepatitis: An under-recognized disease feature associated with autoantibodies to ribosomal P. *Am. J. Med.* **1995**, *99*, 465–472. [CrossRef]
19. Miller, M.H.; Urowitz, M.B.; Gladman, D.D.; Blendis, L.M. The liver in systemic lupus erythematosus. *Q. J. Med.* **1984**, *53*, 401–409.
20. Zheng, R.H.; Wang, J.H.; Wang, S.B.; Chen, J.; Guan, W.M.; Chen, M.H. Clinical and immunopathological features of patients with lupus hepatitis. *Chin. Med. J.* **2013**, *126*, 260–266.
21. Mathsson, L.; Ahlin, E.; Sjowall, C.; Skogh, T.; Ronnelid, J. Cytokine induction by circulating immune complexes and signs of in-vivo complement activation in systemic lupus erythematosus are associated with the occurrence of anti-Sjogren's syndrome A antibodies. *Clin. Exp. Immunol.* **2007**, *147*, 513–520. [CrossRef]
22. Akhter, E.; Burlingame, R.W.; Seaman, A.L.; Magder, L.; Petri, M. Anti-C1q antibodies have higher correlation with flares of lupus nephritis than other serum markers. *Lupus* **2011**, *20*, 1267–1274. [CrossRef]
23. Trendelenburg, M. Autoantibodies against complement component C1q in systemic lupus erythematosus. *Clin. Transl. Immunol.* **2021**, *10*, e1279. [CrossRef] [PubMed]
24. Ohira, H.; Takiguchi, J.; Rai, T.; Abe, K.; Yokokawa, J.; Sato, Y.; Takeda, I.; Kanno, T. High frequency of anti-ribosomal P antibody in patients with systemic lupus erythematosus-associated hepatitis. *Hepatol. Res.* **2004**, *28*, 137–139. [CrossRef] [PubMed]
25. Calich, A.L.; Viana, V.S.; Cancado, E.; Tustumi, F.; Terrabuio, D.R.; Leon, E.P.; Silva, C.A.; Borba, E.F.; Bonfa, E. Anti-ribosomal P protein: A novel antibody in autoimmune hepatitis. *Liver Int.* **2013**, *33*, 909–913. [CrossRef] [PubMed]
26. Adiga, A.; Nugent, K. Lupus Hepatitis and Autoimmune Hepatitis (Lupoid Hepatitis). *Am. J. Med. Sci.* **2017**, *353*, 329–335. [CrossRef]

27. Danielsson Borssen, A.; Marschall, H.U.; Bergquist, A.; Rorsman, F.; Weiland, O.; Kechagias, S.; Nyhlin, N.; Verbaan, H.; Nilsson, E.; Werner, M. Epidemiology and causes of death in a Swedish cohort of patients with autoimmune hepatitis. *Scand. J. Gastroenterol.* **2017**, *52*, 1022–1028. [CrossRef] [PubMed]

28. Gronbaek, L.; Vilstrup, H.; Jepsen, P. Autoimmune hepatitis in Denmark: Incidence, prevalence, prognosis, and causes of death. A nationwide registry-based cohort study. *J. Hepatol.* **2014**, *60*, 612–617. [CrossRef] [PubMed]

29. Ngu, J.H.; Bechly, K.; Chapman, B.A.; Burt, M.J.; Barclay, M.L.; Gearry, R.B.; Stedman, C.A. Population-based epidemiology study of autoimmune hepatitis: A disease of older women? *J. Gastroenterol. Hepatol.* **2010**, *25*, 1681–1686. [CrossRef]

30. Primo, J.; Maroto, N.; Martinez, M.; Anton, M.D.; Zaragoza, A.; Giner, R.; Devesa, F.; Merino, C.; del Olmo, J.A. Incidence of adult form of autoimmune hepatitis in Valencia (Spain). *Acta Gastroenterol. Belg.* **2009**, *72*, 402–406.

31. Gonzalez-Regueiro, J.A.; Cruz-Contreras, M.; Merayo-Chalico, J.; Barrera-Vargas, A.; Ruiz-Margain, A.; Campos-Murguia, A.; Espin-Nasser, M.; Martinez-Benitez, B.; Mendez-Cano, V.H.; Macias-Rodriguez, R.U. Hepatic manifestations in systemic lupus erythematosus. *Lupus* **2020**, *29*, 813–824. [CrossRef]

32. Carey, E.J.; Ali, A.H.; Lindor, K.D. Primary biliary cirrhosis. *Lancet* **2015**, *386*, 1565–1575. [CrossRef]

33. Danielsson, A.; Boqvist, L.; Uddenfeldt, P. Epidemiology of primary biliary cirrhosis in a defined rural population in the northern part of Sweden. *Hepatology* **1990**, *11*, 458–464. [CrossRef]

34. European Association for the Study of the Liver. EASL Clinical Practice Guidelines: The Diagnosis and Management of Patients with Primary Biliary Cholangitis. *J. Hepatol.* **2017**, *67*, 145–172. [CrossRef]

35. Invernizzi, P.; Floreani, A.; Carbone, M.; Marzioni, M.; Craxi, A.; Muratori, L.; Vespasiani Gentilucci, U.; Gardini, I.; Gasbarrini, A.; Kruger, P.; et al. Primary biliary cholangitis: Advances in management and treatment of the disease. *Dig. Liver Dis.* **2017**, *49*, 841–846. [CrossRef]

36. Tsianos, E.V.; Hoofnagle, J.H.; Fox, P.C.; Alspaugh, M.; Jones, E.A.; Schafer, D.F.; Moutsopoulos, H.M. Sjogren's syndrome in patients with primary biliary cirrhosis. *Hepatology* **1990**, *11*, 730–734. [CrossRef]

37. Hatzis, G.S.; Fragoulis, G.E.; Karatzaferis, A.; Delladetsima, I.; Barbatis, C.; Moutsopoulos, H.M. Prevalence and longterm course of primary biliary cirrhosis in primary Sjogren's syndrome. *J. Rheumatol.* **2008**, *35*, 2012–2016. [PubMed]

38. Shizuma, T. Clinical characteristics of concomitant systemic lupus erythematosus and primary biliary cirrhosis: A literature review. *J. Immunol. Res.* **2015**, *2015*, 713728. [CrossRef] [PubMed]

39. Ahmad, A.; Heijke, R.; Eriksson, P.; Wirestam, L.; Kechagias, S.; Dahle, C.; Sjowall, C. Autoantibodies associated with primary biliary cholangitis are common among patients with systemic lupus erythematosus even in the absence of elevated liver enzymes. *Clin. Exp. Immunol.* **2021**, *203*, 22–31. [CrossRef] [PubMed]

40. Frodlund, M.; Dahlstrom, O.; Kastbom, A.; Skogh, T.; Sjowall, C. Associations between antinuclear antibody staining patterns and clinical features of systemic lupus erythematosus: Analysis of a regional Swedish register. *BMJ Open* **2013**, *3*, e003608. [CrossRef] [PubMed]

41. Fries, J.F.; Holman, H.R. Systemic lupus erythematosus: A clinical analysis. In *Major problems in internal medicine*; Smith, L.H., Jr., Ed.; W.B. Saunders: Philadelphia, NY, USA, 1975; pp. 8–20.

42. Vitali, C.; Bombardieri, S.; Jonsson, R.; Moutsopoulos, H.M.; Alexander, E.L.; Carsons, S.E.; Daniels, T.E.; Fox, P.C.; Fox, R.I.; Kassan, S.S.; et al. Classification criteria for Sjogren's syndrome: A revised version of the European criteria proposed by the American-European Consensus Group. *Ann. Rheum. Dis.* **2002**, *61*, 554–558. [CrossRef] [PubMed]

43. Miyakis, S.; Lockshin, M.D.; Atsumi, T.; Branch, D.W.; Brey, R.L.; Cervera, R.; Derksen, R.H.; De Groot, P.G.; Koike, T.; Meroni, P.L.; et al. International consensus statement on an update of the classification criteria for definite antiphospholipid syndrome (APS). *J. Thromb. Haemost.* **2006**, *4*, 295–306. [CrossRef] [PubMed]

44. Hennes, E.M.; Zeniya, M.; Czaja, A.J.; Pares, A.; Dalekos, G.N.; Krawitt, E.L.; Bittencourt, P.L.; Porta, G.; Boberg, K.M.; Hofer, H.; et al. Simplified criteria for the diagnosis of autoimmune hepatitis. *Hepatology* **2008**, *48*, 169–176. [CrossRef] [PubMed]

45. Lindor, K.D.; Bowlus, C.L.; Boyer, J.; Levy, C.; Mayo, M. Primary biliary cholangitis: 2018 practice guidance from the American Association for the Study of Liver Diseases. *Hepatology* **2019**, *69*, 394–419. [CrossRef]

46. Enocsson, H.; Wirestam, L.; Dahle, C.; Padyukov, L.; Jonsen, A.; Urowitz, M.B.; Gladman, D.D.; Romero-Diaz, J.; Bae, S.C.; Fortin, P.R.; et al. Soluble urokinase plasminogen activator receptor (suPAR) levels predict damage accrual in patients with recent-onset systemic lupus erythematosus. *J. Autoimmun.* **2020**, *106*, 102340. [CrossRef]

47. Frodlund, M.; Wettero, J.; Dahle, C.; Dahlstrom, O.; Skogh, T.; Ronnelid, J.; Sjowall, C. Longitudinal anti-nuclear antibody (ANA) seroconversion in systemic lupus erythematosus: A prospective study of Swedish cases with recent-onset disease. *Clin. Exp. Immunol.* **2020**, *199*, 245–254. [CrossRef]

48. Sjowall, C.; Mandl, T.; Skattum, L.; Olsson, M.; Mohammad, A.J. Epidemiology of hypocomplementaemic urticarial vasculitis (anti-C1q vasculitis). *Rheumatology* **2018**, *57*, 1400–1407. [CrossRef]

49. West, M.; Jasin, H.E.; Medhekar, S. The development of connective tissue diseases in patients with autoimmune hepatitis: A case series. *Semin. Arthritis Rheum.* **2006**, *35*, 344–348. [CrossRef]

50. Olafsson, S.; Stridh, P.; Bos, S.D.; Ingason, A.; Euesden, J.; Sulem, P.; Thorleifsson, G.; Gustafsson, O.; Johannesson, A.; Geirsson, A.J.; et al. Fourteen sequence variants that associate with multiple sclerosis discovered by meta-analysis informed by genetic correlations. *NPJ Genom. Med.* **2017**, *2*, 24. [CrossRef]

51. Tizaoui, K.; Terrazzino, S.; Cargnin, S.; Lee, K.H.; Gauckler, P.; Li, H.; Shin, J.I.; Kronbichler, A. The role of PTPN22 in the pathogenesis of autoimmune diseases: A comprehensive review. *Semin. Arthritis Rheum.* **2021**, *51*, 513–522. [CrossRef]

52. Lim, D.H.; Kim, Y.G.; Lee, D.; Min Ahn, S.; Hong, S.; Lee, C.K.; Yoo, B. Immunoglobulin G levels as a prognostic factor for autoimmune hepatitis combined with systemic lupus erythematosus. *Arthritis Care Res.* **2016**, *68*, 995–1002. [CrossRef]
53. Miyake, Y.; Iwasaki, Y.; Terada, R.; Okamaoto, R.; Ikeda, H.; Makino, Y.; Kobashi, H.; Takaguchi, K.; Sakaguchi, K.; Shiratori, Y. Persistent elevation of serum alanine aminotransferase levels leads to poor survival and hepatocellular carcinoma development in type 1 autoimmune hepatitis. *Aliment. Pharmacol. Ther.* **2006**, *24*, 1197–1205. [CrossRef] [PubMed]
54. Verma, S.; Gunuwan, B.; Mendler, M.; Govindrajan, S.; Redeker, A. Factors predicting relapse and poor outcome in type I autoimmune hepatitis: Role of cirrhosis development, patterns of transaminases during remission and plasma cell activity in the liver biopsy. *Am. J. Gastroenterol.* **2004**, *99*, 1510–1516. [CrossRef] [PubMed]
55. Rustad, P.; Felding, P.; Franzson, L.; Kairisto, V.; Lahti, A.; Martensson, A.; Hyltoft Petersen, P.; Simonsson, P.; Steensland, H.; Uldall, A. The Nordic Reference Interval Project 2000: Recommended reference intervals for 25 common biochemical properties. *Scand. J. Clin. Lab. Investig.* **2004**, *64*, 271–284. [CrossRef] [PubMed]
56. Arnaud, L.; Tektonidou, M.G. Long-term outcomes in systemic lupus erythematosus: Trends over time and major contributors. *Rheumatology* **2020**, *59* (Suppl. 5), v29–v38. [CrossRef] [PubMed]

Journal of
*Clinical Medicine*

*Review*

# Patient-Reported Outcomes for Quality of Life in SLE: Essential in Clinical Trials and Ready for Routine Care

Matthew H. Nguyen [1,2,†], Frank F. Huang [3,†] and Sean G. O'Neill [3,4,*]

1    Liverpool Hospital, Liverpool, NSW 2170, Australia; Matthew.nguyen4@gmail.com
2    Pathology Department, School of Medical Sciences, Faculty of Medicine, University of New South Wales, Kensington, NSW 2052, Australia
3    Rheumatology Department, Royal North Shore Hospital, St Leonards, NSW 2065, Australia; fhuangfrank@gmail.com
4    Northern Clinical School, Faculty of Medicine and Health, University of Sydney, St Leonards, NSW 2065, Australia
*    Correspondence: Sean.ONeill@sydney.edu.au; Tel.: +61-02-94631890
†    Co-first authorship.

**Abstract:** Patient-reported outcome (PRO) instruments are widely used to assess quality of life in Systemic Lupus Erythematosus (SLE) research, and there is growing evidence for their use in clinical care. In this review, we evaluate the current evidence for their use in assessing quality of life in SLE in both research and clinical settings and examine the different characteristics of the commonly used PRO tools. There are now several well-validated generic and SLE-specific tools that have demonstrated utility in clinical trials and several tools that complement activity and damage measures in the clinical setting. PRO tools may help overcome physician–patient discordance in SLE and are valuable in the assessment of fibromyalgia and type 2 symptoms such as widespread pain and fatigue. Future work will identify optimal PRO tools for different settings but, despite current limitations, they are ready to be incorporated into patient care.

**Keywords:** systemic lupus erythematosus; health-related quality of life; patient-reported outcomes; clinical follow-up; outcome measures

**Citation:** Nguyen, M.H.; Huang, F.F.; O'Neill, S.G. Patient-Reported Outcomes for Quality of Life in SLE: Essential in Clinical Trials and Ready for Routine Care. *J. Clin. Med.* **2021**, *10*, 3754. https://doi.org/10.3390/jcm10163754

Academic Editors: Christopher Sjöwall and Ioannis Parodis

Received: 28 July 2021
Accepted: 17 August 2021
Published: 23 August 2021

**Publisher's Note:** MDPI stays neutral with regard to jurisdictional claims in published maps and institutional affiliations.

## 1. Introduction

Systemic Lupus Erythematosus (SLE) is a chronic autoimmune condition that can lead to inflammatory damage of multiple organ systems with clinical manifestations varying from patient to patient [1]. Many patients experience a significantly reduced health-related quality of life (HRQoL), citing fatigue, widespread pain and depression among the most common and debilitating features of the disease [2,3]. It is now well recognised that focusing on disease activity and damage does not allow a physician to adequately quantify or address the patient experience of the disease [3,4]. Good disease control does not guarantee improved quality of life in SLE patients, and failure to address this concern may contribute to treatment non-adherence and/or interruptions [5,6]. Numerous groups now advocate for the measurement of HRQoL and the use of patient-reported outcome (PRO) measures in SLE clinical trials and increasingly in routine care, including the American College of Rheumatology (ACR) [7], the European Alliance of Associations for Rheumatology (EULAR) [8], the Outcome Measures in Rheumatology Clinical Trials (OMERACT) [9] and the World Health Organisation (WHO) [10].

Despite the discordance between physician and patient assessment of SLE, literature on HRQoL is still sparse compared to that on disease activity, organ damage and immunotherapeutics in SLE [11]. A major reason behind the suboptimal focus on HRQoL in SLE management is the fact that HRQoL is a multi-dimensional concept without a concrete definition [12]. If clinicians were to enquire about the factors that patients perceive to be most relevant to their HRQoL, these would be significantly variable. However, overarching themes pertaining to HRQoL have been established, including the perceived impacts of disease and its treatment on physical, emotional and social functioning [10,12]. A variety of PRO tools have been used around the world in both clinical trials and routine care to help clinicians assess and monitor HRQoL in SLE patients [13,14]. Lack of agreement on which PRO to utilise along with real and perceived difficulties in their use in the clinical setting has seen a poor uptake of these tools, even though their routine use was recommended by OMERACT more than 20 years ago [9,15,16].

This focused review will consider the role of various generic and SLE-specific HRQoL tools and their utility in both clinical trials and routine care. To date, SLE-specific tools include Lupus Quality of Life (LupusQoL), Lupus Patient-Reported Outcome (LupusPRO), SLE-Specific Quality of Life (SLEQoL), Lupus Quality of Life (L-QoL) and Lupus Impact Tracker (LIT). All these tools have been validated for use in SLE patients but differ in terms of their item numbers and the domains they encompass [13,14]. Conversely, generic PRO tools were not designed to measure HRQoL in any specific disease population. However, some of these tools including the 36-item Short-Form Health Survey (SF-36) and the EuroQoL-5D (EQ-5D) have been widely utilised in SLE research due to their domains aligning with those relevant to Lupus-related QoL. Generic PRO tools also have the advantage of enabling comparison with other disease states. Currently, there are no clear guidelines or evidence to help clinicians determine what is an optimal PRO, as well as in which specific contexts [10,13].

Recently, two generic HRQoL instruments, Multi-Dimensional Health Assessment Questionnaire (MDHAQ) and Patient-Reported Outcomes Measurement Information System (PROMIS) are gaining traction, with increasing studies demonstrating their validity and utility in measuring HRQoL within the SLE population [13]. As they are not disease-specific, these tools have greater potential in allowing for comparisons with other disease populations or sub-cohorts within the SLE population such as patients with concomitant fibromyalgia or newly termed features of "type 2 SLE". This is clinically useful, as patients with these symptoms are less responsive to traditional immunosuppressive agents [17]. In addition, these tools have multiple benefits for clinicians beyond measuring HRQoL. For example, MDHAQ can be used to screen for Fibromyalgia (FAST3/FAST4 score) [18] and assess patient's flare status (RAPID3 score) [19,20]. Specific PRO tools have also been used to assess SLE disease activity and damage [8,15]. In this review, we will provide a summary of the major PRO tools that have been explored for the assessment of HRQoL in SLE research and clinical practice.

## 2. SLE-Specific PRO Tools

### 2.1. LupusQoL

LupusQoL is an SLE-specific instrument measuring HRQoL that has undergone extensive validation in the UK and has been widely adapted to other cohorts [21]. It has been validated in a US sample of SLE patients [22] and also cross-culturally with cohorts from Spain [23], Iran [24], Turkey [25], Italy [26], France [27], Venezuela [28] and China [29]. LupusQoL is a 34-item questionnaire that covers eight domains including physical health, pain, planning, intimate relationships, burden to others, emotional health, body image and fatigue. The recall period for each item is the preceding four weeks, and responses are given on a 5-point Likert scale. The summary score is reported on a scale from 0 to 100, with higher values indicating overall better HRQoL [21,30].

Draft items of this instrument were generated through the identification of recurring themes in qualitative interviews with 30 SLE patients, alongside input from clinical experts [30]. These were then re-assessed by 20 SLE patients whose feedback was incorporated to form the current questionnaire. Although information on readability was not provided, the tool was shown to have good internal consistency, test–retest reliability, concurrent validity and responsiveness to changes with patient-reported deterioration or improvement in health status [31]. However, only six of eight LupusQoL domains were found to be sensitive to improvement of disease activity, and none to deterioration. Floor effects (an inability of the PRO tool to detect true differences in HRQoL at the low end of the scale and below) and ceiling effects (an inability of the instrument to identify true differences in HRQoL at the high end of the scale and above) were mostly acceptable, aside from the intimate relationships and planning domains [30].

Two studies (McElhone [32] and Devilliers [33]) have established definitions for minimal clinically important difference (MCID) for LupusQoL domains. Using anchor-based analysis, McElhone's study determined that domain MCIDs ranged from −2.4 to −8.7 for deteriorations and from 3.5 to 7.3 for improvements [32]. Devilliers' study used a similar approach and reported MCIDs ranging from −0.5 to −6.4 for deteriorations and from 1.1 to 9.2 for improvements [33]. Nantes and colleagues showed in a prospective study comprised of 78 disease-active SLE patients that the percentages of patients reporting changes (improvements or deteriorations) across domains varied between MCID definitions, with percentages for most domains being greater using Devilliers' definition [34].

To date, LupusQoL has been used in five randomised–controlled clinical trials (RCTs). Two Phase III trials assessing the efficacy and safety of Epratuzumab in SLE patients with moderate-to-severe disease found no significant differences in various disease activity scores or LupusQoL scores between the placebo and the treatment groups, at 48 weeks [35]. Moreover, a Phase 4 multi-centre RCT examining the efficacy and safety of Acthar Gel in persistently active SLE patients demonstrated significant and clinically meaningful improvements in LupusQoL scores for the pain, planning, and fatigue domains in those who had higher disease activity levels [36]. Lastly, the remaining two clinical trials found that upper limb exercises [37] and a digital therapeutic plus telehealth coaching intervention [38] led to significant and clinically meaningful improvements in HRQoL as measured by LupusQoL scores. LupusQoL was not designed for use in the clinical setting and has not been studied as a PRO for use in routine care.

### 2.2. LupusPRO

LupusPRO is another SLE-specific tool that was developed in the United States to account for the ethnically diverse population of SLE patients within this demographic [39]. It also incorporated feedback from both genders in its development and is written in gender-neutral language. The tool itself is a 43-item questionnaire that encompasses not only HRQoL domains such as lupus symptoms, cognition and body image but also non-HRQoL domains including desires–goals, coping, social support and satisfaction with care. For each item, patients respond using a 5-point Likert scale, and the total score will range from 0 (worst QoL) to 100 (best QoL). Developers of LupusPRO have proposed that, beyond its ability to assess QoL longitudinally, it is a useful screening tool to help clinicians in determining important aspects of QoL (both health- and non-health related) which could be addressed through initiating discussion or making appropriate referrals. Another advantage is that it was created using recurring themes through patient feedback, and thus the questionnaire is fairly SLE-specific and simple to comprehend [39]. However, the feasibility of using this tool has not been formally evaluated, but this questionnaire would likely be less favourable for use in busy clinics due to its relatively higher number of items [15].

In an inception cohort comprising 323 SLE patients, adequate internal consistency and reliability were found in all domains except for the lupus medication domain [39]. Test–retest reliability was overall fair but particularly lower in some non-HRQoL domains and the procreation domain. Construct validity of the LupusPRO was established through its strong correlations with domains of SF-36, and criterion validity was demonstrated through its correlations with various disease activity and damage measures [39]. It has been validated in several different languages including Tagalog (Philippines) [40], Turkish (Turkey) [41], Spanish (Spain) [42], French (Canada) [43], Italian (Italy) [44], Japanese (Japan) [45], Hindi (India) [46], Arabic (Egypt) [47] and Chinese (Hong Kong) [48]. LupusPRO was found to be valid and reliable within these populations and showed measurement equivalence.

Interestingly, LupusPRO was shown to perform similarly across two differing samples of SLE patients, one of which included an ethnically diverse urban cohort from Southern California, whilst the other comprised an ethnically homogeneous rural cohort from the Philippines using confirmatory factor analysis [49]. This demonstrates measurement equivalence for LupusPRO across ethnically diverse and homogeneous populations.

Another advantage of LupusPRO is that it included patients with concomitant fibromyalgia in its design to improve generalisability for the fatigue domain [39]. Recently, an updated version of LupusPRO (LupusPRO v1.8) [50] was developed, in which the Pain–Vitality domain was separated into three domains including sleep, pain and vitality and was captured through the addition of six further items. This updated version demonstrates acceptable face, content, convergent, discriminant and criterion validity with acceptable internal consistency and reliability in all domains, except procreation and coping. LupusPRO has only been used in two clinical trials to date [51,52]. One examined the efficacy of an online training program focused on development of pain-coping skills (PainTRAINER) in SLE Patients and reported meaningful improvements in LupusPRO HRQoL scores in patients who received the intervention compared to those in the wait-list control group at 9 weeks [51]. The other trial demonstrated significant improvements in LupusPRO body image (BI) scores in patients with cutaneous involvement who received a novel BI intervention that used a cognitive-based therapy approach compared to those who did not. There was also significant improvement in scores of other HRQoL domains including pain–vitality, cognition and lupus symptoms post-intervention compared to baseline within the intervention group [52].

### 2.3. SLEQoL

SLEQoL is a 40-item questionnaire that was developed and validated in an English-speaking cohort of SLE patients in Singapore [21,53]. Items were originally generated by rheumatology experts and then modified according to feedback from 100 patients for content validity. Responses for each item are given on a 7-point response scale and capture the patient's experience over the preceding week. Items encompass six domains including physical functioning, activities, symptoms, treatment, mood and self-image. The summary score is derived from the sum of all responses across these domains and ranges from 40 to 280, with higher scores denoting worse HRQoL. Regarding its psychometric value, SLEQoL has been shown to have good internal consistency, test–retest reliability, construct validity and responsiveness. However, it is limited by its significant floor effects, whereby patients reported good perceived QoL beyond the instrument's measurement capabilities. Authors suggested that this could be addressed through co-administering another validated PRO tool such as SF-36, although this would impose a higher time burden on both clinicians and patients [53]. To date, cross-cultural adaptation and validation of SLEQoL has been performed in Arabic (Egypt) [54], Thai (Thailand) [55], Chinese (China) [56] and Brazilian Portuguese (Brazil) [57].

### 2.4. L-QoL

L-QoL is another tool that serves to assess quality-of-life in SLE patients but on a needs-based approach [21,58]. It was originally developed and validated in the United Kingdom in 2008 and since then has only been translated and validated in a Turkish SLE cohort [59]. The tool comprises 25 questions which are answered in a "true/not true" response format. Summary scores range from 0 to 25, with higher scores indicating worse QoL. Content validity was achieved through items being generated via patient interviews and being predominantly phrased in their own words. In the original study, L-QoL also demonstrated excellent construct validity, internal consistency and test–retest reliability. However, construct validity was examined against non-validated self-reported measures of disease activity and severity. No validated physician assessments of these parameters were employed, and thus further studies including these will be required to further clarify the construct and discriminant validities of this tool [21,58]. Moreover, L-QoL has not been used in any clinical studies or research trials thus far, and its utility in patients with more severe disease phenotypes is to be explored [21].

*2.5. LIT*

The Lupus Impact Tracker (LIT) is a 10-item PRO tool that was designed with the aims of producing a simple but reliable PRO instrument to monitor the impact of SLE on the lives of patients over time [60,61]. The questions cover seven key concepts including cognition, lupus medications, physical health, pain/fatigue impact, emotional health, body image and planning desires and goals. The questions were generated using a multi-step approach that ultimately filtered out items with the highest psychometric value and strongest correlation with overall wellbeing/disease activity/damage scores and that ranked most importantly to patients from the 43-items of the LupusPRO. LIT was shown to have good internal consistency, responsiveness and test–retest reliability, but there are no data on floor and ceiling effects [14,60,61]. Cross-cultural validation has been displayed in Canada [62], southeastern US [63] and five European countries including Germany, Italy, Spain, Sweden and France [64]. In addition, LIT was found to be a valid PRO tool in a multicultural Australian cohort and could distinguish between groups of patients with active or inactive disease [16]. Similar findings were also recently observed in a single-centre but ethnically diverse cohort of paediatric SLE patients [65]. In this paediatric study, LIT was deemed to be highly patient-feasible, given all patients (100%, $n = 46$) had completed their forms in all visits (115 in total) with accurate self-scoring. In correspondence, developers of the LIT reported that, among patients (pts) and physicians (phs) across 20 different centres, more than half agreed that LIT was not burdensome (>80% pts, >60% phs), helped foster better communication (>75% pts, >50% physicians) and facilitated discussion about the impact of SLE on QoL (>80% pts, >70% phs) [61].

## 3. GENERIC PRO Instruments

*3.1. SF-36*

The Medical Outcomes Study Short Form 36 (SF-36) is one of the most widely used generic HRQoL measurements in SLE. It consists of 36 question items grouped across 8 domains (physical functioning, general health, mental health, vitality, role physical, role emotional, bodily pain and social functioning) and can be expressed as two summary scores (physical and mental health component, PCS and MCS, respectively). Individual domain scores are then transformed into a scale ranging from 0 (worst) to 100 (best). As a generic questionnaire, it has the advantage of enabling comparisons with healthy population norms and other chronic diseases. This is particularly important, as patients with SLE have a significantly reduced QoL across all health domains when compared to other conditions [66].

The SF-36 has been extensively validated in various SLE populations over the last 25 years with good results [34,67–70]. As such, it has essentially become a gold standard among HRQoL instruments for validating other generic PRO tools such as PROMIS and MDHAQ. The SF-36 has been incorporated into clinical trials and randomised controlled trials (RCTs) to assess the efficacy of new therapies for SLE. For example, SF-36 scores were used as a major secondary end point in the BLISS trials, which facilitated the approval of belimumab after establishing its efficacy and safety in active SLE patients [71]. The SF-36 has remained the most widely used generic PRO instrument for clinical trials involving biologics, including rituximab [72,73], abatacept [74,75], cyclophosphamide [76], eprutuzumab [77] and sirukumab [78].

The need for a HRQoL tool as a key indicator in the routine clinical monitoring of SLE is becoming increasingly recognised in the literature [79,80]. Although not designed for clinical use, SF-36 remains an option, given its widespread use and international validation for a range of chronic diseases [79]. However, there are limited data regarding the actual use of SF-36 in routine clinical practice for SLE. A Canadian SLE clinic reported only minimal change in SF-36 scores over an 8-year period [81]. Despite the extensive validation and use in research settings, the use of the SF-36 in a purely clinical context remains limited.

Further studies examining the longitudinal responsiveness of SF-36 have yielded conflicting results [33,69,70,82]. One study demonstrated that social functioning and MCS scores were minimally responsive in patients with worsening disease damage [70]. Devilliers' study in 2015 also showed that the LupusQoL was more responsive to changes in QoL than the SF-36 [33]. Given the complex multifaceted nature of SLE, it is not surprising that social and emotional nuances are perhaps more accurately captured and tracked by disease-specific or at least rheumatology-specific PRO tools. Furthermore, the SF-36 is relatively time-consuming to complete, and the scoring system is difficult in a busy clinical setting, requiring computer programming software. The SF-36 is likely to remain more appropriate in short- to medium-term clinical studies than in routine clinical use.

### 3.2. EQ-5D

The EuroQoL five-dimensional (EQ-5D) is a simple and standardised questionnaire which can yield clinical and economic data. It tackles the five domains of mobility, self-care, usual activities, pain and discomfort, and anxiety and depression using three-point response scales, which are then converted to a summary score from 0 (worst) to 1.0 (best). There is an additional visual analogue scale measuring the patient's health perception from 0 (worst) to 100 (best). The EQ-5D has demonstrated favourable psychometric properties and exhibits satisfactory criterion validity, convergent validity and sensitivity to self-reported change in health [83]. Construct validity was also proven against equivalent domains in disease-specific PROs in a cohort of 240 patients; however, the same study also reported significant ceiling effects [84]. The EQ-5D has been used alongside the SF-36 in multiple clinical studies [85,86], as well as for comparison between rheumatic groups [87,88].

The EQ-5D has utility in forming economic appraisals rather than in just simply measuring HRQoL. Different utility values are generated from different health outcomes to calculate quality-adjusted life-years (QALYs) using the time trade-off method (TTO). This method has been validated against direct utility instruments in a cohort of 245 consecutive SLE patients in China [89]. As such, the EQ-5D has enabled economic appraisals and cost studies involving SLE patients [90–92]. For example, one Italian study demonstrated belimumab to be cost-effective (32,859 euros per QALY) [91]. However, it is unlikely that these benefits can be translated to a routine, patient-focused clinical setting.

### 3.3. PROMIS

PROMIS (Patient-Reported Outcomes Measurement Information System) is a relatively recent initiative developed by the NIH (National Institutes of Health) aimed at measuring PROs across various medical conditions [93]. It consists of question items from the eight core domains examining fatigue, pain intensity, pain interference, physical function, sleep disturbance, anxiety, depression and ability to participate in social roles and activities. Unlike conventional PRO tools, PROMIS also enables the application of the item response theory (IRT) and computerised adaptive testing (CAT) in order to develop calibrated item banks for more precise and efficient outcome measures [94].

PROMIS is not disease-specific, and as such, its content relevance to SLE needs to be examined. A study comprised of multi-ethnic English-speaking Asian individuals demonstrated that the eight core PROMIS domains largely aligned with the pertinent issues faced by patients with SLE [95]. However, this study also identified content gaps such as family burden, stigma and discrimination, although this may have been influenced by the demographics of the study cohort. There has been a paucity of further studies specifically analysing the content relevance of PROMIS to SLE in different populations.

One of the first studies to evaluate the validity of PROMIS in SLE was conducted in a childhood-onset SLE population. Most notably, it found that PROMIS demonstrated internal consistency and construct validity, despite taking less than five minutes to complete [96].

Since then, the body of literature has grown, with further studies in the adult SLE population showing similar results [97,98]. The California Lupus Epidemiology Study (CLUES) consisted of a racially and linguistically diverse cohort of 431 individuals. In this cohort, the PROMIS was able to demonstrate consistent reliability across racial/ethnic/language groups and was able to correlate well with the SF-36 [97]. Floor effects were minimal, and it was noted that ceiling effects were prevalent, especially in Social Health measures, which could adversely affect longitudinal effectiveness. The PROMIS has been shown to be sensitive to change in patient-reported improvement or worsening (effect size $> |0.27|$); however, this was only examined across physical and mental health domains [99]. Further studies are required to investigate the responsiveness to change in the social health domain.

One of the advantages of PROMIS is that it encompasses a wide variety of domains, despite placing a reduced burden on the patient. As such, PROMIS has been increasingly used in studies to investigate a range of SLE symptoms including pain [51], fatigue [100,101], depression [102], quality of life [103], cognitive impairment [104] and sleep quality [105]. Most notably, some of these symptoms such as cognitive impairment and sleep resonate strongly with SLE patients yet are known to be content gaps in other generic questionnaires [106]. Furthermore, the standardised metric of PROMIS also enables direct comparisons between SLE and other rheumatological or chronic conditions. Interestingly, the validity of PROMIS in fibromyalgia was found to be markedly lower when compared to OA, RA and SLE in a rheumatology cohort [107]. No studies to date have examined the use of PROMIS in concomitant fibromyalgia in SLE specifically. Further studies of PROMIS, including in RCTs and routine clinical care, are anticipated.

### 3.4. MDHAQ

The MDHAQ is a double-sided one-page questionnaire developed in rheumatology practice and contains six core measures. Pain, patient global and fatigue are scored on a 0–10 VAS, whereas function, joint count and symptom checklist are scored between 0 and 10, 0 and 48 and 0 and 60, respectively. Various scores can be calculated from the MDHAQ, including RAPID3, an index that incorporates three of the MDHAQ items, i.e., function, pain and patient global. The MDHAQ and RAPID3 are well validated in rheumatoid arthritis and other rheumatic diseases, including several studies supporting their utility in SLE. An American study supported the use of MDHAQ/RAPID3 in a cohort of 161 SLE patients in routine care [20]. The study reported robust internal consistency reliability (Cronbach's alpha 0.88), validity and responsiveness to change for MDHAQ items and the RAPID3. However, the study noted significant floor effects compared to similar studies in the rheumatoid arthritis population. The RAPID3 will inherently align more strongly with the RA phenotype, where painful joints can significantly impact function, whereas the multisystem complexity of SLE may not translate as effectively to RAPID3 scores.

MDHAQ is rheumatology-specific rather than generic or SLE-specific, which gives it a unique advantage in examining the interplay between different rheumatic diseases. For example, it has been shown to be able to provide clues of concomitant fibromyalgia in SLE [18]. This is particularly important in the context of SLE, where there may be a high prevalence of non-inflammatory symptoms, characteristic of type 2 SLE. Recently, the concept of type 1 and type 2 SLE was proposed [108], whereby type 1 entails autoimmune and organ damage, whilst type 2 SLE is driven by symptoms typically observed in fibromyalgia. Type 2 symptoms of fatigue, myalgia, mood disturbance and cognitive function are typically not responsive to immunosuppression and thus crucial to recognise to avoid over-treatment and guide appropriate management.

The precise role of the MDHAQ in SLE is still yet to be fully established. The MDHAQ can be given to all patients in the waiting room of a rheumatology clinic regardless of their precise diagnosis, making it feasible in busy clinical settings. It is quick to complete and interpret in the clinic. It differs from typical HRQoL tools in that it has also been shown to reflect inflammatory disease activity and clinical improvement [19,109]. The utility of

MDHAQ is multifaceted, much like SLE itself, as long as scores are interpreted in the context of traditional patient workups.

## 4. Discussion

There are numerous well-validated tools for measuring HRQoL in people with SLE, both disease-specific and generic (Table 1). These instruments have become essential in clinical trials of SLE, acknowledging that disease activity and damage are insufficient measures of the patient experience of living with the disease [13,14]. The use of an HRQoL tool has long been recommended by leading rheumatology groups including ACR [7], EULAR [8] and OMERACT [9] and indeed would seem mandatory for regulatory approval of new therapeutics. Despite their importance, there is no single instrument that is universally accepted as the gold standard for capturing every aspect of HRQoL in people with SLE. In general, the SF-36 and EQ-5D have been widely used due to their broad acceptance and application to numerous populations and diseases. Several SLE-specific tools have been used in the assessment of therapeutic strategies, with some evidence for increased sensitivity to change (when compared with SF36), making them appealing options in the clinical trial setting [13,14]. Further research is needed to determine the optimal instrument for assessing HRQoL in SLE.

Despite their widespread use in the research setting, measures of HRQoL remain underutilised in many clinical settings [11,15]. There are several real and perceived challenges to their use, including the complexity of the instruments, as well as the time and expertise needed to administer the instruments, calculate and interpret the results. These challenges have led to the development of more clinically focused tools, (such as MDHAQ [19,20]) and computerised adaptive questioning (PROMIS [93–96,103]) that sacrifice comprehensive assessment for the sake of practicality. Again, there is no single measure that is appropriate to all circumstances, though we would argue that despite this limitation, some form of measurement is better than none. Given that disease activity and damage may not capture the most prevalent and concerning symptoms that matter to people with SLE [3,4], it would seem prudent to attempt documentation of the patient's concerns. More work is needed to determine which instruments best capture this and are sensitive to change whilst remaining practical and convenient for both the patient and the clinician.

The concept of 'type 1' and 'type 2' lupus symptoms has recently been proposed as a method of categorising symptoms and acknowledging the disparity between physician and patient assessment of SLE [17,108]. This proposal essentially acknowledges that 'type 1' symptoms (often considered inflammatory) are different from 'type 2' symptoms (fatigue, widespread pain, sleep disorders, depression and anxiety, frequently considered as fibromyalgia) that are prevalent in SLE. Incorporating PRO tools into patient management allows for documentation and validation of these symptoms, which may help bridge the gap between physician and patient assessment of the disease. Patient-reported instruments, interpreted by clinicians experienced in the care of people with SLE, aid in the recognition of fibromyalgia and type 2 symptoms, which may in turn allow for more accurate assessment of disease activity and inform treatment decisions. Whilst challenges remain in determining whether these measures accurately quantify severity and are sensitive to change, they seem sensible additions to patient-centred care.

In this review, we have highlighted the evolving role of PROs in the assessment of HRQoL in people with SLE, in both research and clinical settings. Many options now exist that have been validated in the trial setting, with increasing evidence for several PROMs in clinical practice. Further work is anticipated to better define the optimal tool for various clinical settings. Despite this limitation, integrating PROMs into clinical practice complements disease activity and damage measures and enhances patient-centred care.

Table 1. Summary table of PRO instruments validated for use in SLE.

| | HRQoL Tool | Country | Year | Purpose and Content | Time Burden | Number of Items | Response | Summary Score | Recall Period | Strengths | Weaknesses |
|---|---|---|---|---|---|---|---|---|---|---|---|
| Generic | **SF-36** [67] | US | 1990 | Self-report measure QOL in various populations. Eight domains (physical functioning, general health, mental health, vitality, role physical, role emotional, bodily pain, and social functioning) and two summary scores (physical and mental health component) | Respondent: <10 min. Administration: <10 min. | 36 | 3–6 point response scale | 0–100 (higher scores = better QoL) | 4 weeks | Most extensively studied and validated generic PRO tool in SLE. Has been used in multiple clinical trials. Facilitates comparison with other diseases. | Time-consuming and difficult scoring system. Questionable longitudinal responsiveness. |
| | **EQ-5D** [83] | UK | 1994 | Generic questionnaire for clinical and economic appraisal. Five dimensions of health: mobility, self-care, usual activities, pain and discomfort, anxiety and depression | Respondent: <5 min. Administration: <5 min | 6 | 3-point response scale | 0–100 (higher scores = better QoL) | 1 day | Allows for economic evaluation. Has been used in clinical trials. | Encompasses fewer domains. |
| | **MDHAQ** [20] | US | 1999 | Assessment and monitoring of patients with rheumatic diseases | Respondent: <10 min. Administration: <5 min | 4 core items | Variable—Visual analogue scores, checklists, 4-item response scales | 0–10 for individual items (Function, Pain, Patient Global) Also utilises composite indices, e.g., RAPID3 | Variable—Day (joint count) to week (function) to month (systems review) | Rheumatology-specific, allowing for analyses of specific sub-cohorts, e.g., concomitant fibromyalgia in SLE. A single double-sided page. Simple to calculate. | Minimally studied and validated in SLE cohorts. |
| | **PROMIS** [95,103] | US | 2004 | Standardised HRQoL measure. PROMIS29 covers eight domains (fatigue, pain intensity, pain interference, physical function, sleep disturbance, anxiety, depression and ability to participate in social roles and activities) | Respondent: <10 min. Administration: <10 min | Variable: 10 (PROMIS10), 29 (PROMIS29) | 5-item response scale | T-score compared to US general population | 1 week | Enables comparisons across a wide range of domains. Allows for item response theory and computer adaptive tests. | Significant ceiling effects. Difficult to score. |

**Table 1.** *Cont.*

| | HRQoL Tool | Country | Year | Purpose and Content | Time Burden | Number of Items | Response | Summary Score | Recall Period | Strengths | Weaknesses |
|---|---|---|---|---|---|---|---|---|---|---|---|
| **SLE-Specific** | **LupusQOL** [30] | UK | 2017 | To measure disease-specific HRQoL in adult patients with SLE. It covers eight domains including physical health, emotional health, body image, pain, planning, fatigue, intimate relationships and burden to others | Respondent: <10 min. Administration: <5 min | 34 | 5-point Likert response scale | 0–100 (higher score = better QoL) | 4 weeks | Most extensively studied and validated SLE-specific PRO Tool. Has been used in multiple clinical trials and studies. Has been translated and validated in the most languages of PRO tools. | Can be time-consuming. |
| | **LupusPRO** [39] | US | 2012 | To assess HRQoL in ethnically heterogeneous SLE populations and provide a gender-neutral PRO tool. Includes both HRQoL domains (lupus symptoms, lupus medications, physical health, emotional health, pain-vitality, procreation, cognition, body image) and non-HRQoL domains (desires/goals, social support, coping, satisfaction with care) | Respondent: Not reported. Administration: Not reported | 43 | 5-point Likert response scale | 0–100 (higher score = better QoL) | 4 weeks | Was designed to accommodate ethnically heterogenous populations of SLE and also provides a gender-balanced PRO tool for SLE patients. Also encompasses non-HRQoL domains. | Has the most items of all SLE-Specific PRO tools and is thus time-consuming. Has not been widely used in clinical trials to date. |
| | **SLEQoL** [53] | Singapore | 2005 | To assess HRQoL in SLE patients. It covers six domains including physical functioning, activities, symptoms, treatment, mood and self-image. | Respondent: <5 min. Administration: Not reported | 40 | 7-point response scale | 40–280 (higher score = worst QoL) | 1 week | Has good internal consistency, test–retest reliability, construct validity and responsiveness to change. | Significant floor effects requires a second PRO tool to address this and is thus both time-consuming and less feasible. Has not been validated in more ethnically diverse SLE populations. |

**Table 1.** *Cont.*

| HRQoL Tool | Country | Year | Purpose and Content | Time Burden | Number of Items | Response | Summary Score | Recall Period | Strengths | Weaknesses |
|---|---|---|---|---|---|---|---|---|---|---|
| L-QoL [58] | UK | 2008 | To assess HRQoL in SLE patients on a needs-based approach. | Respondent: <5 min. Administration: Not reported | 25 | Dichotomous. "True/Not True" | 0–25 (higher score = worst QoL) | Nil (needs-based model) | Relatively simple and feasible to complete, given the dichotomous responses and lower item numbers. Constructed on a needs-based model. | Construct and discriminant validities need to be further explored. Has not been used in any clinical trials or SLE cohort studies to date. |
| LIT [60] | US | 2014 | Short, simple and feasible PRO tool that monitors the impact of SLE over time | Respondent: <5 min. Administration:<5 min | 10 | 5-point Likert response scale | 0–100 (higher score = worst QoL) | 4 weeks | Shortest, simplest and most feasible SLE-specific PRO tool that has been extensively validated. Demonstrates responsiveness to disease activity and can be used in paediatric SLE. | May miss certain aspects of HRQoL in SLE patients due to its brevity. |

Abbreviations: EQ-5D; European Quality of Life Five-Dimensional, LIT; Lupus Impact Tracker, L-QoL; Lupus Quality of Life Questionnaire, LupusPRO; Lupus Patient-Reported Outcome, LupusQOL; Lupus Quality of Life, MDHAQ; Multi-Dimensional Health Assessment Questionnaire, PRO; Patient-Reported Outcomes, PROMIS; Patient-Reported Outcomes Measurement Information System, QOL; Quality of Life, SF-36; 36-item Short-Form Health Survey, SLE; Systemic Lupus Erythematosus, SLEQoL, Systemic Lupus Erythematosus Quality of Life.

**Author Contributions:** Conceptualization, S.G.O.; Search strategy and review of publications, M.H.N., F.F.H. and S.G.O.; Draft preparation, review and editing, M.H.N., F.F.H., S.G.O. All authors have read and agreed to the published version of the manuscript.

**Funding:** This research received no external funding.

**Institutional Review Board Statement:** Not applicable.

**Informed Consent Statement:** Not applicable.

**Conflicts of Interest:** The authors declare no conflict of interest.

## Abbreviations

| | |
|---|---|
| ACR | American College of Rheumatology |
| EULAR | European Alliance of Associations for Rheumatology |
| EQ-5D | European Quality of Life Five-Dimensional |
| LIT | Lupus Impact Tracker |
| L-QoL | Lupus Quality of Life Questionnaire |
| LupusPRO | Lupus Patient-Reported Outcome |
| LupusQOL | Lupus Quality of Life |
| MDHAQ | Multi-Dimensional Health Assessment Questionnaire |
| OMERACT | Outcome Measures in Rheumatology Clinical Trials |
| PROMIS | Patient-Reported Outcomes Measurement Information System |
| SF-36 | 36-item Short-Form Health Survey |
| SLE | Systemic Lupus Erythematosus |
| SLEQoL | Systemic Lupus Erythematosus Quality of Life |
| WHO | World Health Organisation |

## References

1. Maidhof, W.; Hilas, O. Lupus: An overview of the disease and management options. *Pharm. Ther.* **2012**, *37*, 240.
2. McElhone, K.; Abbott, J.; Gray, J.; Williams, A.; Teh, L.S. Patient perspective of systemic lupus erythematosus in relation to health-related quality of life concepts: A qualitative study. *Lupus* **2010**, *19*, 1640–1647. [CrossRef] [PubMed]
3. Golder, V.; Ooi, J.J.; Antony, A.S.; Ko, T.; Morton, S.; Kandane-Rathnayake, R.; Morand, E.F.; Hoi, A.Y. Discordance of patient and physician health status concerns in systemic lupus erythematosus. *Lupus* **2018**, *27*, 501–506. [CrossRef] [PubMed]
4. Elera-Fitzcarrald, C.; Vega, K.; Gamboa-Cárdenas, R.V.; Zúñiga, K.; Medina, M.; Pimentel-Quiroz, V.; Pastor-Asurza, C.; Perich-Campos, R.; Bellido, Z.R.; Griffin, R.; et al. Discrepant perception of lupus disease activity: A comparison between patients' and physicians' disease activity scores. *J. Clin. Rheumatol.* **2020**, *26*, 165–169. [CrossRef] [PubMed]
5. Julian, L.J.; Yelin, E.; Yazdany, J.; Panopalis, P.; Trupin, L.; Criswell, L.A.; Katz, P. Depression, medication adherence, and service utilization in systemic lupus erythematosus. *Arthritis Care Res.* **2009**, *61*, 240–246. [CrossRef]
6. Bennett, J.K.; Fuertes, J.N.; Keitel, M.; Phillips, R. The role of patient attachment and working alliance on patient adherence, satisfaction, and health-related quality of life in lupus treatment. *Patient Educ. Couns.* **2011**, *85*, 53–59. [CrossRef]
7. Ad Hoc Committee on Systemic Lupus Erythematosus Response Criteria for Fatigue. Measurement of fatigue in systemic lupus erythematosus: A systematic review. *Arthritis Care Res.* **2007**, *57*, 1348–1357. [CrossRef]
8. Castrejon, I.; Carmona, L.; Agrinier, N.; Andres, M.; Briot, K.; Caron, M.; Christensen, R.; Consolaro, A.; Curbelo, R.; Ferrer, M.; et al. The EULAR Outcome Measures Library: Development and an example from a systematic review for systemic lupus erythematous instruments. *Clin. Exp. Rheumatol.* **2015**, *33*, 910–916.
9. Strand, V.; Gladman, D.; Isenberg, D.; Petri, M.; Smolen, J.; Tugwell, P. Endpoints: Consensus recommendations from OMERACT IV. *Lupus* **2000**, *9*, 322–327. [CrossRef]
10. Kwan, A.; Strand, V.; Touma, Z. The role of patient-reported outcomes in systemic lupus erythematosus. *Curr. Treat. Options Rheumatol.* **2017**, *3*, 308–321. [CrossRef]
11. Dua, A.B.; Touma, Z.; Toloza, S.; Jolly, M. Top 10 recent developments in health-related quality of life in patients with systemic lupus erythematosus. *Curr. Rheumatol. Rep.* **2013**, *15*, 380. [CrossRef]
12. Thumboo, J.; Strand, V. Health-related quality of life in patients with systemic lupus erythematosus: An update. *Ann. Acad. Med. Singap.* **2007**, *36*, 115.
13. Mahieu, M.; Yount, S.; Ramsey-Goldman, R. Patient-reported outcomes in systemic lupus erythematosus. *Rheum. Dis. Clin.* **2016**, *42*, 253–263. [CrossRef]
14. Izadi, Z.; Gandrup, J.; Katz, P.P.; Yazdany, J. Patient-reported outcome measures for use in clinical trials of SLE: A review. *Lupus Sci. Med.* **2018**, *5*, 279. [CrossRef]

15. Castrejón, I.; Tani, C.; Jolly, M.; Huang, A.; Mosca, M. Indices to assess patients with systemic lupus erythematosus in clinical trials, long-term observational studies, and clinical care. *Clin. Exp. Rheumatol.* **2014**, *32*, 85–95.

16. Antony, A.; Kandane-Rathnayake, R.K.; Ko, T.; Boulos, D.; Hoi, A.Y.; Jolly, M.; Morand, E.F. Validation of the Lupus Impact Tracker in an Australian patient cohort. *Lupus* **2017**, *26*, 98–105. [CrossRef]

17. Rogers, J.L.; Eudy, A.M.; Pisetsky, D.; Criscione-Schreiber, L.G.; Sun, K.; Doss, J.; Clowse, M.E. Using Clinical Characteristics and Patient-Reported Outcome Measures to Categorize Systemic Lupus Erythematosus Subtypes. *Arthritis Care Res.* **2021**, *73*, 386–393. [CrossRef]

18. Huang, F.F.; Fang, R.; Nguyen, M.H.; Bryant, K.; Gibson, K.A.; O'Neill, S.G. Identifying co-morbid fibromyalgia in patients with systemic lupus erythematosus using the Multi-Dimensional Health Assessment Questionnaire. *Lupus* **2020**, *29*, 1404–1411. [CrossRef] [PubMed]

19. Askanase, A.D.; Castrejón, I.; Pincus, T. Quantitative data for care of patients with systemic lupus erythematosus in usual clinical settings: A patient Multidimensional Health Assessment Questionnaire and physician estimate of noninflammatory symptoms. *J. Rheumatol.* **2011**, *38*, 1309–1316. [CrossRef]

20. Annapureddy, N.; Giangreco, D.; Devilliers, H.; Block, J.A.; Jolly, M. Psychometric properties of MDHAQ/RAPID3 in patients with systemic lupus erythematosus. *Lupus* **2018**, *27*, 982–990. [CrossRef]

21. Yazdany, J. Health-related quality of life measurement in systemic lupus erythematosus: The LupusQoL, SLEQoL, and L-QoL. *Arthritis Care Res.* **2011**, *63*, 413. [CrossRef] [PubMed]

22. Jolly, M.; Pickard, A.S.; Wilke, C.; Mikolaitis, R.A.; Teh, L.S.; Mcelhone, K.; Fogg, L.; Block, J. Lupus-specific health outcome measure for US patients: The LupusQoL-US version. *Ann. Rheum. Dis.* **2010**, *69*, 29–33. [CrossRef] [PubMed]

23. González-Rodríguez, V.; Peralta-Ramírez, M.I.; Navarrete-Navarrete, N.; Callejas-Rubio, J.L.; Ruiz, S.A.M.; Khamashta, M. Adaptation and validation of the Spanish version of a disease-specific quality of life measure in patients with systemic lupus erythematosus: The Lupus quality of life. *Med. Clin.* **2009**, *134*, 13–16. [CrossRef]

24. Hosseini, N.; Bonakdar, Z.S.; Gholamrezaei, A.; Mirbagher, L. Linguistic validation of the LupusQoL for the assessment of quality of life in Iranian patients with systemic lupus erythematosus. *Int. J. Rheumatol.* **2014**, *2014*. [CrossRef]

25. Pamuk, O.N.; Onat, A.M.; Donmez, S.; Mengüs, C.; Kisacik, B. Validity and reliability of the Lupus QoL index in Turkish systemic lupus erythematosus patients. *Lupus* **2015**, *24*, 816–821. [CrossRef] [PubMed]

26. Conti, F.; Perricone, C.; Reboldi, G.; Gawlicki, M.; Bartosiewicz, I.; Pacucci, V.A.; Massaro, L.; Miranda, F.; Truglia, S.; Alessandri, C.; et al. Validation of a disease-specific health-related quality of life measure in adult Italian patients with systemic lupus erythematosus: LupusQoL-IT. *Lupus* **2014**, *23*, 743–751. [CrossRef]

27. Devilliers, H.; Amoura, Z.; Besancenot, J.F.; Bonnotte, B.; Pasquali, J.L.; Wahl, D.; Maurier, F.; Kaminsky, P.; Pennaforte, J.L.; Magy-Bertrand, N.; et al. LupusQoL-FR is valid to assess quality of life in patients with systemic lupus erythematosus. *Rheumatology* **2012**, *51*, 1906–1915. [CrossRef]

28. Carrión-Nessi, F.S.; Marcano-Rojas, M.V.; Freitas-DeNobrega, D.C.; Arocha, S.R.R.; Antuarez-Magallanes, A.W.; Fuentes-Silva, Y.J. Validation of the LupusQoL in Venezuela: A Specific Measurement of Quality of Life in Patients with Systemic Lupus Erythematosus. *Reumatol. Clin.* **2021**, in press. [CrossRef]

29. Wang, S.L.; Wu, B.; Leng, L.; Bucala, R.; Lu, L.J. Validity of LupusQoL-China for the assessment of health related quality of life in Chinese patients with systemic lupus erythematosus. *PLoS ONE* **2013**, *8*, e63795. [CrossRef]

30. McElhone, K.; Abbott, J.; Shelmerdine, J.; Bruce, I.N.; Ahmad, Y.; Gordon, C.; Peers, K.; Isenberg, D.; Ferenkeh-Koroma, A.; Griffiths, B.; et al. Development and validation of a disease-specific health-related quality of life measure, the LupusQoL, for adults with systemic lupus erythematosus. *Arthritis Care Res.* **2007**, *57*, 972–979. [CrossRef]

31. Louthrenoo, W.; Kasitanon, N.; Morand, E.; Kandane-Rathnayake, R. Comparison of performance of specific (SLEQOL) and generic (SF36) health-related quality of life questionnaires and their associations with disease status of systemic lupus erythematosus: A longitudinal study. *Arthritis Res. Ther.* **2020**, *22*, 8. [CrossRef] [PubMed]

32. McElhone, K.; Abbott, J.; Sutton, C.; Mullen, M.; Lanyon, P.; Rahman, A.; Yee, C.S.; Akil, M.; Bruce, I.N.; Ahmad, Y.; et al. Sensitivity to change and minimal important differences of the LupusQoL in patients with systemic lupus erythematosus. *Arthritis Care Res.* **2016**, *68*, 1505–1513. [CrossRef]

33. Devilliers, H.; Amoura, Z.; Besancenot, J.F.; Bonnotte, B.; Pasquali, J.L.; Wahl, D.; Maurier, F.; Kaminsky, P.; Pennaforte, J.L.; Magy-Bertrand, N.; et al. Responsiveness of the 36-item Short Form Health Survey and the Lupus Quality of Life questionnaire in SLE. *Rheumatology* **2015**, *54*, 940–949. [CrossRef] [PubMed]

34. Nantes, S.G.; Strand, V.; Su, J.; Touma, Z. Comparison of the Sensitivity to Change of the 36-Item Short Form Health Survey and the Lupus Quality of Life Measure Using Various Definitions of Minimum Clinically Important Differences in Patients with Active Systemic Lupus Erythematosus. *Arthritis Care Res.* **2018**, *70*, 125–133. [CrossRef] [PubMed]

35. Clowse, M.E.; Wallace, D.J.; Furie, R.A.; Petri, M.A.; Pike, M.C.; Leszczyński, P.; Neuwelt, C.M.; Hobbs, K.; Keiserman, M.; Duca, L.; et al. Efficacy and safety of epratuzumab in moderately to severely active systemic lupus erythematosus: Results from two phase III randomized, double-blind, placebo-controlled trials. *Arthritis Rheumatol.* **2017**, *69*, 362–375. [CrossRef] [PubMed]

36. Askanase, A.D.; Wan, G.J.; Panaccio, M.P.; Zhao, E.; Zhu, J.; Bilyk, R.; Furie, R.A. Patient-Reported Outcomes from a Phase 4, Multicenter, Randomized, Double-Blind, Placebo-Controlled Trial of Repository Corticotropin Injection (Acthar® Gel) for Persistently Active Systemic Lupus Erythematosus. *Rheumatol. Ther.* **2021**, *8*, 573–584. [CrossRef] [PubMed]

37. Keramiotou, K.; Anagnostou, C.; Kataxaki, E.; Galanos, A.; Sfikakis, P.P.; Tektonidou, M.G. The impact of upper limb exercise on function, daily activities and quality of life in systemic lupus erythematosus: A pilot randomised controlled trial. *RMD Open.* **2020**, *6*, 1141. [CrossRef] [PubMed]
38. Khan, F.; Granville, N.; Malkani, R.; Chathampally, Y. Health-Related Quality of Life Improvements in Systemic Lupus Erythematosus Derived from a Digital Therapeutic Plus Tele-Health Coaching Intervention: Randomized Controlled Pilot Trial. *J. Med. Internet Res.* **2020**, *22*, 23868. [CrossRef]
39. Jolly, M.; Pickard, A.S.; Block, J.A.; Kumar, R.B.; Mikolaitis, R.A.; Wilke, C.T.; Rodby, R.A.; Fogg, L.; Sequeira, W.; Utset, T.O.; et al. Disease-specific patient reported outcome tools for systemic lupus erythematosus. *Semin. Arthritis Rheum.* **2012**, *42*, 56–65. [CrossRef] [PubMed]
40. Navarra, S.; Tanangunan, R.M.; Mikolaitis-Preuss, R.A.; Kosinski, M.; Block, J.A.; Jolly, M. Cross-cultural validation of a disease-specific patient-reported outcome measure for lupus in Philippines. *Lupus* **2013**, *22*, 262–267. [CrossRef] [PubMed]
41. Kaya, A.; Goker, B.; Cura, E.S.; Tezcan, M.E.; Tufan, A.; Mercan, R.; Bitik, B.; Haznedaroglu, S.; Ozturk, M.A.; Mikolaitis-Preuss, R.A.; et al. Turkish lupusPRO: Cross-cultural validation study for Lupus. *Clin. Rheumatol.* **2014**, *33*, 1079–1084. [CrossRef]
42. Jolly, M.; Toloza, S.; Block, J.; Mikolaitis, R.; Kosinski, M.; Wallace, D.; Durran-Barragan, S.; Bertoli, A.; Blazevic, I.; Vilá, L.; et al. Spanish LupusPRO: Cross-cultural validation study for lupus. *Lupus* **2013**, *22*, 431–436. [CrossRef]
43. Bourré-Tessier, J.; Clarke, A.E.; Kosinski, M.; Mikolaitis-Preuss, R.A.; Bernatsky, S.; Block, J.A.; Jolly, M. The French-Canadian validation of a disease-specific, patient-reported outcome measure for lupus. *Lupus* **2014**, *23*, 1452–1459. [CrossRef]
44. Jolly, M.; Mazzoni, D.; Cicognani, E. Measurement properties of Web based Lupuspro, a disease targeted Outcome Tool, among Italian Patients with Lupus. *Value Health* **2015**, *18*, 302. [CrossRef]
45. Inoue, M.; Shiozawa, K.; Yoshihara, R.; Yamane, T.; Shima, Y.; Hirano, T.; Jolly, M.; Makimoto, K. The Japanese LupusPRO: A cross-cultural validation of an outcome measure for lupus. *Lupus* **2017**, *26*, 849–856. [CrossRef] [PubMed]
46. Pinto, B.; Jolly, M.; Dhooria, A.; Grover, S.; Raj, J.M.; Devilliers, H.; Sharma, A. Hindi LupusPRO: Cross cultural validation of disease specific patient reported outcome measure of lupus. *Lupus* **2019**, *28*, 1534–1540. [CrossRef] [PubMed]
47. Elkaraly, N.E.; Nasef, S.I.; Omar, A.S.; Fouad, A.M.; Jolly, M.; Mohamed, A.E. The Arabic LupusPRO: A Cross-Cultural Validation of a Disease-Specific Patient-Reported Outcome Tool for Quality of Life in Lupus Patients. *Lupus* **2020**, *29*, 1727–1735. [CrossRef]
48. Mok, C.C.; Kosinski, M.; Ho, L.Y.; Chan, K.L.; Jolly, M. Validation of the LupusPRO in Chinese patients from Hong Kong with systemic lupus erythematosus. *Arthritis Care Res.* **2015**, *67*, 297–304. [CrossRef]
49. Azizoddin, D.R.; Olmstead, R.; Cost, C.; Jolly, M.; Ayeroff, J.; Racaza, G.; Sumner, L.A.; Ormseth, S.; Weisman, M.; Nicassio, P.M. A multi-group confirmatory factor analyses of the LupusPRO between southern California and Filipino samples of patients with systemic lupus erythematosus. *Lupus* **2017**, *26*, 967–974. [CrossRef] [PubMed]
50. Azizoddin, D.R.; Weinberg, S.; Gandhi, N.; Arora, S.; Block, J.A.; Sequeira, W.; Jolly, M. Validation of the LupusPRO version 1.8: An update to a disease-specific patient-reported outcome tool for systemic lupus erythematosus. *Lupus* **2018**, *27*, 728–737. [CrossRef]
51. Allen, K.D.; Beauchamp, T.; Rini, C.; Keefe, F.J.; Bennell, K.L.; Cleveland, R.J.; Grimm, K.; Huffman, K.; Hu, D.G.; Santana, A.; et al. Pilot study of an internet-based pain coping skills training program for patients with systemic Lupus Erythematosus. *BMC Rheumatol.* **2021**, *5*, 1–10. [CrossRef]
52. Jolly, M.; Peters, K.F.; Mikolaitis, R.; Evans-Raoul, K.; Block, J.A. Body image intervention to improve health outcomes in lupus: A pilot study. *JCR J. Clin. Rheumatol.* **2014**, *20*, 403–410. [CrossRef] [PubMed]
53. Leong, K.P.; Kong, K.O.; Thong, B.Y.; Koh, E.T.; Lian, T.Y.; Teh, C.L.; Cheng, Y.K.; Chng, H.H.; Badsha, H.; Law, W.G.; et al. Development and preliminary validation of a systemic lupus erythematosus-specific quality-of-life instrument (SLEQOL). *Rheumatology* **2005**, *44*, 1267–1276. [CrossRef]
54. Aziz, M.M.; Galal, M.A.; Elzohri, M.H.; El-Nouby, F.; Leong, K. Cross-cultural adaptation and validation of Systemic Lupus Erythematosus Quality of Life questionnaire into Arabic. *Lupus* **2018**, *27*, 780–787. [CrossRef] [PubMed]
55. Kasitanon, N.; Wangkaew, S.; Puntana, S.; Sukitawut, W.; Leong, K.P.; Tan Tock Seng Hospital Lupus Study Group; Louthrenoo, W. The reliability, validity and responsiveness of the Thai version of Systemic Lupus Erythematosus Quality of Life (SLEQOL-TH) instrument. *Lupus* **2013**, *22*, 289–296. [CrossRef] [PubMed]
56. Jiang, H.Z.; Lin, Z.G.; Li, H.J.; Wang, S.Y.; Guan, S.Q.; Mei, Y.F. The Chinese version of the SLEQOL is a reliable assessment of health-related quality of life in Han Chinese patients with systemic lupus erythematosus. *Clin. Rheumatol.* **2018**, *37*, 151–160. [CrossRef]
57. Freire, E.A.; Bruscato, A.; Leite, D.R.; Sousa, T.T.; Ciconelli, R.M. Translation into Brazilian Portuguese, cultural adaptation and validatation of the systemic lupus erythematosus quality of life questionnaire (SLEQOL). *Acta Reumatol. Port.* **2010**, *35*, 334–339.
58. Doward, L.C.; McKenna, S.P.; Whalley, D.; Tennant, A.; Griffiths, B.; Emery, P.; Veale, D.J. The development of the L-QoL: A quality-of-life instrument specific to systemic lupus erythematosus. *Ann. Rheum. Dis.* **2009**, *68*, 196–200. [CrossRef]
59. Duruöz, M.T.; Unal, C.; Toprak, C.S.; Sezer, I.; Yilmaz, F.; Ulutatar, F.; Atagündüz, P.; Baklacioglu, H.S. The validity and reliability of Systemic Lupus Erythematosus Quality of Life Questionnaire (L-QoL) in a Turkish population. *Lupus* **2017**, *26*, 1528–1533. [CrossRef]
60. Jolly, M.; Garris, C.P.; Mikolaitis, R.A.; Jhingran, P.M.; Dennis, G.; Wallace, D.J.; Clarke, A.; Dooley, M.A.; Parke, A.; Strand, V.; et al. Development and validation of the Lupus Impact Tracker: A patient-completed tool for clinical practice to assess and monitor the impact of systemic lupus erythematosus. *Arthritis Care Res.* **2014**, *66*, 1542–1550. [CrossRef]

61. Jolly, M.; Kosinski, M.; Garris, C.P.; Oglesby, A.K. Prospective validation of the Lupus Impact Tracker: A patient-completed tool for clinical practice to evaluate the impact of systemic lupus erythematosus. *Arthritis Rheumatol.* **2016**, *68*, 1422–1431. [CrossRef]
62. Bourré-Tessier, J.; Clarke, A.E.; Mikolaitis-Preuss, R.A.; Kosinski, M.; Bernatsky, S.; Block, J.A.; Jolly, M. Cross-cultural validation of a disease-specific patient-reported outcome measure for systemic lupus erythematosus in Canada. *J. Rheumatol.* **2013**, *40*, 1327–1333. [CrossRef]
63. Brandt, J.E.; Drenkard, C.; Kan, H.; Bao, G.; Dunlop-Thomas, C.; Pobiner, B.; Chang, D.J.; Jolly, M.; Lim, S.S. External validation of the lupus impact tracker in a southeastern US longitudinal cohort with systemic lupus erythematosus. *Arthritis Care Res.* **2017**, *69*, 842–848. [CrossRef]
64. Schneider, M.; Mosca, M.; Pego-Reigosa, J.M.; Gunnarsson, I.; Maurel, F.; Garofano, A.; Perna, A.; Porcasi, R.; Devilliers, H. Cross-cultural validation of Lupus Impact Tracker in five European clinical practice settings. *Rheumatology* **2017**, *56*, 818–828. [CrossRef] [PubMed]
65. Ganguli, S.K.; Hui-Yuen, J.S.; Jolly, M.; Cerise, J.; Eberhard, B.A. Performance and psychometric properties of lupus impact tracker in assessing patient-reported outcomes in pediatric lupus: Report from a pilot study. *Lupus* **2020**, *29*, 1781–1789. [CrossRef] [PubMed]
66. Jolly, M. How does quality of life of patients with systemic lupus erythematosus compare with that of other common chronic illnesses? *J. Rheumatol.* **2005**, *32*, 1706–1708.
67. Stoll, T.; Gordon, C.; Seifert, B.; Richardson, K.; Malik, J.; Bacon, P.A.; Isenberg, D.A. Consistency and validity of patient administered assessment of quality of life by the MOS SF-36; its association with disease activity and damage in patients with systemic lupus erythematosus. *J. Rheumatol.* **1997**, *24*, 1608–1614.
68. Thumboo, J.; Fong, K.Y.; Ng, T.P.; Leong, K.H.; Feng, P.H.; Thio, S.T.; Boey, M.L. Validation of the MOS SF-36 for quality of life assessment of patients with systemic lupus erythematosus in Singapore. *J. Rheumatol.* **1999**, *26*, 97–102. [PubMed]
69. Touma, Z.; Gladman, D.D.; Ibañez, D.; Urowitz, M.B. Is there an advantage over SF-36 with a quality of life measure that is specific to systemic lupus erythematosus? *J. Rheumatol.* **2011**, *38*, 1898–1905. [CrossRef] [PubMed]
70. Baba, S.; Katsumata, Y.; Okamoto, Y.; Kawaguchi, Y.; Hanaoka, M.; Kawasumi, H.; Yamanaka, H. Reliability of the SF-36 in Japanese patients with systemic lupus erythematosus and its associations with disease activity and damage: A two-consecutive year prospective study. *Lupus* **2018**, *27*, 407–416. [CrossRef] [PubMed]
71. Furie, R.; Petri, M.; Zamani, O.; Cervera, R.; Wallace, D.J.; Tegzová, D.; Sanchez-Guerrero, J.; Schwarting, A.; Merrill, J.T.; Chatham, W.W.; et al. A phase III, randomized, placebo-controlled study of belimumab, a monoclonal antibody that inhibits B lymphocyte stimulator, in patients with systemic lupus erythematosus. *Arthritis Rheum.* **2011**, *63*, 3918–3930. [CrossRef] [PubMed]
72. Merrill, J.T.; Neuwelt, C.M.; Wallace, D.J.; Shanahan, J.C.; Latinis, K.M.; Oates, J.C.; Utset, T.O.; Gordon, C.; Isenberg, D.A.; Hsieh, H.J.; et al. Efficacy and safety of rituximab in moderately-to-severely active systemic lupus erythematosus: The randomized, double-blind, phase II/III systemic lupus erythematosus evaluation of rituximab trial. *Arthritis Rheum.* **2010**, *62*, 222–233. [CrossRef] [PubMed]
73. Rovin, B.H.; Furie, R.; Latinis, K.; Looney, R.J.; Fervenza, F.C.; Sanchez-Guerrero, J.; Maciuca, R.; Zhang, D.; Garg, J.P.; Brunetta, P.; et al. Efficacy and safety of rituximab in patients with active proliferative lupus nephritis: The Lupus Nephritis Assessment with Rituximab study. *Arthritis Rheum.* **2012**, *64*, 1215–1226. [CrossRef] [PubMed]
74. Merrill, J.T.; Burgos-Vargas, R.; Westhovens, R.; Chalmers, A.; D'cruz, D.; Wallace, D.J.; Bae, S.C.; Sigal, L.; Becker, J.C.; Kelly, S.; et al. The efficacy and safety of abatacept in patients with non–life-threatening manifestations of systemic lupus erythematosus: Results of a twelve-month, multicenter, exploratory, phase IIb, randomized, double-blind, placebo-controlled trial. *Arthritis Rheum.* **2010**, *62*, 3077–3087. [CrossRef] [PubMed]
75. Furie, R.; Nicholls, K.; Cheng, T.T.; Houssiau, F.; Burgos-Vargas, R.; Chen, S.L.; Hillson, J.L.; Meadows-Shropshire, S.; Kinaszczuk, M.; Merrill, J.T. Efficacy and safety of abatacept in lupus nephritis: A twelve-month, randomized, double-blind study. *Arthritis Rheumatol.* **2014**, *66*, 379–389. [CrossRef]
76. ACCESS Trial Group. Treatment of lupus nephritis with abatacept: The abatacept and cyclophosphamide combination efficacy and safety study. *Arthritis Rheumatol.* **2014**, *66*, 3096–3104. [CrossRef]
77. Wallace, D.J.; Kalunian, K.; Petri, M.A.; Strand, V.; Houssiau, F.A.; Pike, M.; Kilgallen, B.; Bongardt, S.; Barry, A.; Kelley, L.; et al. Efficacy and safety of epratuzumab in patients with moderate/severe active systemic lupus erythematosus: Results from EMBLEM, a phase IIb, randomised, double-blind, placebo-controlled, multicentre study. *Ann. Rheum. Dis.* **2014**, *73*, 183–190. [CrossRef]
78. Thorne, C.; Takeuchi, T.; Karpouzas, G.A.; Sheng, S.; Kurrasch, R.; Fei, K.; Hsu, B. Investigating sirukumab for rheumatoid arthritis: 2-year results from the phase III SIRROUND-D study. *RMD Open* **2018**, *4*, 731. [CrossRef]
79. Fernando, M.M.; Isenberg, D.A. How to monitor SLE in routine clinical practice. *Ann. Rheum. Dis.* **2005**, *64*, 524–527. [CrossRef]
80. Mosca, M.; Tani, C.; Aringer, M.; Bombardieri, S.; Boumpas, D.; Cervera, R.; Doria, A.; Jayne, D.; Khamashta, M.A.; Kuhn, A.; et al. Development of quality indicators to evaluate the monitoring of SLE patients in routine clinical practice. *Autoimmun. Rev.* **2011**, *10*, 383–388. [CrossRef]
81. Kuriya, B.; Gladman, D.D.; Ibañez, D.; Urowitz, M.B. Quality of life over time in patients with systemic lupus erythematosus. *Arthritis Care Res.* **2008**, *59*, 181–185. [CrossRef] [PubMed]

82. Hanly, J.G.; Urowitz, M.B.; Jackson, D.; Bae, S.C.; Gordon, C.; Wallace, D.J.; Clarke, A.; Bernatsky, S.; Vasudevan, A.; Isenberg, D.; et al. SF-36 summary and subscale scores are reliable outcomes of neuropsychiatric events in systemic lupus erythematosus. *Ann. Rheum. Dis.* **2011**, *70*, 961–967. [CrossRef] [PubMed]

83. Aggarwal, R.; Wilke, C.T.; Pickard, A.S.; Vats, V.; Mikolaitis, R.; Fogg, L.; Block, J.A.; Jolly, M. Psychometric properties of the EuroQol-5D and Short Form-6D in patients with systemic lupus erythematosus. *J. Rheumatol.* **2009**, *36*, 1209–1216. [CrossRef] [PubMed]

84. Wang, S.L.; Wu, B.; Zhu, L.A.; Leng, L.; Bucala, R.; Lu, L.J. Construct and criterion validity of the Euro Qol-5D in patients with systemic lupus erythematosus. *PLoS ONE* **2014**, *9*, e98883. [CrossRef] [PubMed]

85. Parodis, I.; Lopez Benavides, A.H.; Zickert, A.; Pettersson, S.; Möller, S.; Welin Henriksson, E.; Voss, A.; Gunnarsson, I. The Impact of Belimumab and Rituximab on Health-Related Quality of Life in Patients with Systemic Lupus Erythematosus. *Arthritis Care Res.* **2019**, *71*, 811–821. [CrossRef]

86. Shi, Y.; Li, M.; Liu, L.; Wang, Z.; Wang, Y.; Zhao, J.; Wang, Q.; Tian, X.; Li, M.; Zeng, X. Relationship between disease activity, organ damage and health-related quality of life in patients with systemic lupus erythematosus: A systemic review and meta-analysis. *Autoimmun. Rev.* **2020**, *20*, 102691. [CrossRef]

87. Park, E.H.; Strand, V.; Oh, Y.J.; Song, Y.W.; Lee, E.B. Health-related quality of life in systemic sclerosis compared with other rheumatic diseases: A cross-sectional study. *Arthritis Res. Ther.* **2019**, *21*, 1–10. [CrossRef]

88. Chen, H.H.; Chen, D.Y.; Chen, Y.M.; Lai, K.L. Health-related quality of life and utility: Comparison of ankylosing spondylitis, rheumatoid arthritis, and systemic lupus erythematosus patients in Taiwan. *Clin. Rheumatol.* **2017**, *36*, 133–142. [CrossRef]

89. Wang, S.L.; Hsieh, E.; Zhu, L.A.; Wu, B.; Lu, L.J. Comparative assessment of different health utility measures in systemic lupus erythematosus. *Sci. Rep.* **2015**, *5*, 13297. [CrossRef] [PubMed]

90. Ivorra, J.R.; Fernández-Llanio-Comella, N.; San-Martín-Álvarez, A.; Vela-Casasempere, P.; Saurí-Ferrer, I.; González-de-Julián, S.; Vivas-Consuelo, D. Health-related quality of life in patients with systemic lupus erythematosus: A Spanish study based on patient reports. *Clin. Rheumatol.* **2019**, *38*, 1857–1864. [CrossRef] [PubMed]

91. Pierotti, F.; Palla, I.; Treur, M.; Pippo, L.; Turchetti, G. Assessment of the economic impact of belimumab for the treatment of systemic lupus erythematosus in the Italian setting: A cost-effectiveness analysis. *PLoS ONE* **2015**, *10*, e0140843. [CrossRef]

92. Bexelius, C.; Wachtmeister, K.; Skare, P.; Jönsson, L.; Vollenhoven, R.V. Drivers of cost and health-related quality of life in patients with systemic lupus erythematosus (SLE): A Swedish nationwide study based on patient reports. *Lupus* **2013**, *22*, 793–801. [CrossRef]

93. Cella, D.; Yount, S.; Rothrock, N.; Gershon, R.; Cook, K.; Reeve, B.; Ader, D.; Fries, J.F.; Bruce, B.; Rose, M. The Patient-Reported Outcomes Measurement Information System (PROMIS): Progress of an NIH Roadmap cooperative group during its first two years. *Med. Care* **2007**, *45*, 3. [CrossRef] [PubMed]

94. Fries, J.F.; Witter, J.; Rose, M.; Cella, D.; Khanna, D.; Morgan-DeWitt, E. Item response theory, computerized adaptive testing, and PROMIS: Assessment of physical function. *J. Rheumatol.* **2014**, *41*, 153–158. [CrossRef] [PubMed]

95. Ow, Y.L.; Thumboo, J.; Cella, D.; Cheung, Y.B.; Yong Fong, K.; Wee, H.L. Domains of health-related quality of life important and relevant to multiethnic English-speaking Asian systemic lupus erythematosus patients: A focus group study. *Arthritis Care Res.* **2011**, *63*, 899–908. [CrossRef] [PubMed]

96. Jones, J.T.; Carle, A.C.; Wootton, J.; Liberio, B.; Lee, J.; Schanberg, L.E.; Ying, J.; Morgan DeWitt, E.; Brunner, H.I. Validation of Patient-Reported Outcomes Measurement Information System Short Forms for use in childhood-onset systemic lupus erythematosus. *Arthritis Care Res.* **2017**, *69*, 133–142. [CrossRef] [PubMed]

97. Katz, P.; Yazdany, J.; Trupin, L.; Rush, S.; Helmick, C.G.; Murphy, L.B.; Lanata, C.; Criswell, L.A.; Dall'Era, M. Psychometric evaluation of the National Institutes of Health Patient-Reported Outcomes Measurement Information System in a multiracial, multiethnic systemic lupus erythematosus cohort. *Arthritis Care Res.* **2019**, *71*, 1630–1639. [CrossRef] [PubMed]

98. Kasturi, S.; Burket, J.C.; Berman, J.R.; Kirou, K.A.; Levine, A.B.; Sammaritano, L.R.; Mandl, L.A. Feasibility of Patient-Reported Outcomes Measurement Information System (PROMIS®) computerized adaptive tests in systemic lupus erythematosus outpatients. *Lupus* **2018**, *27*, 1591–1599. [CrossRef]

99. Kasturi, S.; Szymonifka, J.; Berman, J.R.; Kirou, K.A.; Levine, A.B.; Sammaritano, L.R.; Mandl, L.A. Responsiveness of the Patient-Reported Outcomes Measurement Information System Global Health Short Form in Outpatients with Systemic Lupus Erythematosus. *Arthritis Care Res.* **2020**, *72*, 1282–1288. [CrossRef]

100. Mahieu, M.A.; Ahn, G.E.; Chmiel, J.S.; Dunlop, D.D.; Helenowski, I.B.; Semanik, P.; Song, J.; Yount, S.; Chang, R.W.; Ramsey-Goldman, R. Serum adipokine levels and associations with patient-reported fatigue in systemic lupus erythematosus. *Rheumatol. Int.* **2018**, *38*, 1053–1061. [CrossRef]

101. Mahieu, M.A.; Ahn, G.E.; Chmiel, J.S.; Dunlop, D.D.; Helenowski, I.B.; Semanik, P.; Song, J.; Yount, S.; Chang, R.W.; Ramsey-Goldman, R. Fatigue, patient reported outcomes, and objective measurement of physical activity in systemic lupus erythematosus. *Lupus* **2016**, *25*, 1190–1199. [CrossRef]

102. Dietz, B.; Katz, P.; Dall'Era, M.; Murphy, L.B.; Lanata, C.; Trupin, L.; Criswell, L.A.; Yazdany, J. Major Depression and Adverse Patient-Reported Outcomes in Systemic Lupus Erythematosus: Results from a Prospective Longitudinal Cohort. *Arthritis Care Res.* **2021**, *73*, 48–54. [CrossRef]

103. Lai, J.S.; Beaumont, J.L.; Jensen, S.E.; Kaiser, K.; Van Brunt, D.L.; Kao, A.H.; Chen, S.Y. An evaluation of health-related quality of life in patients with systemic lupus erythematosus using PROMIS and Neuro-QoL. *Clin. Rheumatol.* **2017**, *36*, 555–562. [CrossRef] [PubMed]
104. Kim, M.Y.; Sen, D.; Drummond, R.R.; Brandenburg, M.C.; Biesanz, K.L.; Kim, A.H.; Eisen, S.A.; Baum, C.M.; Foster, E.R. Cognitive dysfunction among people with systemic lupus erythematosus is associated with reduced participation in daily life. *Lupus* **2021**, *30*, 1100–1107. [CrossRef] [PubMed]
105. Charoenwoodhipong, P.; Harlow, S.D.; Marder, W.; Hassett, A.L.; McCune, W.J.; Gordon, C.; Helmick, C.G.; Barbour, K.E.; Wang, L.; Mancuso, P.; et al. Dietary Omega Polyunsaturated Fatty Acid Intake and Patient-Reported Outcomes in Systemic Lupus Erythematosus: The Michigan Lupus Epidemiology and Surveillance Program. *Arthritis Care Res.* **2020**, *72*, 874–881. [CrossRef] [PubMed]
106. Stamm, T.A.; Bauernfeind, B.; Coenen, M.; Feierl, E.; Mathis, M.; Stucki, G.; Smolen, J.S.; Machold, K.P.; Aringer, M. Concepts important to persons with systemic lupus erythematosus and their coverage by standard measures of disease activity and health status. *Arthritis Care Res.* **2007**, *57*, 1287–1295. [CrossRef]
107. Katz, P.; Pedro, S.; Michaud, K. Performance of the patient-reported outcomes measurement information system 29-item profile in rheumatoid arthritis, osteoarthritis, fibromyalgia, and systemic lupus erythematosus. *Arthritis Care Res.* **2017**, *69*, 1312–1321. [CrossRef] [PubMed]
108. Pisetsky, D.S.; Clowse, M.E.; Criscione-Schreiber, L.G.; Rogers, J.L. A novel system to categorize the symptoms of systemic lupus erythematosus. *Arthritis Care Res.* **2019**, *71*, 735–741. [CrossRef]
109. Castrejón, I.; Bergman, M.J.; Pincus, T. MDHAQ/RAPID3 to recognize improvement over 2 months in usual care of patients with osteoarthritis, systemic lupus erythematosus, spondyloarthropathy, and gout, as well as rheumatoid arthritis. *J. Clin. Rheumatol.* **2013**, *19*, 169–174. [CrossRef]

Journal of
*Clinical Medicine*

*Article*

# A Comparison of the Correlation of Systemic Lupus Erythematosus Disease Activity Index 2000 (SLEDAI-2K) and Systemic Lupus Erythematosus Disease Activity Score (SLE-DAS) with Health-Related Quality of Life

Ning-Sheng Lai [1,2,†], Ming-Chi Lu [1,2,†], Hsiu-Hua Chang [3], Hui-Chin Lo [3], Chia-Wen Hsu [3], Kuang-Yung Huang [1,2], Chien-Hsueh Tung [1,2], Bao-Bao Hsu [1], Cheng-Han Wu [1] and Malcolm Koo [4,5,*]

[1]  Division of Allergy, Immunology and Rheumatology, Dalin Tzu Chi Hospital, Buddhist Tzu Chi Medical Foundation, Dalin, Chiayi 622401, Taiwan; tzuchilai@gmail.com (N.-S.L.); e360187@yahoo.com.tw (M.-C.L.); hky0919@yahoo.com.tw (K.-Y.H.); dr5188@yahoo.com.tw (C.-H.T.); baobaohsu100@yahoo.com.tw (B.-B.H.); wickham_wu@yahoo.com.tw (C.-H.W.)

[2]  School of Medicine, Tzu Chi University, Hualien City, Hualien 97004, Taiwan

[3]  Department of Medical Research, Dalin Tzu Chi Hospital, Buddhist Tzu Chi Medical Foundation, Dalin, Chiayi 622401, Taiwan; df274760@tzuchi.com.tw (H.-H.C.); df289469@tzuchi.com.tw (H.-C.L.); chiawen0114@yahoo.com.tw (C.-W.H.)

[4]  Graduate Institute of Long-Term Care, Tzu Chi University of Science and Technology, Hualien City, Hualien 973302, Taiwan

[5]  Dalla Lana School of Public Health, University of Toronto, Toronto, ON M5T 3M7, Canada

*   Correspondence: m.koo@utoronto.ca

†   Ning-Sheng Lai and Ming-Chi Lu contributed equally to this work.

**Citation:** Lai, N.-S.; Lu, M.-C.; Chang, H.-H.; Lo, H.-C.; Hsu, C.-W.; Huang, K.-Y.; Tung, C.-H.; Hsu, B.-B.; Wu, C.-H.; Koo, M. A Comparison of the Correlation of Systemic Lupus Erythematosus Disease Activity Index 2000 (SLEDAI-2K) and Systemic Lupus Erythematosus Disease Activity Score (SLE-DAS) with Health-Related Quality of Life. *J. Clin. Med.* **2021**, *10*, 2137. https://doi.org/10.3390/jcm10102137

Academic Editors: Christopher Sjöwall and Ioannis Parodis

Received: 20 March 2021
Accepted: 11 May 2021
Published: 15 May 2021

**Abstract:** Background and Aim: The aim of this study was to compare the correlation of a recently developed systemic lupus erythematosus disease activity score (SLE-DAS) with the SLE disease activity index 2000 (SLEDAI-2K) with the Lupus Quality of Life questionnaire (LupusQoL) in Taiwanese patients with SLE. Methods: A cross-sectional study was conducted in a regional teaching hospital in Taiwan from April to August 2019. Adult patients with a clinician-confirmed diagnosis of SLE based on the 1997 American College of Rheumatology revised criteria or the 2012 Systemic Lupus International Collaborating Clinics Classification Criteria were recruited. SLE disease activity was measured with both SLEDAI-2K and SLE-DAS. Disease-specific quality of life was assessed using the LupusQoL. Results: Of the 333 patients with SLE in this study, 90.4% were female and 40% were between the ages of 20 and 39 years. The median SLEDAI-2K score was 4.00 (interquartile range [IQR] 2.00–7.50) and the median SLE-DAS score was 2.08 (IQR 1.12–8.24) in our patients with SLE. After adjusting for sex and age intervals, both SLEDAI-2k and SLE-DAS were significantly and inversely associated with all eight domains of LupusQoL. The magnitudes of the mean absolute error, root mean square error, Akaike Information Criterion, Bayesian Information Criterion, and coefficient of determination were comparable between SLEDAI-2K and SLE-DAS. Conclusions: There were no clear differences in the use of SLE-DAS over SLEDAI-2K in assessing HRQoL in patients with SLE. We suggest that, in this aspect, both SLEDAI-2K and SLE-DAS are effective tools for measuring disease activity in patients with SLE.

**Keywords:** systemic lupus erythematosus; quality of life; cross-sectional studies; surveys and questionnaires

## 1. Introduction

Systemic lupus erythematosus (SLE) is a chronic systemic autoimmune disease involving multiple organ systems, such as the skin, kidneys, blood, joints, and brain [1]. The disease predominantly affects women of childbearing age, with female-to-male ratio of 9 to 1. The clinical course of SLE is highly variable with recurrent relapses and exacerbations.

Despite the advancements in therapeutic options and the improvement in the survival rate for SLE [2], a high proportion of patients living with SLE have a poor health-related quality of life (HRQoL) compared with healthy individuals as well as patients with other chronic diseases, such as diabetes, hypertension, and even heart failure [3]. Fatigue, pain, and musculoskeletal distress associated with SLE have been reported to be the main predictors of poor HRQoL [4]. Older age, poverty, lower educational level, behavioral issues, some clinical manifestations, and comorbidities could also have an impact on HRQoL [5]. In addition, disease activity status has been suggested to adversely affect HRQoL in patients with SLE [6–11].

One of the most commonly used measures for the global disease activity of SLE is the Systemic Lupus Erythematosus Disease Activity Index 2000 (SLEDAI-2K) introduced in 2002. It is a modification of the original Systemic Lupus Erythematosus Disease Activity Index (SLEDAI) developed by consensus of a group of experienced clinicians in the field of lupus research [12]. The SLEDAI-2K was validated against SLEDAI in a cohort of 960 patients and a high correlation of 0.97 between the two indices was reported [13]. More recently, a new 17-item Systemic Lupus Erythematosus Disease Activity Score (SLE-DAS) with improved sensitivity to changes in SLE disease activity as compared with SLEDAI was proposed. In a study of 520 patients with SLE, the SLE-DAS showed a significantly better performance than SLEDAI-2K in identifying clinically meaningful changes in disease activity and in predicting damage accrual [14]. The scale was subsequently validated in an independent cohort of 227 Latin American patients with Mexican Mestizo ethnicity. Nevertheless, the authors concluded that SLE-DAS did not add an advantage over the existing SLEDAI-2K score, particularly regarding its suboptimal performance in patients with high disease activity [15]. In addition, the choice of outcome measures for the musculoskeletal component in SLE-DAS has been challenged by a study that reanalyzed the data with SLE-DAS obtained from a longitudinal study of patients with SLE [16]. Furthermore, another study retrospectively calculated SLE-DAS for 41 patients with lupus nephritis and revealed that the performance of SLE-DAS among patients of high disease activity might not be robust. The authors concluded that there might be no added advantage over the existing SLEDAI-2K score in the current state of SLE-DAS [17].

Given that measuring SLE disease activity remains a challenging and complex task, it is clear that a broader evaluation of the new SLE-DAS is needed, particularly, in diverse populations across a spectrum of severity and types of clinical manifestations of SLE [18]. At present, no studies have yet attempted to compare the correlation of these two indices in predicting HRQoL in patients with SLE. Therefore, the aim of this cross-sectional study was to compare the correlation of SLEDAI-2K and SLE-DAS with a disease-specific HRQoL, the Lupus Quality of Life questionnaire (LupusQoL) [19], in patients with SLE.

## 2. Materials and Methods

### 2.1. Study Design and Study Population

This cross-sectional study was conducted at the rheumatology outpatient clinic in a regional hospital in southern Taiwan from April to August 2019. Patients were consecutively enrolled and all participants signed informed consent under a study protocol approved by the institutional review board of the Dalin Tzu Chi Hospital, Buddhist Tzu Chi Medical Foundation (No. B10801017). The study was carried out in accordance with the Declaration of Helsinki.

Patients aged 20 years and older, with a clinician-confirmed diagnosis of SLE based on the 1997 American College of Rheumatology (ACR) revised criteria [20] or the 2012 Systemic Lupus International Collaborating Clinics Classification Criteria [21] were recruited. The exclusion criteria included patients who had previously been diagnosed with other major systemic diseases, including rheumatoid arthritis, polymyositis, dermatomyositis, systemic sclerosis, spondyloarthritis, and juvenile idiopathic arthritis.

### 2.2. Measurement of Disease Activity

SLE disease activity was assessed using rheumatologist-scored SLEDAI-2K [13] and SLE-DAS [14]. The SLEDAI-2K consists of 24 items covering nine organ systems. The recall period for disease activity is the previous 10 days. The score ranges from 0 to 105 points, with higher values signifying greater disease activity.

The SLE-DAS consists of 17 items and has all disease manifestations in the 24-item SLEDAI-2K with added items for hemolytic anemia, cardiopulmonary, and gastrointestinal involvement. The SLE-DAS is a continuous disease activity score with higher values signifying greater disease activity [14].

### 2.3. Measurement of Disease-Specific Quality of Life and Other Variables

Demographic and clinical information of the patients was collected using a paper-based questionnaire consisting of questions on sex, age interval, body mass index, educational level, marital status, job change due to SLE, employment status, self-perceived health status, duration of SLE, age of diagnosis of SLE, alcohol use, smoking, betel nut chewing, regular exercise, and sleep duration. The questionnaire was administered by two experienced research nurses of the rheumatology outpatient clinic.

The LupusQoL, which is one of the most validated measures of disease-specific HRQoL in patients with SLE, was used in this study [22]. The original LupusQoL was developed from qualitative interviews with patients with SLE and expert panel agreement followed by psychometric evaluation [19]. The LupusQoL consists of 34 items grouped in eight domains of HRQoL, including physical health (8 items), emotional health (6 items), body image (5 items), pain (3 items), planning (3 items), fatigue (4 items), intimate relationships (2 items), and burden to others (3 items). The recall period is the previous four weeks. The response scale was a five-point Likert format, where 0 = all of the time, 1 = most of the time, 2 = a good bit of the time, 3 = occasionally, and 4 = never. For each domain, the mean domain score is obtained by dividing the total score by the number of items in that domain. The mean domain score is rescaled to a final score ranging from 0 to 100 by dividing by 4 (the number of Likert responses minus 1) and then multiplying by 100. A non-applicable response is available in six of the items, and it is treated as unanswered. A higher score in a domain indicates a better health-related quality of life for that particular domain. The validity of the original English version of LupusQoL has been demonstrated in patients with SLE in the United Kingdom [19] and the United States [23]. In this study, we used the official Chinese for Taiwan version of the LupusQoL, which was obtained from RWS Life Sciences with permission for use in this study. A study in China on 208 patients with SLE, using the LupusQoL-China culturally adapted from the Chinese for Taiwan version, demonstrated evidence of construct validity when compared with equivalent domains on the EQ-5D. In addition, the internal consistency reliability Cronbach's $\alpha$ ranged from 0.81 to 0.96 with the test–retest reliability ranging from 0.84 to 0.97 across the different domains for the LupusQoL-China [24].

### 2.4. Statistical Analysis

All statistical analyses were performed using SAS software release 9.4 (SAS Institute, Inc., Cary, NC, USA). Continuous variables were summarized as mean with standard deviation (SD) and median with interquartile range, as appropriate. Categorical variables were presented as frequencies and percentages. Separate linear regression analyses for each of the eight domains of LupusQoL were performed with SLEDAI-2K and SLE-DAS as independent variables. Because sex differences were observed in HRQoL in patients with SLE [25], linear regression models were fitted with and without adjusting for sex and age interval.

The correlations of SLEDAI-2K and SLE-DAS with LupusQoL were assessed using five regression model accuracy metrics, including mean absolute error (MAE), root mean square error (RMSE), Akaike Information Criterion (AIC), Bayesian Information Criterion (BIC), and coefficient of determination ($R^2$). The MAE is the average of the absolute differences between prediction and actual observation with all individual differences has equal weight. The RMSE is the square root of the average of squared differences between prediction and actual observation, and therefore it gives relatively high weight to large errors. A smaller value in MAE and RMSE indicates better model performance. Similarly, a lower AIC or BIC value indicates a better model fit. Conversely, because $R^2$ is the proportion of variation in the outcome that is explained by the predictor variables; therefore the higher the $R^2$, the better the model [26]. The differences in the MAE and RMSE between SLEDAI-2K and SLE-DAS were compared using the paired *t*-test. In addition, the correlation between SLEDAI-2K and SLE-DAS was determined using Pearson's correlation coefficient and Spearman's rank correlation coefficient.

## 3. Results

The demographic and clinical information of the 333 patients with SLE are shown in Table 1. In brief, 90.4% were female and 40% were between the ages of 20 and 39 years. Approximately 54% of the patients had a normal body mass index, and 50% had an educational level of college or above. About 29% had to change their jobs due to SLE, and 73% rated their own health as average or below. In addition, 64% of the patients had SLE for more than nine years. In addition, 55.3% patients with SLE had low complement levels and 35.1% had increased anti-double strain DNA antibody titer. Clinically, 61.6% patients with SLE had Raynaud's phenomenon and 51.7% had photosensitivity.

**Table 1.** Demographic data of patients with systemic lupus erythematosus ($N$ = 333).

| Variable | $n$ | (%) |
|---|---|---|
| Sex | | |
| female | 301 | (90.4) |
| male | 32 | (9.6) |
| Age interval (years) | | |
| 20–29 | 40 | (12.0) |
| 30–39 | 94 | (28.2) |
| 40–49 | 78 | (23.4) |
| 50–59 | 64 | (19.2) |
| $\geq$60 | 57 | (17.1) |
| Body mass index (kg/m$^2$) | | |
| normal ($\geq$18.5 and <24.0) | 179 | (53.8) |
| other | 154 | (46.2) |
| Educational level | | |
| high school or below | 165 | (49.5) |
| college or above | 168 | (50.5) |
| Marital status | | |
| single | 111 | (33.3) |
| married, widowed, divorced | 222 | (66.7) |
| Change job due to SLE | | |
| no | 237 | (71.2) |
| yes | 96 | (28.8) |
| Employment status | | |
| unemployed | 119 | (35.7) |
| employed | 214 | (64.3) |
| Self-perceived health status | | |
| good or very good | 90 | (27.0) |
| average | 189 | (56.8) |
| poor or very poor | 54 | (16.2) |
| Disease duration, years | | |
| $\leq$9 | 121 | (36.3) |
| >9 | 212 | (63.7) |
| Age at diagnosis of SLE, years | | |
| <29 | 177 | (53.2) |
| $\geq$30 | 156 | (46.8) |

**Table 1.** *Cont.*

| Variable | *n* | (%) |
|---|---|---|
| Alcohol use | | |
| no | 255 | (76.6) |
| yes | 78 | (23.4) |
| Smoking | | |
| no | 301 | (90.4) |
| yes | 32 | (9.6) |
| Betel nut chewing | | |
| no | 326 | (97.9) |
| yes/ever | 7 | (2.1) |
| Regular exercise | | |
| no | 57 | (17.1) |
| yes | 276 | (82.9) |
| Sleep duration, hours | | |
| 0–7 | 268 | (80.5) |
| $\geq$8 | 65 | (19.5) |
| Low complement level | 184 | (55.3) |
| Increased anti-dsDNA antibody titer | 117 | (35.1) |
| Thrombocytopenia ($<100{,}000/mm^3$) | 12 | (3.6) |
| Leukopenia ($<3000/mm^3$) | 17 | (5.1) |
| Anemia | 138 | (41.4) |
| Raynaud's phenomenon | 205 | (61.6) |
| Photosensitivity | 172 | (51.7) |
| Sjögren's syndrome | 93 | (27.9) |
| Arthritis | 72 | (21.6) |
| Renal involvement | 47 | (14.1) |

Summary statistics of SLEDAI-2K, SLE-DAS, and individual domains of LupusQoL are also presented in Table 2. The median SLEDAI-2K and SLE-DAS was 4.00 (interquartile range [IQR] 2.00–7.50) and 2.08 (IQR 1.12–8.24), respectively. Figure 1 shows a scatter plot of SLEDAI-2K and SLE-DAS. There was a moderate correlation between SLEDAI-2K and SLE-DAS (Pearson's r = 0.66; 95% CI 0.60, 0.72; $p < 0.001$; Spearman's $\rho$ = 0.78; 95% CI 0.71, 0.83; $p < 0.001$).

**Table 2.** Systemic Lupus Erythematosus Disease Activity Score (SLE-DAS), Systemic Lupus Erythematosus Disease Activity Index 2000 (SLEDAI-2K), and Lupus Quality of Life (LupusQoL) of patients with systemic lupus erythematosus (*N* = 333).

| Variable | Mean | Standard Deviation | Median | Interquartile Range |
|---|---|---|---|---|
| SLEDAI-2K | 4.87 | (4.42) | 4.00 | (2.00, 7.50) |
| SLE-DAS | 5.43 | (6.97) | 2.08 | (1.12, 8.24) |
| Domain of LupusQoL | | | | |
| Physical health | 81.2 | (20.0) | 87.5 | (75.0, 96.9) |
| Emotional health | 83.0 | (20.1) | 87.5 | (75.0, 100.0) |
| Body image | 82.4 | (23.4) | 90.0 | (70.0, 100.0) |
| Pain | 80.0 | (26.9) | 91.7 | (75.0, 100.0) |
| Planning | 81.2 | (26.0) | 91.7 | (75.0, 100.0) |
| Fatigue | 72.0 | (23.8) | 75.0 | (56.2, 93.8) |
| Intimate relationships | 73.8 | (33.4) | 87.5 | (62.5, 100.0) |
| Burden to others | 72.1 | (30.3) | 75.00 | (58.3, 100.0) |

3.0% (*N* = 10) in the body image domain and 22.2% (*N* = 74) in the intimate relationships of the responses were missing because items were reported as not applicable by the patients.

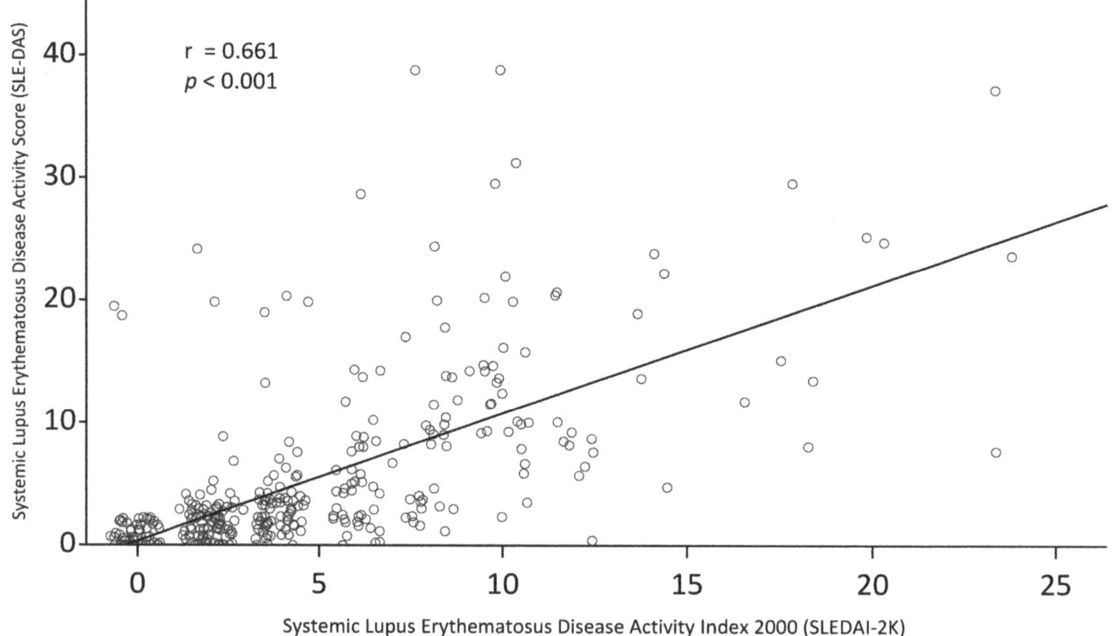


**Figure 1.** Scatter plot of Systemic Lupus Erythematosus Disease Activity Index 2000 (SLEDAI-2K) and Systemic Lupus Erythematosus Disease Activity Score (SLE-DAS).

Tables 3 and 4 show the association of the eight domains of LupusQoL with SLEDAI-2K and SLE-DAS, respectively. In Table 3, SLEDAI-2K was significantly and inversely associated with five domains of LupusQoL, namely, emotional health ($p = 0.036$), body image ($p = 0.033$), pain ($p = 0.033$), fatigue ($p = 0.003$), and burden to others ($p < 0.001$). When adjusting for sex and age interval, SLEDAI-2K became significantly and inversely associated with all eight domains of LupusQoL. The standardized beta coefficients for the eight domains ranged from the highest at $-0.238$ in burden to others to the lowest at $-0.123$ in planning. The three domains with the highest standardized beta coefficients were burden to others ($-0.238$), followed by pain ($-0.196$) and physical health ($-0.192$).

**Table 3.** Linear regression analyses of the eight domains of Lupus Quality of Life (LupusQoL) with the Systemic Lupus Erythematosus Disease Activity Index 2000 (SLEDAI-2K) in patients with systemic lupus erythematosus.

| Domain of LupusQoL | Simple Linear Regression Analysis | | | | Linear Regression Analysis Adjusted for Sex and Age Interval | | | |
|---|---|---|---|---|---|---|---|---|
| | β | (95% CI) | Std. β | *p* | β | (95% CI) | Std. β | *p* |
| Physical health | −0.459 | (−0.945, 0.028) | −0.101 | 0.065 | −0.871 | (−1.347, −0.394) | −0.192 | <0.001 |
| Emotional health | −0.523 | (−1.012, −0.034) | −0.115 | 0.036 | −0.670 | (−1.181, −0.159) | −0.147 | 0.010 |
| Body image | −0.628 | (−1.205, −0.051) | −0.119 | 0.033 | −0.673 | (−1.276, −0.069) | −0.127 | 0.029 |
| Pain | −0.715 | (−1.370, −0.059) | −0.117 | 0.033 | −1.196 | (−1.859, −0.532) | −0.196 | <0.001 |
| Planning | −0.627 | (−1.262, 0.007) | −0.106 | 0.053 | −0.728 | (−1.392, −0.063) | −0.123 | 0.032 |
| Fatigue | −0.866 | (−1.441, −0.290) | −0.160 | 0.003 | −0.997 | (−1.598, −0.397) | −0.185 | 0.001 |
| Intimate relationships | −0.228 | (−1.156, 0.699) | −0.030 | 0.628 | −1.297 | (−2.175, −0.419) | −0.172 | 0.004 |
| Burden to others | −1.663 | (−2.383, −0.944) | −0.243 | <0.001 | −1.633 | (−2.389, −0.877) | −0.238 | <0.001 |

CI: confidence interval; std: standardized. 3.0% ($N = 10$) in the body image domain and 22.2% ($N = 74$) in the intimate relationships of the responses were missing because these items were reported as not applicable.

In Table 4, SLE-DAS was significantly and inversely associated with six domains of LupusQoL, namely, physical health ($p = 0.003$), emotional health ($p = 0.007$), pain ($p = 0.002$), fatigue ($p = 0.001$), intimate relationships ($p = 0.022$), and burden to others ($p < 0.001$). When adjusting for sex and age interval, SLE-DAS also became significantly and inversely associated with all eight domains of LupusQoL. The standardized beta coefficients for the eight domains ranged from the highest at $-0.217$ in physical health to the two lowest at $-0.115$ in planning and body image. The three domains with the highest standardized beta coefficients were physical health ($-0.217$), followed by burden to others ($-0.216$), and pain ($-0.203$).

**Table 4.** Linear regression analyses of the eight domains of Lupus Quality of Life (LupusQoL) with the Systemic Lupus Erythematosus Disease Activity Score (SLE-DAS) in patients with systemic lupus erythematosus.

| Domain of LupusQoL | Simple Linear Regression Analysis | | | | Linear Regression Analysis Adjusted for Sex and Age Interval | | | |
|---|---|---|---|---|---|---|---|---|
| | $\beta$ | (95% CI) | Std. $\beta$ | $p$ | $\beta$ | (95% CI) | Std. $\beta$ | $p$ |
| Physical health | $-0.469$ | $(-0.775, -0.164)$ | $-0.164$ | 0.003 | $-0.623$ | $(-0.913, -0.332)$ | $-0.217$ | <0.001 |
| Emotional health | $-0.423$ | $(-0.731, -0.115)$ | $-0.147$ | 0.007 | $-0.469$ | $(-0.782, -0.156)$ | $-0.163$ | 0.003 |
| Body image | $-0.356$ | $(-0.720, 0.007)$ | $-0.107$ | 0.055 | $-0.385$ | $(-0.755, -0.015)$ | $-0.115$ | 0.042 |
| Pain | $-0.642$ | $(-1.054, -0.230)$ | $-0.166$ | 0.002 | $-0.784$ | $(-1.191, -0.377)$ | $-0.203$ | <0.001 |
| Planning | $-0.381$ | $(-0.783, 0.021)$ | $-0.102$ | 0.063 | $-0.429$ | $(-0.838, -0.021)$ | $-0.115$ | 0.039 |
| Fatigue | $-0.594$ | $(-0.957, -0.230)$ | $-0.174$ | 0.001 | $-0.638$ | $(-1.006, -0.269)$ | $-0.187$ | <0.001 |
| Intimate relationships | $-0.664$ | $(-1.230, -0.098)$ | $-0.041$ | 0.022 | $-0.915$ | $(-1.434, -0.396)$ | $-0.197$ | 0.001 |
| Burden to others | $-0.943$ | $(-1.401, -0.485)$ | $-0.217$ | <0.001 | $-0.936$ | $(-1.402, -0.470)$ | $-0.216$ | <0.001 |

CI: confidence interval; std: standardized. 3.0% ($N = 10$) in the body image domain and 22.2% ($N = 74$) in the intimate relationships of the responses were missing because these items were reported as not applicable.

Correlations of SLEDAI-2K and SLE-DAS with LupusQoL were evaluated by comparing five regression model accuracy metrics (Table 5). The magnitudes of MAE, RMSE, AIC, BIC, and $R^2$ were comparable between SLEDAI-2K and SLE-DAS. In addition, MAE and RMSE obtained from SLEDAI-2K and SLE-DAS were not significantly different for all eight domains of LupusQoL.

**Table 5.** Regression model accuracy metrics of the eight domains of Lupus Quality of Life (LupusQoL) with Systemic Lupus Erythematosus Disease Activity Score (SLE-DAS) and Systemic Lupus Erythematosus Disease Activity Index 2000 (SLEDAI-2K) adjusted for age and sex.

| Domain of LupusQoL | SLE-DAS | | | | | SLEDAI-2K | | | | | $p$ | |
|---|---|---|---|---|---|---|---|---|---|---|---|---|
| | MAE | RMSE | AIC | BIC | $R^2$ | MAE | RMSE | AIC | BIC | $R^2$ | MAE | RMSE |
| Physical health | 13.31 | 18.28 | 2904.7 | 2923.8 | 0.159 | 13.47 | 18.41 | 2905.9 | 2925.0 | 0.147 | 0.370 | 0.578 |
| Emotional health | 14.71 | 19.68 | 2942.9 | 2961.9 | 0.037 | 14.68 | 19.74 | 2944.7 | 2963.8 | 0.032 | 0.859 | 0.722 |
| Body image | 17.58 | 23.15 | 2958.7 | 2977.6 | 0.021 | 17.60 | 23.13 | 2957.7 | 2976.6 | 0.023 | 0.899 | 0.864 |
| Pain | 19.29 | 25.60 | 3116.9 | 3135.9 | 0.094 | 19.51 | 25.67 | 3118.0 | 3137.0 | 0.089 | 0.322 | 0.758 |
| Planning | 19.34 | 25.70 | 3120.2 | 3139.2 | 0.024 | 19.31 | 25.68 | 3119.1 | 3138.2 | 0.025 | 0.829 | 0.909 |
| Fatigue | 18.89 | 23.18 | 3051.8 | 3070.8 | 0.050 | 18.89 | 23.21 | 3052.0 | 3071.0 | 0.048 | 0.965 | 0.865 |
| Intimate relationships | 23.27 | 29.52 | 2504.9 | 2522.7 | 0.216 | 23.43 | 29.72 | 2506.7 | 2524.5 | 0.205 | 0.583 | 0.548 |
| Burden to others | 23.48 | 29.34 | 3208.2 | 3227.2 | 0.058 | 23.33 | 29.23 | 3205.8 | 3224.9 | 0.064 | 0.580 | 0.702 |

AIC: Akaike Information Criterion; BIC: Bayesian Information Criterion; MAE: Mean Absolute Error; RMSE: Root Mean Square Error. 3.0% ($N = 10$) in the body image domain and 22.2% ($N = 74$) in the intimate relationships of the responses were missing because these items were reported as not applicable.

In Tables 6 and 7, correlations of SLEDAI-2K and SLE-DAS with LupusQoL in patients with or without renal involvement were evaluated by comparing five regression model accuracy metrics. The magnitudes of MAE, RMSE, AIC, BIC, and $R^2$ were comparable between SLEDAI-2K and SLE-DAS. In addition, MAE and RMSE obtained from SLEDAI-

2K and SLE-DAS were not significantly different for all eight domains of LupusQoL in patients with SLE with renal involvement or not.

**Table 6.** Regression model accuracy metrics of the eight domains of Lupus Quality of Life (LupusQoL) with Systemic Lupus Erythematosus Disease Activity Score (SLE-DAS) and Systemic Lupus Erythematosus Disease Activity Index 2000 (SLEDAI-2K) adjusted for age and sex in SLE patients with renal involvement ($N = 47$).

| Domain of LupusQoL | SLE-DAS | | | | | SLEDAI-2K | | | | | $p$ | |
|---|---|---|---|---|---|---|---|---|---|---|---|---|
| | MAE | RMSE | AIC | BIC | $R^2$ | MAE | RMSE | AIC | BIC | $R^2$ | MAE | RMSE |
| Physical health | 12.17 | 17.81 | 419.4 | 428.7 | 0.191 | 10.70 | 15.76 | 408.3 | 417.5 | 0.367 | 0.252 | 0.178 |
| Emotional health | 12.84 | 17.08 | 419.4 | 428.6 | 0.198 | 12.68 | 16.90 | 418.7 | 428.0 | 0.216 | 0.765 | 0.594 |
| Body image | 18.04 | 25.05 | 448.4 | 457.7 | 0.091 | 18.17 | 24.22 | 446.5 | 455.8 | 0.150 | 0.890 | 0.428 |
| Pain | 16.43 | 22.47 | 442.2 | 451.4 | 0.290 | 16.59 | 21.50 | 437.9 | 447.2 | 0.350 | 0.900 | 0.436 |
| Planning | 19.12 | 25.82 | 450.6 | 459.9 | 0.100 | 18.28 | 24.01 | 443.0 | 452.2 | 0.222 | 0.547 | 0.144 |
| Fatigue | 16.97 | 20.60 | 434.5 | 443.7 | 0.155 | 16.64 | 20.53 | 434.4 | 443.6 | 0.161 | 0.575 | 0.883 |
| Intimate relationships ($n = 37$) | 21.17 | 26.61 | 363.4 | 371.4 | 0.372 | 21.32 | 27.13 | 364.1 | 372.1 | 0.347 | 0.867 | 0.654 |
| Burden to others | 23.18 | 28.65 | 466.1 | 475.3 | 0.191 | 22.82 | 27.96 | 462.4 | 471.7 | 0.230 | 0.790 | 0.602 |

AIC: Akaike Information Criterion; BIC: Bayesian Information Criterion; MAE: Mean Absolute Error; RMSE: Root Mean Square Error; 22.3% ($N = 10$) in intimate relationships of the responses were missing because these items were reported as not applicable.

**Table 7.** Regression model accuracy metrics of the eight domains of Lupus Quality of Life (LupusQoL) with Systemic Lupus Erythematosus Disease Activity Score (SLE-DAS) and Systemic Lupus Erythematosus Disease Activity Index 2000 (SLEDAI-2K) adjusted for age and sex SLE patients without renal involvement ($N = 286$).

| Domain of LupusQoL | SLE-DAS | | | | | SLEDAI-2K | | | | | $p$ | |
|---|---|---|---|---|---|---|---|---|---|---|---|---|
| | MAE | RMSE | AIC | BIC | $R^2$ | MAE | RMSE | AIC | BIC | $R^2$ | MAE | RMSE |
| Physical health | 13.19 | 18.00 | 2487.8 | 2506.1 | 0.186 | 13.51 | 18.28 | 2492.6 | 2510.8 | 0.160 | 0.107 | 0.331 |
| Emotional health | 14.92 | 19.83 | 2532.2 | 2550.5 | 0.037 | 14.88 | 19.92 | 2535.2 | 2553.4 | 0.027 | 0.772 | 0.645 |
| Body image | 17.38 | 22.69 | 2517.6 | 2535.7 | 0.013 | 17.37 | 22.69 | 2517.4 | 2535.5 | 0.013 | 0.879 | 0.996 |
| Pain | 19.56 | 25.79 | 2682.0 | 2700.2 | 0.082 | 19.78 | 25.89 | 2683.9 | 2702.2 | 0.075 | 0.305 | 0.659 |
| Planning | 19.18 | 25.55 | 2677.3 | 2695.6 | 0.018 | 19.20 | 25.59 | 2677.7 | 2696.0 | 0.015 | 0.865 | 0.720 |
| Fatigue | 19.01 | 23.35 | 2625.1 | 2643.4 | 0.052 | 19.05 | 23.39 | 2625.7 | 2644.0 | 0.049 | 0.878 | 0.880 |
| Intimate relationships | 23.49 | 29.62 | 2149.4 | 2166.4 | 0.208 | 23.67 | 29.84 | 2151.5 | 2168.5 | 0.197 | 0.607 | 0.580 |
| Burden to others | 23.09 | 29.15 | 2751.3 | 2769.6 | 0.043 | 23.07 | 29.09 | 2750.8 | 2769.1 | 0.047 | 0.938 | 0.837 |

AIC: Akaike Information Criterion; BIC: Bayesian Information Criterion; MAE: Mean Absolute Error; RMSE: Root Mean Square Error; 3.5% ($N = 10$) in the body image domain and 22.4% ($N = 64$) in the intimate relationships of the responses were missing because these items were reported as not applicable.

## 4. Discussion

Measuring disease activity in patients with SLE is important but complex. In this study on 333 patients with SLE, a commonly used SLEDAI-2K was compared with a more recently developed SLE-DAS scoring tool. Overall, we found that the correlations between SLEDAI-2K and SLE-DAS with HRQoL, as measured by LupusQoL, were similar in our patients with SLE. We used five regression model accuracy metrics to assess the performance of the two disease activity measures, and no clear advantages were observed with the newer SLE-DAS over the SLEDAI-2K with respect to their associations with HRQoL. In addition, while there were small differences in the magnitude of the $R^2$ between the SLEDAI-2K and SLE-DAS, the differences were not in the same direction for the eight domains of LupusQoL. Furthermore, the magnitudes of the $R^2$ ranged from 0.023 to 0.205 in SLEDAI-2K and 0.021 to 0.216 in SLE-DAS support the view that HRQoL is a different entity from disease activity. Reduced disease activity as a result of treatment may not correlate with improved HRQoL because of the side effects of the medication [27]. Therefore, both of these entities need to be measured for a more complete clinical picture.

The agreement between SLEDAI-2K and SLE-DAS was evaluated using Spearman's correlation coefficient. In the original SLE-DAS study, SLE-DAS was shown to be strongly correlated with SLEDAI-2K measured at the last follow-up visit of the external validation cohort, with a $\rho = 0.94$ [14]. In our study, a $\rho$ of 0.78 was observed between SLEDAI-2K and SLE-DAS, which is similar to the 0.70 in a study of 41 Indian patients with lupus nephritis [17]. The low correlation could be attributed to a difference in the distribution of the disease activity between the studies. In a study of 227 Latin American patients with SLE, the authors pointed out that the correlation appeared to depend on the level of the disease activity, with a stronger correlation observed in patients with quiescence or low disease activity [15].

Regarding the associations with various domains of the LupusQoL, SLEDAI-2K and SLE-DAS were similar. When adjusting for sex and age interval, both SLEDAI-2K and SLE-DAS were significantly and inversely associated with all eight domains of LupusQoL. In terms of the magnitude of the standardized beta coefficients of SLEDAI-2K and SLE-DAS, while their rankings were not identical, they were in general agreement. Burden to others, pain, and physical health were the top three domains, whereas emotional health, body image, and planning were the bottom three domains. Several previous studies on patients with SLE from different cultural and ethnic groups have shown varying degrees of association between disease activity and HRQoL. Some studies showed that all the domains were significantly associated with active disease status, whereas some did not. In a study assessing the psychometric properties of LupusQoL in 208 Chinese patients with SLE, the Chinese version of LupusQoL could discriminate patients with active disease activity, defined as a SLEDAI score >4, in all domains except for body image [24]. In addition, a study on 132 Turkish patients with SLE found that all domains except planning of the Turkish version of LupusQoL were able to discriminate between active and inactive SLE groups [28]. Moreover, a study on 78 Iranian patients with SLE showed that active disease, assessed by SLEDAI-2K, was significantly associated with planning, emotional health, and body image domains of the Persian version of the LupusQoL [29]. Furthermore, a cohort study of 182 French patients with SLE showed that the French version of LupusQoL was significantly lower only for physical health, pain, and intimate relationship in patients with SLEDAI >4 [30]. Conversely, no significant differences in any domains of an Argentine version of LupusQoL were observed between 147 patients with a SLEDAI score of <4 and ≥4 [31]. The heterogeneity of the findings from the abovementioned studies might be explained by differences in ethnic composition, cultural setting, and healthcare infrastructure, which could affect the perception of HRQoL in patients with SLE [32].

Our study has some limitations that deserve mention. First, our patients were enrolled from our outpatient clinic, and therefore, the disease activities were relatively mild. Correlations of SLEDAI-2k and SLE-DAS and LupusQoL in patients with more severe disease activity should be investigated in future studies. Second, we did not measure other variables that might potentially affect HRQoL. Nevertheless, we adjusted the association between the two indexes and HRQoL for age and sex, which are likely to be the two most notable potential confounders of the associations. Despite these limitations, to the best of our knowledge, this is the first study to compare the association of HRQoL between SLEDAI-2k and SLE-DAS. The large sample size is also a strength of this study.

In conclusion, findings from this study showed that there were no clear differences in the use of SLE-DAS over SLEDAI-2K in assessing various domains of HRQoL in patients with SLE. We suggest that, in this aspect, both SLEDAI-2K and SLE-DAS are comparable in their associations with disease activity in patients with SLE.

**Author Contributions:** N.-S.L. and M.-C.L. contributed to the conception and methodology of the study. H.-C.L. and H.-H.C. collected data and administrated the project. C.-W.H. and M.K. performed statistical analysis. N.-S.L., M.-C.L., K.-Y.H., C.-H.T., B.-B.H., and C.-H.W. assisted in data collection and interpretation. M.-C.L. and M.K. wrote the manuscript. All authors have read and agreed to the published version of the manuscript.

**Funding:** This work was supported by grants from the Dalin Tzu Chi Hospital, Buddhist Tzu Chi Medical Foundation, Taiwan (No. DTCRD109-I-21 and DTCRD109-I-23) and the Buddhist Tzu Chi Medical Foundation, Taiwan (No. TCMF-A 108-05).

**Institutional Review Board Statement:** The study was conducted according to the guidelines of the Declaration of Helsinki, and approved by the Institutional Review Board of the Dalin Tzu Chi Hospital, Buddhist Tzu Chi Medical Foundation (No. B10801017, date of approval 29 March 2019).

**Informed Consent Statement:** All participants signed informed consent.

**Data Availability Statement:** The data presented in this study are available on request from the corresponding author.

**Conflicts of Interest:** The authors declare that they have no competing interest.

# References


1. Tsokos, G.C. Systemic lupus erythematosus. *N. Engl. J. Med.* **2011**, *365*, 2110–2121. [CrossRef] [PubMed]
2. Ippolito, A.; Petri, M. An update on mortality in systemic lupus erythematosus. *Clin. Exp. Rheumatol.* **2008**, *26*, 72–79.
3. Jolly, M. How does quality of life of patients with systemic lupus erythematosus compare with that of other common chronic illnesses? *J. Rheumatol.* **2005**, *32*, 1706–1708. [PubMed]
4. Pettersson, S.; Lövgren, M.; Eriksson, L.; Moberg, C.; Svenungsson, E.; Gunnarsson, I.; Henriksson, E.W. An exploration of patient-reported symptoms in systemic lupus erythematosus and the relationship to health-related quality of life. *Scand. J. Rheumatol.* **2012**, *41*, 383–390. [CrossRef]
5. Elera-Fitzcarrald, C.; Fuentes, A.; González, L.A.; Burgos, P.I.; Alarcón, G.S.; Ugarte-Gil, M.F. Factors affecting quality of life in patients with systemic lupus erythematosus: Important considerations and potential interventions. *Expert Rev. Clin. Immunol.* **2018**, *14*, 915–931. [CrossRef]
6. Mok, C.C.; Ho, L.Y.; Cheung, M.Y.; Yu, K.L.; To, C.H. Effect of disease activity and damage on quality of life in patients with systemic lupus erythematosus: A 2-year prospective study. *Scand. J. Rheumatol.* **2009**, *38*, 121–127. [CrossRef]
7. Ivorra, J.A.R.; Fernández-Llanio-Comella, N.; San-Martín-Álvarez, A.; Vela-Casasempere, P.; Saurí-Ferrer, I.; González-De-Julián, S.; Vivas-Consuelo, D. Health-related quality of life in patients with systemic lupus erythematosus: A Spanish study based on patient reports. *Clin. Rheumatol.* **2019**, *38*, 1857–1864. [CrossRef] [PubMed]
8. Bexelius, C.; Wachtmeister, K.; Skare, P.; Jönsson, L.; Van Vollenhoven, R. Drivers of cost and health-related quality of life in patients with systemic lupus erythematosus (SLE): A Swedish nationwide study based on patient reports. *Lupus* **2013**, *22*, 793–801. [CrossRef]
9. Poomsalood, N.; Narongroeknawin, P.; Chaiamnuay, S.; Asavatanabodee, P.; Pakchotanon, R. Prolonged clinical remission and low disease activity statuses are associated with better quality of life in systemic lupus erythematosus. *Lupus* **2019**, *28*, 1189–1196. [CrossRef]
10. Chaigne, B.; Chizzolini, C.; Perneger, T.; Trendelenburg, M.; Huynh-Do, U.; Dayer, E.; Stoll, T.; von Kempis, J.; Ribi, C.; Swiss Systemic Lupus Erythematosus Cohort Study Group. Impact of disease activity on health-related quality of life in sys-temic lupus erythematosus-across-sectional analysis of the Swiss Systemic Lupus Erythematosus Cohort Study (SSCS). *BMC Immunol.* **2017**, *18*, 17. [CrossRef]
11. Shen, B.; Feng, G.; Tang, W.; Huang, X.; Yan, H.; He, Y.; Chen, W.; Da, Z.; Liu, H.; Gu, Z. The quality of life in Chinese pa-tients with systemic lupus erythematosus is associated with disease activity and psychiatric disorders: A path analysis. *Clin. Exp. Rheumatol.* **2014**, *32*, 101–107. [PubMed]
12. Bombardier, C.; Gladman, D.D.; Urowitz, M.B.; Caron, D.; Chang, C.H.; Austin, A.; Bell, A.; Bloch, D.A.; Corey, P.N.; Decker, J.L.; et al. Derivation of the sledai. A disease activity index for lupus patients. *Arthritis Rheum.* **1992**, *35*, 630–640. [CrossRef]
13. Gladman, D.D.; Ibañez, M.; Urowitz, M.B. Systemic lupus erythematosus disease activity index. *J. Rheumatol.* **2002**, *29*, 288–291.
14. Jesus, D.; Matos, A.; Henriques, C.; Zen, M.; Larosa, M.; Iaccarino, L.; Da Silva, J.; Doria, A.; Inês, L.S. Derivation and vali-dation of the SLE Disease Activity Score (SLE-DAS): A new SLE continuous measure with high sensitivity for changes in dis-ease activity. *Ann. Rheum. Dis.* **2019**, *78*, 365–371. [CrossRef] [PubMed]
15. Rodríguez-González, M.G.; A Valero-Gaona, G.; Vargas-Aguirre, T.; Guerra, L.M.A. Performance of the systemic lupus ery-thematosus disease activity score (SLE-DAS) in a Latin American population. *Ann. Rheum. Dis.* **2020**, *79*, e158. [CrossRef] [PubMed]
16. Hassan, S.-U.; Mahmoud, K.; Vital, E.M. Assessment of responsiveness of the musculoskeletal component of SLE-DAS in an independent cohort. *Ann. Rheum. Dis.* **2020**, *79*, e51. [CrossRef] [PubMed]
17. Mathew, A.; Chengappa, K.G.; Shah, S.; Negi, V.S. SLE-DAS: Ready for routine use? *Ann. Rheum. Dis.* **2019**, *79*, e116. [CrossRef] [PubMed]
18. Askanase, A.D.; Merrill, J.T. Measuring disease activity in SLE is an ongoing struggle. *Nat. Rev. Rheumatol.* **2019**, *15*, 194–195. [CrossRef]
19. McElhone, K.; Abbott, J.; Shelmerdine, J.; Bruce, I.N.; Ahmad, Y.; Gordon, C.; Peers, K.; Isenberg, D.; Ferenkeh-Koroma, A.; Griffiths, B.; et al. Development and validation of a disease-specific health-related quality of life measure, the LupusQol, for adults with systemic lupus erythematosus. *Arthritis Rheum.* **2007**, *57*, 972–979. [CrossRef]
20. Hochberg, M.C. Updating the American College of Rheumatology revised criteria for the classification of systemic lupus ery-thematosus. *Arthritis Rheum.* **1997**, *40*, 1725. [CrossRef] [PubMed]

21. Petri, M.; Orbai, A.-M.; Alarcón, G.S.; Gordon, C.; Merrill, J.T.; Fortin, P.R.; Bruce, I.N.; Isenberg, D.; Wallace, D.J.; Nived, O.; et al. Derivation and validation of the Systemic Lupus International Collaborating Clinics classification criteria for systemic lupus erythematosus. *Arthritis Rheum.* **2012**, *64*, 2677–2686. [CrossRef] [PubMed]
22. Yazdany, J. Health-related quality of life measurement in adult systemic lupus erythematosus: Lupus Quality of Life (Lu-pusQoL), Systemic Lupus Erythematosus-Specific Quality of Life Questionnaire (SLEQOL), and Systemic Lupus Erythemato-sus Quality of Life Questionnaire (L-QoL). *Arthritis Care. Res.* **2011**, *63* (Suppl. 11), S413–S419.
23. Jolly, M.; Pickard, A.S.; Wilke, C.; Mikolaitis, R.A.; Teh, L.S.; McElhone, K.; Fogg, L.; Block, J. Lupus-specific health outcome measure for US patients: The LupusQoL-US version. *Ann. Rheum. Dis.* **2010**, *69*, 29–33. [CrossRef]
24. Wang, S.L.; Wu, B.; Leng, L.; Bucala, R.; Lu, L.J. Validity of LupusQoL-China for the assessment of health related quality of life in Chinese patients with systemic lupus erythematosus. *PLoS ONE* **2013**, *8*, e63795. [CrossRef]
25. Jolly, M.; Sequeira, W.; Block, J.A.; Toloza, S.; Bertoli, A.; Blazevic, I.; Vila, L.M.; Moldovan, I.; Torralba, K.D.; Mazzoni, D.; et al. Sex Differences in Quality of Life in Patients With Systemic Lupus Erythematosus. *Arthritis Rheum.* **2019**, *71*, 1647–1652. [CrossRef]
26. Pham, H. A New Criterion for Model Selection. *Mathematics* **2019**, *7*, 1215. [CrossRef]
27. McElhone, K.; Abbott, J.; Teh, L.-S. A review of health related quality of life in systemic lupus erythematosus. *Lupus* **2006**, *15*, 633–643. [CrossRef] [PubMed]
28. Pamuk, O.N.; Onat, A.M.; Donmez, S.; Mengüs, C.; Kisacik, B. Validity and reliability of the Lupus QoL index in Turkish systemic lupus erythematosus patients. *Lupus* **2014**, *24*, 816–821. [CrossRef]
29. Hosseini, N.; Bonakdar, Z.S.; Gholamrezaei, A.; Mirbagher, L. Linguistic Validation of the LupusQoL for the Assessment of Quality of Life in Iranian Patients with Systemic Lupus Erythematosus. *Int. J. Rheumatol.* **2014**, *2014*, 1–5. [CrossRef]
30. Devilliers, H.; Amoura, Z.; Besancenot, J.F.; Bonnotte, B.; Pasquali, J.L.; Wahl, D.; Maurier, F.; Kaminsky, P.; Pennaforte, J.L.; Magy-Bertrand, N.; et al. LupusQoL-FR is valid to assess quality of life in patients with sys-temic lupus erythematosus. *Rheumatology* **2012**, *51*, 1906–1915. [CrossRef] [PubMed]
31. Escobar, M.A.M.; Yacuzzi, M.S.; Martinez, R.N.; Lucero, L.G.; I Bellomio, V.; Santana, M.; Galindo, L.; Mayer, M.M.; Barreira, J.C.; Sarano, J.; et al. Validation of an Argentine version of Lupus Quality of Life questionnaire. *Lupus* **2016**, *25*, 1615–1622. [CrossRef] [PubMed]
32. Toloza, S.M.A.; Jolly, M.; Alarcón, G.S. Quality-of-Life Measurements in Multiethnic Patients with Systemic Lupus Erythematosus: Cross-Cultural Issues. *Curr. Rheumatol. Rep.* **2010**, *12*, 237–249. [CrossRef] [PubMed]

Journal of
*Clinical Medicine*

*Review*

# Fatigue in Systemic Lupus Erythematosus: An Update on Its Impact, Determinants and Therapeutic Management

Lou Kawka [1,2], Aurélien Schlencker [1,2], Philippe Mertz [1,2], Thierry Martin [2,3] and Laurent Arnaud [1,2,*]

1   Department of Rheumatology, INSERM UMR-S1109, Hôpitaux Universitaires de Strasbourg, 67000 Strasbourg, France; lou.kawka@chru-strasbourg.fr (L.K.); aurelien.schlencker@chru-strasbourg.fr (A.S.); Philippe.mertz@chru-strasbourg.fr (P.M.)
2   Centre National de Référence des Maladies Auto-Immunes Systémiques Rares Est Sud-Ouest (RESO), 67000 Strasbourg, France; Thierry.martin@chru-strasbourg.fr
3   Department of Clinical Immunology, Hôpitaux Universitaires de Strasbourg, 67000 Strasbourg, France
*   Correspondence: laurent.arnaud@chru-strasbourg.fr

**Abstract:** Fatigue is a complex and multifactorial phenomenon which is often neglected by clinicians. The aim of this review was to analyze the impact, determinants and management of fatigue in patients with Systemic Lupus Erythematosus (SLE). Fatigue is one of the most prevalent symptoms in SLE, reported by 67% to 90% of patients. It is also described as the most bothersome symptom, considering that it may impair key aspects of health-related quality of life, while also leading to employment disability. It is a multifactorial phenomenon involving psychological factors, pain, lifestyle factors such as reduced physical activity, whereas the contribution of disease activity remains controversial. The management of fatigue in patients with SLE should rely upon a person-centered approach, with targeted interventions. Some pharmacological treatments used to control disease activity have demonstrated beneficial effects upon fatigue and non-pharmacological therapies such as psychological interventions, pain reduction and lifestyle changes, and each of these should be incorporated into fatigue management in SLE.

**Keywords:** systemic lupus erythematosus; fatigue; quality of life

**Citation:** Kawka, L.; Schlencker, A.; Mertz, P.; Martin, T.; Arnaud, L. Fatigue in Systemic Lupus Erythematosus: An Update on Its Impact, Determinants and Therapeutic Management. *J. Clin. Med.* **2021**, *10*, 3996. https://doi.org/10.3390/jcm10173996

Academic Editors: Christopher Sjöwall, Ioannis Parodis and Chrong-Reen Wang

Received: 21 July 2021
Accepted: 2 September 2021
Published: 3 September 2021

**Publisher's Note:** MDPI stays neutral with regard to jurisdictional claims in published maps and institutional affiliations.



## 1. Introduction

Fatigue is a universal symptom experienced by nearly everyone in the general population. However, we lack a consensual definition of fatigue. Fatigue can be described as a subjective unpleasant sensation of exhaustion with physical and mental components, which interferes with individuals' ability to function at their normal capacity. Fatigue impairs quality of life, and may lead to irritability, inability to concentrate, and poor motivation [1,2]. In chronic conditions such as Systemic Lupus Erythematosus (SLE), but also in other autoimmune diseases such as Sjögren's syndrome or systemic sclerosis, the experience of fatigue seems to differ from 'everyday tiredness', as being more frequent, unpredictable and typically unresolved by rest [3]. This symptom remains a complex, multidimensional and poorly understood concept, often neglected by clinicians who prefer to focus on objective manifestations. The aim of this review was to report upon the impact, determinants and management of fatigue in patients with SLE.

## 2. The Most Frequent and Disabling Symptom

Fatigue is recognized as one of the most prevalent symptoms in SLE, reported by 67% to 90% of patients, depending on the series [4]. In a 2020 survey analyzing the burden of SLE from the patients' perspective in European countries [4], fatigue was described as the most common symptom (affecting 85.3% of the 4375 respondents) and the most bothersome symptom, which is consistent with previous studies. Fatigue is reported as severe in intensity in more than a third of SLE patients [4].

Fatigue may impair several key aspects of the patient's quality of life, with repercussions on both physical and mental health. Indeed, SLE patients report that fatigue has a negative impact on emotions, cognition, work, activities of daily living, leisure activities, social activities and family activities. They describe physical impairment, with walking and exercising difficulties. Emotional consequences of fatigue, such as frustrations and stress due to being unable to accomplish tasks, sadness or loss of motivation, are also common [5–9]. Fatigue in SLE has a significantly negative impact on work ability and work productivity, as it can lead to limitations in workplace activities by affecting endurance, mobility, concentration, or interactions with employees and coworkers. Fatigue in SLE is also associated with a higher risk of absenteeism and unemployment [10]. Altogether, fatigue is an important determinant in the perception of SLE impact upon patients' daily living, even for those in remission.

### 3. A Multifactorial Manifestation

Altogether, fatigue is a highly multifactorial manifestation (Figure 1), caused by a complex interplay between disease itself, psychosocial, behavioral and personal variables. A recent study from our group described 3 main clusters of fatigue in SLE patients: (1) the most frequent profile (67.5% of the patients) was represented by patients with moderate fatigue, low disease activity and low anxiety and depression; (2) a quarter of the patients had very high fatigue, high depression and anxiety but low disease activity; and (3) less than 10% of the patients had high levels of fatigue, with high disease activity, low anxiety and no depression [11]. This suggests that the mental health status is an important predictor of fatigue in SLE, that disease activity plays a weaker role in SLE fatigue, and that, most of the time, other factors contribute to fatigue in SLE.

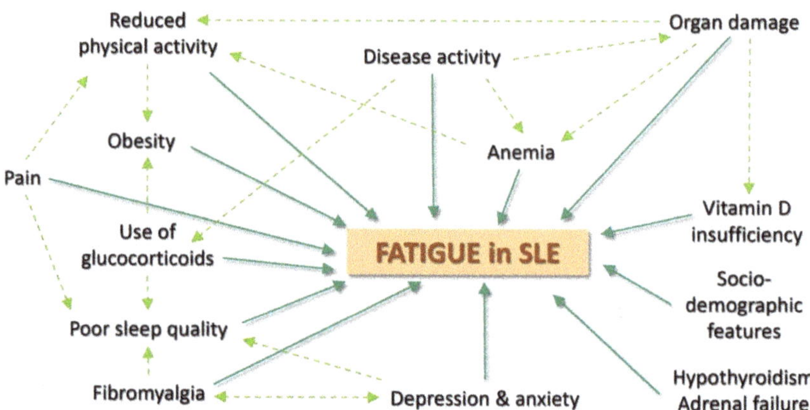


**Figure 1.** Main determinants of fatigue in Systemic Lupus Erythematosus.

*3.1. Lupus-Related Determinants*

The association between fatigue and disease activity in SLE has been widely studied and debated for a long time, with controversial results [11–19]. Type I interferons, which are key cytokines in SLE, are associated with fatigue and may provide a clue towards a pathogenic explanation for fatigue in SLE. Disease activity seems to play a role in the genesis of fatigue but it cannot fully explain fatigue by itself. Indeed, a study based on an inception cohort of adult patients with SLE found that fatigue and disease activity followed distinct trajectories over 10 years [20]. Additionally, in our recent FATILUP studies [21], the association between fatigue and disease activity was significant, but weak (OR: 1.05 (95% CI: 1.00–1.12) per 1 point increase in SELENA-SLEDAI score). Therefore, it is likely that disease activity has a complex and potentially indirect contribution to fatigue, such as by influencing other major determinants of fatigue, for example pain and psychological factors. Some specific organ involvements such as neurological impairment and painful manifestations such as arthritis or oral ulcers have been found to be associated with fatigue in some studies [14,21,22]. Pain has been reported to have a specific role in SLE fatigue, and chronic pain treatment is essential to the management of fatigue in SLE [12,14,17,23]. Organ damage, especially renal or cardiac failure, can also be important causes of fatigue in SLE patients [24–27]. Furthermore, the use of glucocorticoid has been shown to be independently associated with fatigue in SLE [21].

### 3.2. Psychological Determinants

Mental health status, emotional and functional wellbeing have shown to be major determinants of fatigue in SLE patients. Depression and anxiety appear to be among the strongest predictors of fatigue in SLE patients [11–13,21,23,28,29]. There is a clear association between fatigue and depression in general, and scales assessing depression often include fatigue-related items. Depression affects both physical and mental dimensions of fatigue in SLE. Additionally, depression is frequent in SLE patients (between 17 and 75% of patients), and some authors have mentioned that SLE contributes to depression through its neurological involvement, an autoimmune effect, and the emotional consequences of pain and disability [30–32]. Stress, which is a subjective negative perception of life events, which may be influenced by sociological and psychological factors and SLE burden, seems to mediate the relationship between depression and fatigue over time in SLE patients. Decline in stress has been associated with a meaningful improvement in fatigue in SLE [31]. Sleep disorders have also been shown to be common and significant predictors of fatigue, occurring in more than half of SLE patients [28,29,33,34]. SLE may contribute to sleep disorders because of pain and inflammation, and steroid use has been associated with sleep disorders [35]. In addition, helplessness (a state in which a person remains passive in negative situations), coping disability (difficulties in facing problems in an adequate manner) and abnormal illness-related behavior have been associated, although not independently, with fatigue in SLE in some series [14,15,19]. The role of psychological determinants is therefore major in SLE fatigue. Consequently, it is crucial to suggest a thorough psychological assessment of SLE patients reporting severe fatigue, especially for those with no or low disease activity, since mood disorders are frequent in patients with SLE [4,21,30–32] and multifactorial.

### 3.3. Comorbidities

Fibromyalgia is a major predictor of fatigue in SLE [14,18,36]. In a study conducted by Touma et al., trajectories with higher fatigue scores were associated with a higher prevalence of fibromyalgia [20]. Fibromyalgia is common in SLE patients (from 6.2% to 30% of patients) but may be underdiagnosed by physicians [21,37]. Consequently, the role of fibromyalgia should be considered in SLE patients who complain about fatigue and widespread pain. Other frequent comorbidities such as anemia, hypothyroidism, or adrenal failure are risk factors in fatigue. Vitamin D insufficiency was associated with fatigue in SLE in some but not all studies [38,39]. SLE patients have a high risk of vitamin D deficiency because of photoprotection as well as in case of renal failure.

### 3.4. Behavioral and Socio-Demographic Features

Reduced levels of physical activity and aerobic capacity significantly increase fatigue in SLE. SLE patients have many actual and perceived barriers to exercise. It has been shown that, compared to sedentary controls, SLE patients have reduced levels of aerobic fitness, reduced exercise capacity and reduced muscle strength, which further leads to a reduced ability to perform physical activity. Furthermore, SLE patients are limited by arthralgia, anemia, and other SLE organ involvements. For all of those reasons, SLE patients often have limited physical activity and assume a sedentary lifestyle [40–42]. Obesity and smoking are other potential behavioral determinants of fatigue in this population [14,43]. The role of sociodemographic features is contradictory, but some studies found higher levels of fatigue in SLE patients with low annual income, low education level, or difficulty in accessing health care [14,15]. In some studies, a low level of perceived social support was also associated with fatigue [12].

## 4. Interventions to Improve Fatigue

Recently, an increasing number of interventional studies focused on fatigue in SLE, and some pharmacologic and non-pharmacologic therapies have demonstrated beneficial effects on fatigue. Improving disease activity is associated with significant reduction in fatigue in randomized controlled trials of belimumab, blisibimod, and hydroxychloroquine [44–46]. This effect is likely to be observed with any treatment improving disease activity in SLE, although this has not been formally proven. N-acetyl-cysteine has also been shown to improve fatigue in SLE. A double-blind, placebo controlled, randomized trial found that a 2.4 g/day dose of N-acetyl-cysteine is effective for reducing fatigue and improving disease activity, and is safe and well-tolerated [47]. Vitamin D supplementation also seems to have positive effects on fatigue in SLE patients. An observational study found significantly lower fatigue scores after vitamin D supplementation in 80 SLE patients, and a randomized double-blind placebo-controlled trial showed a decrease in fatigue in juvenile-onset SLE patient receiving vitamin D supplementation [48,49]. Physical activities such as supervised training, home training, and appropriately prescribed graded aerobic exercise, have been associated with favorable improvements in patient-reported fatigue in different studies. Importantly, exercise was reported to be safe and well tolerated, with rare adverse effects, and no reported deleterious effects on disease activity or inflammation [50–52]. Physical activity should therefore be generally recommended for the management of fatigue in SLE patients, especially since it also leads to less pain interference, better physical function, cardiovascular risk reduction, and even positive impact on anxiety and depression. A trial conducted by Davies et al. indicates that a low glycemic index diet and a low-calorie diet were both associated with reduction in fatigue in SLE, indicating the role of weight loss in the improvement in fatigue [53].

Different psychosocial interventions have been associated with significant improvement in fatigue in SLE: cognitive behavioral therapy, psychoeducation, psychotherapy, relaxation and self-management. Those interventions focus on coping ability improvement, cognitive restructuring and perceived social supports [52,54–56]. Even if the effect in reducing fatigue has been shown to be weak in most of these studies, such interventions can decrease psychological distress and pain and therefore might be integrated into the general management of SLE patients. Some interventions targeting pain have also shown their ability to improve fatigue in patients with SLE. A randomized trial found a significant decrease in fatigue in SLE patients receiving transcutaneous auricular vagus nerve stimulation [57]. Additionally, a randomized controlled trial indicates benefits of acupuncture in reducing fatigue in patients with SLE [58].

## 5. The Need for a Personalized Management

At this time, there is no validated recommendation for the management of fatigue in SLE. Since fatigue may be influenced by a variety of factors and because of the diverse profiles of fatigue in SLE, the management of fatigue should rely upon an individualized person-centered approach (Figure 2). Women with SLE have reported the need for fatigue acknowledgement by clinicians, as well as conversations about fatigue, with information about coping strategies [59]. Fatigue management in SLE would start with an assessment of the intensity and the characteristics of fatigue using validated scales, enabling an individual follow-up of fatigue over time. In recent years, there has been an increasing interest in using Patient Reported Outcomes (PROs), because they place the patients at the center of their health management and help to establish a trusting physician-patient relationship. The most commonly PROs used to evaluate fatigue in SLE are the Fatigue Severity Scale (FSS), the FACIT-fatigue score, which we use in clinical practice, the Fatigue-VAS, which are unidimensional scales measuring fatigue intensity, and the Multi-dimensional Fatigue

Inventory (MFI), which analyze general fatigue, physical and mental components of fatigue as well as the reduction in activities and motivation [60]. A personalized investigation of fatigue predictors is needed, with evaluation of disease activity, search for intricate causes (major organ damage, chronic pain, anemia … ) and psychosocial factors, assessment of life habits (physical activity, quality of sleep, smoking, obesity … ). Common medical causes of fatigue, such as pregnancy, infections, metabolic diseases or drug-induced fatigue must not be forgotten. Finally, optimal management of fatigue for patients with SLE should be based on providing targeted interventions, according to the patient profile [61].

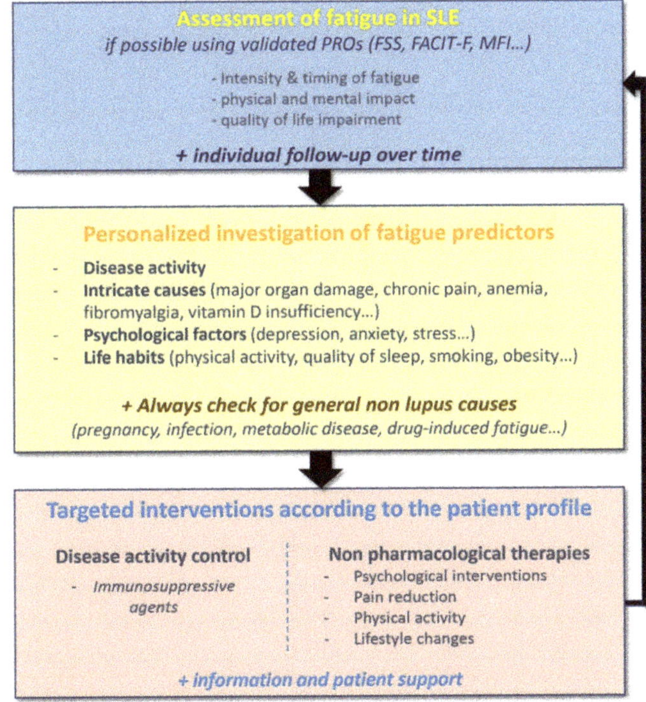


**Figure 2.** Personalized strategy for the assessment of fatigue in Systemic Lupus Erythematosus. FSS: Fatigue Severity Scale; FACIT-F: Functional Assessment of Chronic Illness Therapy—Fatigue; MFI: Multidimensional Fatigue Inventory.

## 6. Conclusions

According to patients, fatigue is the most common and disabling symptom in SLE, and this may impair patients' physical and mental health and reduce patients' quality of life by impacting upon their emotions, work, and daily life activities. Fatigue must therefore be adequately assessed and managed in SLE. It is a complex and multifactorial phenomenon, possessing many patterns. Psychological factors seem to be the most important fatigue predictors in SLE patients. Pain and fibromyalgia are also major fatigue determinants, along with lifestyle, especially reduced physical activity. Disease activity seems to have a complex contribution to fatigue, and its role remains controversial. Consequently, the management of fatigue in patients with SLE should rely upon a person-centered approach, with a personalized assessment, and targeted interventions. Some pharmacological treatments used to control disease activity, such as Belimumab, have demonstrated beneficial effects on fatigue. Non-pharmacological therapies, such as psychological interventions, pain reduction and lifestyle changes should be integrated into fatigue management in SLE. In recent years, the scientific community seems to have increased their understanding of the importance of fatigue management in SLE, and we can hope for a better understanding and treatment of fatigue in patients with SLE in the future.

**Funding:** This study was supported by the EU-funded (ERDF) project INTERREG IV and V Rhin and the Centre National de Référence des Maladies Autoimmunes Systémiques Rares Est Sud-Ouest (RESO).

**Acknowledgments:** The authors wish to thank the European Reference Networks (ERNs) ReCON-NET and RITA.

**Conflicts of Interest:** Laurent ARNAUD has acted as a consultant for Alexion, Amgen, Astra-Zeneca, Biogen, BMS, Boehringer-Ingelheim, GSK, Grifols, Janssen-Cilag, LFB, Lilly, Menarini France, Medac, Novartis, Pfizer, Roche-Chugaï, UCB. Lou KAWKA, Aurélien SCHLENCKER, Philippe MERTZ have no relationship to disclose. Thierry MARTIN has acted as a consultant for GSK. The authors declare no conflict of interest.

## References


1. Ream, E.; Richardson, A. Fatigue: A concept analysis. *Int. J. Nurs. Stud.* **1996**, *33*, 519–529. [CrossRef]
2. Hardy, S.E.; Studenski, S.A. Qualities of Fatigue and Associated Chronic Conditions among Older Adults. *J. Pain Symptom Manag.* **2010**, *39*, 1033–1042. [CrossRef]
3. Finsterer, J.; Mahjoub, S.Z. Fatigue in Healthy and Diseased Individuals. *Am. J. Hosp. Palliat. Med.* **2014**, *31*, 562–575. [CrossRef]
4. Cornet, A.; Andersen, J.; Myllys, K.; Edwards, A.; Arnaud, L. Living with systemic lupus erythematosus in 2020: A European patient survey. *Lupus Sci. Med.* **2021**, *8*, e000469. [CrossRef]
5. Sterling, K.; Gallop, K.; Swinburn, P.; Flood, E.; French, A.; Al Sawah, S.; Iikuni, N.; Naegeli, A.; Nixon, A. Patient-reported fatigue and its impact on patients with systemic lupus erythematosus. *Lupus* **2014**, *23*, 124–132. [CrossRef]
6. Raymond, K.; Park, J.; Joshi, A.V.; White, M.K. Patient Experience with Fatigue and Qualitative Interview-Based Evidence of Content Validation of The FACIT-Fatigue in Systemic Lupus Erythematosus. *Rheumatol. Ther.* **2021**, *8*, 541–554. [CrossRef] [PubMed]
7. Mathias, S.D.; Berry, P.; De Vries, J.; Pascoe, K.; Colwell, H.H.; Chang, D.J.; Askanase, A.D. Patient experience in systemic lupus erythematosus: Development of novel patient-reported symptom and patient-reported impact measures. *J. Patient-Rep. Outcomes* **2017**, *2*, 11. [CrossRef]
8. Elefante, E.; Tani, C.; Stagnaro, C.; Ferro, F.; Parma, A.; Carli, L.; Signorini, V.; Zucchi, D.; Peta, U.; Santoni, A.; et al. Impact of fatigue on health-related quality of life and illness perception in a monocentric cohort of patients with systemic lupus erythematosus. *RMD Open* **2020**, *6*, e001133. [CrossRef]
9. Du, X.; Zhao, Q.; Zhuang, Y.; Chen, H.; Shen, B. Fatigue of systemic lupus erythematosus in China: Contributors and effects on the quality of life. *Patient Prefer. Adherence* **2018**, *12*, 1729–1735. [CrossRef]
10. Basta, F.; Margiotta, D.P.E.; Vadacca, M.; Vernuccio, A.; Mazzuca, C.; Picchianti Diamanti, A.; Afeltra, A. Is fatigue a cause of work disability in systemic lupus erythematosus? Results from a systematic literature review. *Eur. Rev. Med. Pharmacol. Sci.* **2018**, *22*, 4589–4597.
11. Arnaud, L.; Mertz, P.; Amoura, Z.; Voll, R.E.; Schwarting, A.; Maurier, F.; Blaison, G.; Bonnotte, B.; Poindron, V.; Fiehn, C.; et al. Patterns of fatigue and association with disease activity and clinical manifestations in systemic lupus erythematosus. *Rheumatology* **2020**, *60*, 2672–2677. [CrossRef]
12. Jump, R.L.; Robinson, M.E.; Armstrong, A.E.; Barnes, E.V.; Kilbourn, K.M.; Richards, H.B. Fatigue in systemic lupus erythematosus: Contributions of disease activity, pain, depression, and perceived social support. *J. Rheumatol.* **2005**, *32*, 1699–1705.
13. Yilmaz-Oner, S.; Ilhan, B.; Can, M.; Alibaz-Oner, F.; Polat-Korkmaz, O.; Ozen, G.; Mumcu, G.; Kremers, H.M.; Tuglular, S.; Direskeneli, H. Fatigue in systemic lupus erythematosus: Association with disease activity, quality of life and psychosocial factors. *Z. Rheumatol.* **2017**, *76*, 913–919. [CrossRef]
14. Burgos, P.I.; Alarcón, G.S.; McGwin, G.; Crews, K.Q.; Reveille, J.D.; Vilá, L.M. Disease Activity and Damage are not Associated with Increased Levels of Fatigue in Systemic Lupus Erythematosus Patients from LUMINA LXVII, a Multiethnic Cohort. *Arthritis Care Res.* **2009**, *61*, 1179–1186. [CrossRef]
15. Zonana-Nacach, A.; Roseman, J.M.; McGwin, G.; Friedman, A.W.; Baethge, B.A.; Reveille, J.D. Systemic lupus erythematosus in three ethnic groups. VI: Factors associated with fatigue within 5 years of criteria diagnosis. *Lupus* **2000**, *9*, 101–109. [CrossRef]
16. Li, H.J.; Du, Q.; Wang, S.Y.; Guan, S.Q.; Zhan, H.H.; Tian, W.; Shao, Y.X.; Zhang, Z.Y.; Mei, Y.F. The application and influence factors of FACIT Fatigue Scale in SLE patients. *Ann. Rheum. Dis.* **2017**, *97*, 2775–2778.
17. Bruce, I.; Mak, V.; Hallett, D.; Gladman, D.; Urowitz, M. Factors associated with fatigue in patients with systemic lupus erythematosus. *Ann. Rheum. Dis.* **1999**, *58*, 379–381. [CrossRef]
18. Wang, B.; Gladman, D.D.; Urowitz, M.B. Fatigue in lupus is not correlated with disease activity. *J. Rheumatol.* **1998**, *25*, 892–895.
19. Tayer, W.G.; Nicassio, P.M.; Weisman, M.H.; Schuman, C.; Daly, J. Disease status predicts fatigue in systemic lupus erythematosus. *J. Rheumatol.* **2001**, *28*, 1999–2007.

20. Moazzami, M.; Strand, V.; Su, J.; Touma, Z. Dual trajectories of fatigue and disease activity in an inception cohort of adults with systemic lupus erythematosus over 10 years. *Lupus* **2021**, *30*, 578–586. [CrossRef]
21. Arnaud, L.; Gavand, P.E.; Voll, R.; Schwarting, A.; Maurier, F.; Blaison, G.; Magy-Bertrand, N.; Pennaforte, J.-L.; Peter, H.-H.; Kieffer, P.; et al. Predictors of fatigue and severe fatigue in a large international cohort of patients with systemic lupus erythematosus and a systematic review of the literature. *Rheumatology* **2019**, *58*, 987–996. [CrossRef] [PubMed]
22. Harboe, E.; Greve, O.J.; Beyer, M.; Goransson, L.G.; Tjensvoll, A.B.; Maroni, S.; Omdal, R. Fatigue is associated with cerebral white matter hyperintensities in patients with systemic lupus erythematosus. *J. Neurol. Neurosurg. Psychiatry* **2008**, *79*, 199–201. [CrossRef]
23. Moldovan, I.; Cooray, D.; Carr, F.; Katsaros, E.; Torralba, K.; Shinada, S.; Ishimori, M.; Jolly, M.; Wilson, A.; Wallace, D.; et al. Pain and depression predict self-reported fatigue/energy in lupus. *Lupus* **2013**, *22*, 684–689. [CrossRef] [PubMed]
24. Gregg, L.P.; Jain, N.; Carmody, T.; Minhajuddin, A.T.; Rush, A.J.; Trivedi, M.H.; Hedayati, S.S. Fatigue in Nondialysis Chronic Kidney Disease: Correlates and Association with Kidney Outcomes. *Am. J. Nephrol.* **2019**, *50*, 37–47. [CrossRef]
25. Artom, M.; Moss-Morris, R.; Caskey, F.; Chilcot, J. Fatigue in advanced kidney disease. *Kidney Int.* **2014**, *86*, 497–505. [CrossRef]
26. Barnes, S.; Gott, M.; Payne, S.; Parker, C.; Seamark, D.; Gariballa, S.; Small, N. Prevalence of symptoms in a community-based sample of heart failure patients. *J. Pain Symptom Manag.* **2006**, *32*, 208–216. [CrossRef]
27. Conley, S.; Feder, S.; Redeker, N.S. The Relationship between Pain, Fatigue, Depression and Functional Performance in Stable Heart Failure. *Heart Lung. J. Crit. Care* **2015**, *44*, 107–112. [CrossRef]
28. Da Costa, D.; Dritsa, M.; Bernatsky, S.; Pineau, C.; Ménard, H.A.; Dasgupta, K.; Keschani, A.; Rippen, N.; Clarke, A.E. Dimensions of fatigue in systemic lupus erythematosus: Relationship to disease status and behavioral and psychosocial factors. *J. Rheumatol.* **2006**, *33*, 1282–1288.
29. Iaboni, A.; Ibanez, D.; Gladman, D.D.; Urowitz, M.B.; Moldofsky, H. Fatigue in systemic lupus erythematosus: Contributions of disordered sleep, sleepiness, and depression. *J. Rheumatol.* **2006**, *33*, 2453–2457.
30. Figueiredo-Braga, M.; Cornaby, C.; Cortez, A.; Bernardes, M.; Terroso, G.; Figueiredo, M.; Santos Mesquita, C.D.; Costa, L.; Poole, B.D. Depression and anxiety in systemic lupus erythematosus: The crosstalk between immunological, clinical, and psychosocial factors. *Medicine* **2018**, *97*, e11376. [CrossRef]
31. Azizoddin, D.R.; Jolly, M.; Arora, S.; Yelin, E.; Katz, P. Longitudinal Study of Fatigue, Stress, and Depression: Role of Reduction in Stress toward Improvement in Fatigue. *Arthritis Care Res.* **2020**, *72*, 1440–1448. [CrossRef]
32. Palagini, L.; Mosca, M.; Tani, C.; Gemignani, A.; Mauri, M.; Bombardieri, S. Depression and systemic lupus erythematosus: A systematic review. *Lupus* **2013**, *22*, 409–416. [CrossRef]
33. Mckinley, P.S.; Ouellette, S.C.; Winkel, G.H. The contributions of disease activity, sleep patterns, and depression to fatigue in systemic lupus erythematosus. *Arthritis Rheum.* **1995**, *38*, 826–834. [CrossRef]
34. Da Costa, D.; Bernatsky, S.; Dritsa, M.; Clarke, A.E.; Dasgupta, K.; Keshani, A.; Pineau, C. Determinants of sleep quality in women with systemic lupus erythematosus. *Arthritis Rheum.* **2005**, *53*, 272–278. [CrossRef]
35. Palagini, L.; Tani, C.; Mauri, M.; Carli, L.; Vagnani, S.; Bombardieri, S.; Gemignani, A.; Mosca, M. Sleep disorders and systemic lupus erythematosus. *Lupus* **2014**, *23*, 115–123. [CrossRef]
36. Taylor, J.; Skan, J.; Erb, N.; Carruthers, D.; Bowman, S.; Gordon, C.; Isenberg, D. Lupus patients with fatigue—Is there a link with fibromyalgia syndrome? *Rheumatology* **2000**, *39*, 620–623. [CrossRef] [PubMed]
37. Huang, F.F.; Fang, R.; Nguyen, M.H.; Bryant, K.; Gibson, K.A.; O'Neill, S.G. Identifying co-morbid fibromyalgia in patients with systemic lupus erythematosus using the Multi-Dimensional Health Assessment Questionnaire. *Lupus* **2020**, *29*, 1404–1411. [CrossRef] [PubMed]
38. Stockton, K.A.; Kandiah, D.A.; Paratz, J.D.; Bennell, K.L. Fatigue, muscle strength and vitamin D status in women with systemic lupus erythematosus compared with healthy controls. *Lupus* **2012**, *21*, 271–278. [CrossRef]
39. Carrión-Barberà, I.; Salman-Monte, T.C.; Castell, S.; Castro, F.; Ojeda, F.; Carbonell, J. Prevalence and factors associated with fatigue in female patients with systemic lupus erythematosus. *Med. Clin.* **2018**, *151*, 353–358. [CrossRef]
40. Mancuso, C.A.; Perna, M.; Sargent, A.B.; Salmon, J.E. Perceptions and measurements of physical activity in patients with systemic lupus erythematosus. *Lupus* **2011**, *20*, 231–242. [CrossRef]
41. Tench, C.; Bentley, D.; Vleck, V.; McCurdie, I.; White, P.; D'Cruz, D. Aerobic fitness, fatigue, and physical disability in systemic lupus erythematosus. *J. Rheumatol.* **2002**, *29*, 474–481.
42. Balsamo, S.; da Mota, L.M.H.; de Carvalho, J.F.; da Nascimento, D.C.; Tibana, R.A.; de Santana, F.S.; Moreno, R.L.; Gualano, B.; dos Santos-Neto, L. Low dynamic muscle strength and its associations with fatigue, functional performance, and quality of life in premenopausal patients with systemic lupus erythematosus and low disease activity: A case-control study. *BMC Musculoskelet. Disord.* **2013**, *14*, 263. [CrossRef] [PubMed]
43. Patterson, S.L.; Schmajuk, G.; Jafri, K.; Yazdany, J.; Katz, P. Obesity is Independently Associated With Worse Patient-Reported Outcomes in Women with Systemic Lupus Erythematosus. *Arthritis Care Res.* **2019**, *71*, 126–133. [CrossRef]
44. Strand, V.; Berry, P.; Lin, X.; Asukai, Y.; Punwaney, R.; Ramachandran, S. Long-Term Impact of Belimumab on Health-Related Quality of Life and Fatigue in Patients With Systemic Lupus Erythematosus: Six Years of Treatment. *Arthritis Care Res.* **2019**, *71*, 829–838. [CrossRef] [PubMed]
45. Petri, M.A.; Martin, R.S.; Scheinberg, M.A.; Furie, R.A. Assessments of fatigue and disease activity in patients with systemic lupus erythematosus enrolled in the Phase 2 clinical trial with blisibimod. *Lupus* **2017**, *26*, 27–37. [CrossRef] [PubMed]

46. Yokogawa, N.; Eto, H.; Tanikawa, A.; Ikeda, T.; Yamamoto, K.; Takahashi, T.; Mizukami, H.; Sato, T.; Yokota, N.; Furukawa, F. Effects of Hydroxychloroquine in Patients With Cutaneous Lupus Erythematosus: A Multicenter, Double-Blind, Randomized, Parallel-Group Trial. *Arthritis Rheumatol.* **2017**, *69*, 791–799. [CrossRef] [PubMed]

47. Lai, Z.-W.; Hanczko, R.; Bonilla, E.; Caza, T.N.; Clair, B.; Bartos, A.; Miklossy, G.; Jimah, J.; Doherty, E.; Tily, H.; et al. N-acetylcysteine reduces disease activity by blocking mammalian target of rapamycin in T cells from systemic lupus erythematosus patients: A randomized, double-blind, placebo-controlled trial. *Arthritis Rheum.* **2012**, *64*, 2937–2946. [CrossRef] [PubMed]

48. Lima, G.L.; Paupitz, J.; Aikawa, N.E.; Takayama, L.; Bonfa, E.; Pereira, R.M.R. Vitamin D Supplementation in Adolescents and Young Adults With Juvenile Systemic Lupus Erythematosus for Improvement in Disease Activity and Fatigue Scores: A Randomized, Double-Blind, Placebo-Controlled Trial. *Arthritis Care Res.* **2016**, *68*, 91–98. [CrossRef]

49. Ruiz-Irastorza, G.; Gordo, S.; Olivares, N.; Egurbide, M.-V.; Aguirre, C. Changes in vitamin D levels in patients with systemic lupus erythematosus: Effects on fatigue, disease activity, and damage. *Arthritis Care Res.* **2010**, *62*, 1160–1165. [CrossRef]

50. O'Dwyer, T.; Durcan, L.; Wilson, F. Exercise and physical activity in systemic lupus erythematosus: A systematic review with meta-analyses. *Semin. Arthritis Rheum.* **2017**, *47*, 204–215. [CrossRef]

51. Mahieu, M.A.; Ahn, G.E.; Chmiel, J.S.; Dunlop, D.D.; Helenowski, I.B.; Semanik, P.; Song, J.; Yount, S.; Chang, R.W.; Ramsey-Goldman, R. Fatigue, patient reported outcomes, and objective measurement of physical activity in systemic lupus erythematosus. *Lupus* **2016**, *25*, 1190–1199. [CrossRef]

52. Tench, C.M.; McCarthy, J.; McCurdie, I.; White, P.D.; D'Cruz, D.P. Fatigue in systemic lupus erythematosus: A randomized controlled trial of exercise. *Rheumatology* **2003**, *42*, 1050–1054. [CrossRef]

53. Davies, R.J.; Lomer, M.C.E.; Yeo, S.I.; Avloniti, K.; Sangle, S.R.; D'Cruz, D.P. Weight loss and improvements in fatigue in systemic lupus erythematosus: A controlled trial of a low glycaemic index diet versus a calorie restricted diet in patients treated with corticosteroids. *Lupus* **2012**, *21*, 649–655. [CrossRef]

54. Karlson, E.W.; Liang, M.H.; Eaton, H.; Huang, J.; Fitzgerald, L.; Rogers, M.P.; Daltroy, L.H. A randomized clinical trial of a psychoeducational intervention to improve outcomes in systemic lupus erythematosus. *Arthritis Rheum.* **2004**, *50*, 1832–1841. [CrossRef] [PubMed]

55. Health Outcome Improvements in Patients with Systemic Lupus Erythematosus Using Two Telephone Counseling Interventions—Austin—1996—Arthritis & Rheumatism—Wiley Online Library. Available online: https://onlinelibrary.wiley.com/doi/abs/10.1002/1529-0131(199610)9:5%3C391::AID-ANR1790090508%3E3.0.CO;2-V (accessed on 19 July 2021).

56. Sohng, K.Y. Effects of a self-management course for patients with systemic lupus erythematosus. *J. Adv. Nurs.* **2003**, *42*, 479–486. [CrossRef]

57. Aranow, C.; Atish-Fregoso, Y.; Lesser, M.; Mackay, M.; Anderson, E.; Chavan, S.; Zanos, T.P.; Datta-Chaudhuri, T.; Bouton, C.; Tracey, K.J.; et al. Transcutaneous auricular vagus nerve stimulation reduces pain and fatigue in patients with systemic lupus erythematosus: A randomised, double-blind, sham-controlled pilot trial. *Ann. Rheum. Dis.* **2021**, *80*, 203–208. [CrossRef] [PubMed]

58. Greco, C.; Kao, A.; Maksimowicz-McKinnon, K.; Glick, R.; Houze, M.; Sereika, S.; Balk, J.; Manzi, S. Acupuncture for systemic lupus erythematosus: A pilot RCT feasibility and safety study. *Lupus* **2008**, *17*, 1108–1116. [CrossRef] [PubMed]

59. Kier, A.Ø.; Midtgaard, J.; Hougaard, K.S.; Berggreen, A.; Bukh, G.; Hansen, R.B.; Dreyer, L. How do women with lupus manage fatigue? A focus group study. *Clin. Rheumatol.* **2016**, *35*, 1957–1965. [CrossRef] [PubMed]

60. Barbacki, A.; Petri, M.; Aviña-Zubieta, A.; Alarcón, G.S.; Bernatsky, S. Fatigue Measurements in Systemic Lupus Erythematosus. *J. Rheumatol.* **2019**, *46*, 1470–1477. [CrossRef]

61. Mertz, P.; Schlencker, A.; Schneider, M.; Gavand, P.-E.; Martin, T.; Arnaud, L. Towards a practical management of fatigue in systemic lupus erythematosus. *Lupus Sci. Med.* **2020**, *7*, e000441. [CrossRef]

Journal of
*Clinical Medicine*

*Review*

# Fatigue in Systemic Lupus Erythematosus and Rheumatoid Arthritis: A Comparison of Mechanisms, Measures and Management

Mrinalini Dey [1,2,*], Ioannis Parodis [3,4] and Elena Nikiphorou [5,6]

1 Institute of Life Course and Medical Sciences, University of Liverpool, Brownlow Hill, Liverpool L69 3BX, UK
2 Department of Rheumatology, Aintree Hospital, Liverpool University Hospitals NHS Foundation Trust, Lower Lane L9 7AL, UK
3 Division of Rheumatology, Department of Medicine Solna, Karolinska Institutet and Karolinska University Hospital, 171 76 Stockholm, Sweden; ioannis.parodis@ki.se
4 Department of Rheumatology, Faculty of Medicine and Health, Örebro University, 701 85 Örebro, Sweden
5 Rheumatology Department, King's College Hospital, London SE5 9RJ, UK; elena.nikiphorou@kcl.ac.uk
6 Centre for Rheumatic Diseases, King's College London, London SE5 9RJ, UK
* Correspondence: mrinalini.dey@nhs.net

**Abstract:** Fatigue is a common constitutional feature of systemic lupus erythematosus (SLE) and rheumatoid arthritis (RA). While the two diseases share a common mechanism of autoimmunity, they differ in their clinical manifestations and treatment. Fatigue is one of the most commonly reported symptoms in both groups, associated with pain, depression and anxiety, and affecting function, work and quality of life. Fatigue is not easy to assess or conceptualise. It can be linked to disease activity, although it is not always, and is challenging to treat. Several measures have been trialled in RA and SLE; however, none have been adopted into mainstream practice. Despite being a common symptom, fatigue remains poorly managed in both RA and SLE—more so in the latter, where there have been relatively fewer studies. Additionally, comorbidities contribute to fatigue, further complicating its management. Pain, depression and anxiety also need to be addressed, not as separate entities, but together with fatigue in a holistic manner. Here, we describe the similarities and differences between fatigue in patients with RA and SLE, discuss concepts and practices applicable to both conditions and identify areas for further research. Through this review, we aim to highlight the importance of the holistic management of fatigue in SLE.

**Keywords:** systemic lupus erythematosus; rheumatoid arthritis; fatigue; quality of life; pain; psychosocial; disease activity

**Citation:** Dey, M.; Parodis, I.; Nikiphorou, E. Fatigue in Systemic Lupus Erythematosus and Rheumatoid Arthritis: A Comparison of Mechanisms, Measures and Management. *J. Clin. Med.* **2021**, *10*, 3566. https://doi.org/10.3390/jcm10163566

Academic Editor: Mahesh Mohan

Received: 29 July 2021
Accepted: 11 August 2021
Published: 13 August 2021

**Publisher's Note:** MDPI stays neutral with regard to jurisdictional claims in published maps and institutional affiliations.



## 1. Introduction

Fatigue is a subjective symptom of malaise and aversion to activity, comprising both physical and mental aspects [1]. It is often poorly defined in clinical practice and may be reported by patients as "fatigue", "tiredness", "lethargy" or "exhaustion", as well as other descriptors for a lack of energy [1]. It is not easy to assess or conceptualise. Fatigue is one of the most frequent presentations in primary care, affecting up to 20% of the general population, and is twice as common in women than in men [2,3]. In the case of chronic disease, up to 50% of people experience fatigue as part of their condition.

The prevalence of fatigue sharply increases when considering rheumatic diseases. For decades, it has been known that fatigue is one of the most commonly reported symptoms, affecting almost all patients [4]. Rheumatoid arthritis (RA) and systemic lupus erythematosus (SLE) are two rheumatic diseases where fatigue features strongly as one of the predominant symptoms, beyond the articular and connective tissue disease features. Significant fatigue is reported by two-thirds of patients with SLE, and severe fatigue is reported by one-third of these patients, as defined by the Fatigue Scale for Motor and

Cognitive Functions scale (FSMC)—a self-administered questionnaire initially developed for patients with multiple sclerosis [5–8]. Up to 75% of RA patients experience persistently high or worsening levels of fatigue [9]. While the two diseases share a common mechanism of autoimmunity, they differ in their underlying immunopathology, treatment and resulting clinical manifestations, such as organ involvement. Despite these differences, fatigue is one of the most commonly reported symptoms in both patient groups, and is associated with symptoms of pain, depression and anxiety, while impacting function, work and overall quality of life [7,10,11].

Fatigue may be linked to disease activity, although it is not always, and can be challenging to treat. Several scores and measures of fatigue have been trialled in RA and SLE, with variable success, and none have been adopted into mainstream clinical practice. Despite being one of the most troubling symptoms reported by patients, fatigue remains poorly managed in both RA and SLE. It is important to rule out causes of fatigue not related to the primary rheumatological diagnosis, and aim for optimal disease control. Additionally, the comorbidity profile of patients with SLE and RA differs greatly, e.g., renal disease is more common in SLE [12,13]. Comorbidities may be directly or indirectly related to the primary diagnosis and are likely to contribute to the burden of fatigue in both patient groups, further complicating its management. Related factors such as pain, depression and anxiety also need to be addressed, not as separate entities, but together with fatigue to ensure a holistic approach to management.

In this review, we describe the similarities and differences between fatigue and its associations in patients with RA and SLE, discuss the concepts and practices that may be applied in the two conditions, compare and contrast the measures of fatigue and identify areas for further research on fatigue in SLE. Through this review, we wish to highlight the importance of the holistic management of fatigue in SLE, addressing all possible causes, as a symptom that is intertwined with the other aspects of the disease.

## 2. Recognising Fatigue as a Clinical Outcome

Despite its long being recognised as a key symptom in patients with RA and SLE, healthcare professionals and researchers have only recently started to appreciate the clinical relevance of fatigue, its impact on patients (and all aspects of their lives) and the need for appropriate assessment and suitable, discriminative outcome measures. This was largely prompted by a Patient Perspective Workshop at an OMERACT meeting in 2002, with a subsequent formal recommendation in 2006, highlighting the importance of recognising fatigue as a core outcome measure amongst patients with RA [14–16]. Both qualitative and quantitative studies have, subsequently, demonstrated the high prevalence of fatigue amongst patients with RA, encompassing physical, cognitive, social and emotional fatigue, indicating the need to tailor fatigue management to the individual situation of the patient [17,18].

While fatigue is also recognised as highly prevalent amongst patients with SLE, quantifying this and incorporating it into the holistic assessment of the patient has proved challenging, not only due to the multidimensional impacts of fatigue, but also because of the complex aetiology of this multi-system disease. The assessment of fatigue becomes particularly challenging in patients without active disease. The extent to which disease activity plays a role remains uncertain, and there is a clear (albeit complex) association with depression and anxiety, as well as related symptoms of chronic pain and fibromyalgia, and disordered sleep [19].

## 3. The Role of Disease Activity

The aetiology of fatigue and its association with disease activity remains contentious in both RA and SLE (Figure 1). It has been suggested that systemic inflammation may contribute to fatigue. In the case of RA, the primary manifestations of pain, joint problems and functional limitations (which contribute to disease activity scores) may also play a role [20]. Fatigue is a major contributor to patients' global assessment, and it has been suggested, based on trial data, that it is a separate aspect of disease, which may be explored as a treatment target in its own right, separate from disease activity [21,22].

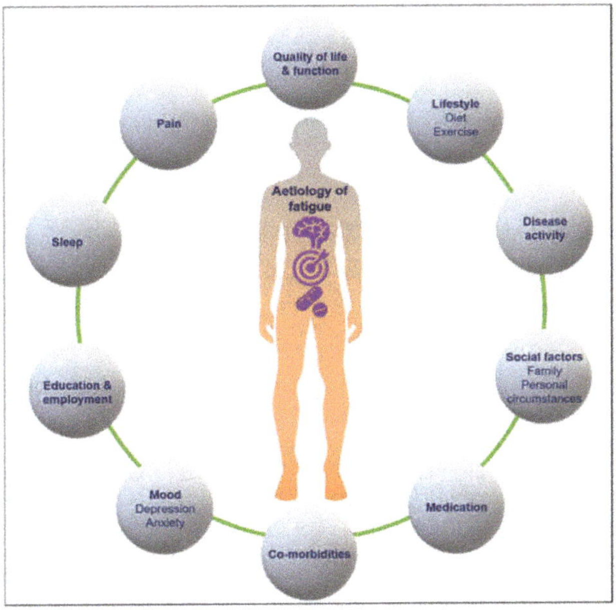

**Figure 1.** Factors associated with fatigue in rheumatoid arthritis and systemic lupus erythematosus. Fatigue is a complex, multi-factorial symptom. Multiple aspects of a patient's biological, social and psychological circumstances contribute; this is not an exhaustive list. Examples of broad contributory factors, such as mood, lifestyle and social factors, are provided below each heading.

Similar findings have been demonstrated in patients with SLE. The FATILUP study was a large observational study assessing the determinants of fatigue in 570 patients with SLE [7]. This study found a significant but limited association between SLE disease activity and fatigue, with an accompanying systematic literature review concluding that there is no major role for disease activity, albeit with some studies reporting a link with neurological involvement [7]. It is notable that, in the FATILUP cohort, arthritis and ulcers showed the strongest associations with fatigue, which, again, suggests a role for pain in its aetiology.

It has been hypothesised that inflammatory molecules (often raised in active RA), such as tumour necrosis factor (TNF), interleukin-6 (IL-6) and C-reactive protein (CRP), may contribute to fatigue symptoms. However, the evidence is inconsistent [20]. Early studies in mouse models demonstrated that high levels of IL-1 and IL-6 induce fatigue and hypersomnia, which can be resolved with the administration of anti-inflammatory drugs [23–25]. However, in human patients, the erythrocyte sedimentation rate (ESR) only poorly correlates with fatigue, with mixed evidence for CRP [26–28]. While these components of disease activity have a significant association with fatigue, one systematic review found this association to be mainly driven by pain [27]. Looking more widely at the other components of disease activity scores, fatigue was only weakly associated with the swollen joint count in one cohort, while a large longitudinal study found a significant, but small, association with both tender and swollen joint counts that did not resolve after improved treatment strategies [27,28]. Furthermore, when considering the evidence of inflammation on joint imaging, such as MRI, patients with greater levels of MRI-detected inflammation are not necessarily those with more severe fatigue, suggesting that fatigue is, at least in part, a separate entity from inflammation [29].

The evidence base is even more sparse when we consider SLE. The FATILUP study found some components of the SLE disease activity score (SLEDAI), arthritis and oral ulcers, to be associated with fatigue, although they are likely to be confounded by pain [7]. A large longitudinal cohort study found no association with disease activity and fatigue, and this has since been repeated in other cohorts [30,31]. However, other studies have shown decreasing fatigue levels following therapy, including signals of a protective role for belimumab against severe fatigue [32–35]. It is important to consider the role of fibromyalgia in these patients, as it is present in up to 25%, and has a strong association with fatigue, as well as related symptoms of depression, anxiety and poor sleep [5,31]. A recent study examining fatigue trajectories in a cohort of SLE patients found no association with disease activity, but identified higher levels of fatigue in those with fibromyalgia. This study also noted an association between higher glucocorticoid use and fatigue, suggesting a role for the side effects of medication, particularly glucocorticoids [36]. Crucially, while many studies have examined the relationship between inflammatory molecules and organ involvement, such as lupus nephritis, very few studies have been performed examining fatigue—in stark contrast to RA.

## 4. Mechanisms of Fatigue and Association with Other Symptoms

In both RA and SLE, fatigue may be correlated with other organ involvement, such as renal, cardiac, respiratory and neurological diseases. This is particularly relevant to SLE. However, something which is important and common to fatigue in both of these conditions is mental health outcomes.

Fatigue, poor sleep and depression are closely intertwined, with a recent study in RA patients demonstrating greater severity and frequency of depressive symptoms with poorer sleep [37]. This has wider psychosocial impacts, such as decreased quality of life, work participation and physical functioning [20,38,39] (Figure 2). These diverse impacts on mental health and quality of life mean the experience of fatigue for each individual patient is unique and requires a personalised approach to management, something which has been highlighted through qualitative work in this area [18]. Mental health outcomes are also strongly associated with fatigue in SLE, affecting approximately 13% of patients with SLE, and are attributed to the disease in 40% of cases [5]. A recent cross-sectional study in the California Lupus Epidemiology cohort found that a quarter of the 326 patients met the criteria for major depression, with these individuals being more likely to have greater levels of fatigue and sleep impairment, negative psychosocial impacts of illness, decreased satisfaction in discretionary social activities and decreased satisfaction in social roles [40]. The FATILUP study found a similarly high prevalence of depression and anxiety. In total, 44.4% of patients with fatigue reported one or both of these conditions, rising to almost 60% in those with severe fatigue, with odds of between four and seven for the association between fatigue or severe fatigue and depression or anxiety [7]. Stress, pain and depression are the largest contributors to fatigue in patients with SLE, indicating a need for specialist assessment and management [5,31]. It is clear that depression and anxiety are highly prevalent in people with either RA or SLE and that they are closely associated with fatigue. However, the management of these symptoms within the context of the patient's disease and wider health and well-being remains poor, impacted by factors such as inadequate access to services (such as psychology) and a continued need for greater awareness amongst clinicians [41].

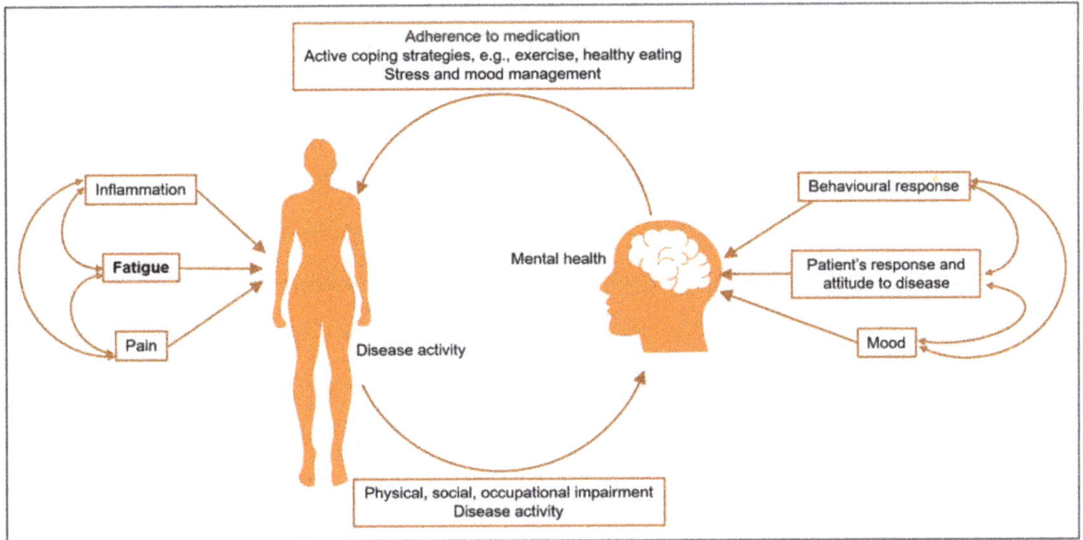



**Figure 2.** The relationship between fatigue and mental health in rheumatic diseases [39].

The similarities between the aetiology and manifestation of fatigue in RA and SLE become less clear when considering multi-organ involvement. Sjögren's syndrome is prevalent in both conditions, affecting 15–30% of both patient groups, and has been associated with fatigue in multiple studies, independent of disease activity or pain conditions such as fibromyalgia [42–45].

Importantly, SLE is characterised by multi-organ involvement. The skin, cardiovascular system and central nervous system are frequently affected, and over half of patients with SLE develop lupus nephritis during their disease course [46]. Most organ involvement in SLE contributes, to some extent, to a patient's overall fatigue, partially explaining its high prevalence. The most frequently self-reported symptom in patients with kidney disease is fatigue, affecting up to two-thirds of non-dialysis patients and being associated with factors such as comorbidity burden, anaemia and the use of anti-depressants [47,48]. The latter may reflect the mental health burden in chronic disease manifesting as fatigue. Anaemia is one plausible explanation for the aetiology of fatigue in these patients, and is often present, even in the absence of kidney disease, in SLE. Anaemia is also present in patients with cardiac failure, a manifestation of SLE (a common comorbidity in those with RA), although few studies have sought to quantify the impact of cardiac failure in either condition [20]. Cardiovascular disease, including ischaemic heart disease, correlates with RA disease activity more than any other comorbidity, including a significant association with patient-reported fatigue [49]. Similar studies on patients with SLE are lacking, signifying a critical gap in the evidence, given the prevalence of heart disease in these patients.

Neurological involvement is seen in both SLE and RA, manifesting in different ways. Common symptoms in RA include neuropathy secondary to atlantoaxial subluxation, mononeuritis multiplex and peripheral neuropathy [50]. Central nervous system involvement is rare but can present as, for example, cerebral rheumatoid vasculitis [51]. Crucially, few studies have considered the effects of these neurological manifestations on fatigue prevalence in RA. The topic remains relatively understudied in SLE, though some evidence is beginning to emerge for certain types of neurological involvement [52]. The chronic inflammatory state in SLE, mediated by pro-inflammatory cytokines, promotes oxidative and nitrosative stress. This results in the production of damage-associated molecular patterns and the engagement of toll-like receptors, which then manifest as fatigue. Interferon-α

(IFNα) is involved in SLE disease pathogenesis and is associated, in a dose-dependent manner, with neuropsychological symptoms, including fatigue, as well as depression and seizures [53,54]. Specifically, increased white matter hyperintensities have been observed in patients with SLE, associated with increased fatigue, similar to patients with multiple sclerosis [55,56].

Other related clinical states, such as vitamin D deficiency, as well as the use of certain medications, also contribute to fatigue, particularly the use of corticosteroids, which are commonly used in both SLE and RA patients [57,58]. The side effects of corticosteroids, particularly insomnia and weight gain, are both independently associated with fatigue. In an individual with active SLE or RA, fatigue is, therefore, often compounded by the use of corticosteroids. Lastly, fatigue in SLE has been shown to be associated with overweight and obesity in a body mass index (BMI)-level-dependent manner [59].

## 5. Measures of Fatigue

Fatigue in both RA and SLE has a complex aetiology and presentation, and is highly subjective. This makes it exceptionally difficult to quantify and compare, even between patients with the same disease. Patient-reported outcomes are crucial. Despite being a required outcome for clinical trials in RA, as mandated by The American College of Rheumatology and European Alliance of Associations for Rheumatology, there are no formally recommended fatigue scores for clinical use. In SLE, no such mandate exists; however, due to the increasing awareness of fatigue as a major symptom in these patients, much work has been done to develop such a tool [5,20,60].

Multiple fatigue scales exist for use in both SLE and RA. Recent reviews have identified 16 such tools in SLE and 23 in RA, with some overlap between the two diseases [5,20,61,62]. Many scales have been created for the purposes of a given study and are, therefore, not necessarily validated (the most commonly used measures are summarised in Table 1). Since these reviews, the Patient-Reported Outcomes Measurement Information System (PROMIS) has been developed for use in RA, although work to identify clinically meaningful results with regard to the magnitude of change in symptoms is ongoing [63–65].

In addition to formally recognising the burden of fatigue in patients, a validated measure enables the clinician to monitor symptom burden and the response to treatment over time. The commonly used scores across both SLE and RA include Visual Analogue Scales (VAS), SF-36 Vitality subscale score and the Functional Assessment of Chronic Illness Therapy–Fatigue (FACIT-Fatigue) score. Others, such as the Bristol Rheumatoid Arthritis Fatigue-Multidimensional Questionnaire (BRAF-MDQ) and Fatigue Severity Scale (FSS) have been favoured for either RA or SLE, respectively [5,66]. The BRAF-MDQ was designed in collaboration with patients, capturing the multiple facets of fatigue, including emotional, psychological and cognitive burden, as well as allowing clinicians to evaluate the impacts of fatigue on the patient's function. Going forward, it is essential to ensure that patients are involved in the development of tools for assessing fatigue, given the variation in experience and impact across patients.

Rather than "reinventing the wheel", it is also essential to look across the rheumatic diseases to assess the performance of each tool. It may be argued that the existence of multiple fatigue scores across SLE and RA alone demonstrates the need for one common measure across the rheumatic diseases, given the prevalence of fatigue in these patients. However, as discussed, there are subtle differences in the aetiology and characteristics of fatigue across patient groups. It may, therefore, be some time before a consensus is reached on this front.

**Table 1.** Summary of the most commonly used measures for assessment of fatigue used in systemic lupus erythematosus (SLE) and rheumatoid arthritis (RA) [5,11].

| SLE and RA | | RA Only | |
|---|---|---|---|
| Fatigue Severity Scale (FSS) | 9-item scale covering psychosocial and cognitive aspects of fatigue. Originally developed for use in multiple sclerosis and SLE. | Bristol Rheumatoid Arthritis Fatigue Multi-Dimensional Questionnaire | 20-item scale assessing the experience and impact fatigue, giving an overall score comprising 4 subscale scores (physical fatigue, living with fatigue, cognitive fatigue, emotional fatigue). |
| Multi-dimensional Fatigue Inventory (MFI) | 20-item scale comprising 5 domains: general fatigue, physical fatigue, mental fatigue, reduced motivation, reduced activity. Significant fatigue is defined depending on age and gender. | Bristol Rheumatoid Arthritis Fatigue Numeric Rating scales | 3 scales, scored 0–10: severity (no fatigue–totally exhausted), effect (no effect–a great deal of effect), coping (not at all well–very well). |
| Visual analogue scale to evaluate fatigue severity (VAS-F) | 18-item scale based on subjective experience of fatigue, using fatigue and energy subscales | Checklist of Individual Strength (CIS20) | 20-item scale giving overall score comprising 4 sub-scores (subjective fatigue, concentration, motivation, physical activity levels). |
| Functional Assessment of Chronic Illness Therapy–Fatigue (FACIT-Fatigue) | 13-item questionnaire on self-reported aspects of physical, mental and functional fatigue, and effect of these on daily living. | Fatigue Severity Inventory | 11-item scale comprising 2 scores: 6 items rating average fatigue in past week, on days with most and least fatigue, number of days with fatigue, duration of fatigue each day and current fatigue levels; 5-item fatigue interference scale. |
| | | Multidimensional Assessment of Fatigue (MAF) | 15-item scale comprising 4 aspects of fatigue (severity, distress, ability to undertake activities of daily living, frequency and change during previous week). |
| | | Profile of Mood States (POMS) | 7-item scale, focussing mainly on mood plus cognitive components and overwhelming fatigue. |
| | | SF-36 (36-Item Short Form Survey) | 4-item score, 2 on energy and 2 on fatigue. |
| | | Patient-Reported Outcomes Measurement Information System (PROMIS) Fatigue scales | Scale ranging from subjective feelings of tiredness to overwhelming exhaustion impacting activities of daily living. |

All scores have been used in the assessment of patients with RA, with four also being used in patients with SLE. No single universally agreed score for the assessment of fatigue within either diagnosis or across SLE and RA exists.

## 6. Management of Fatigue

There is no single treatment for fatigue in either RA or SLE. It may be argued that better disease control may abate symptoms; however, as discussed, the aetiology is complex and remission does not equate to the absence of fatigue [67]. Nonetheless, optimal disease control is essential to reduce the inflammation-driven component of fatigue. A 2016

Cochrane review concluded that biologic DMARDs elicit a moderate reduction in fatigue when measured with FACIT-F and the SF-36 vitality subscale [68,69]. Specifically, the use of TNF inhibitors was found to reduce fatigue by 6.3 units on FACIT-F, or 7.5 units on SF-36, compared to the controls, based on 19 studies with 8946 patients. The use of other biologics produced an even greater effect, leading to a reduction in FACIT-F of 6.9 units or 8.19 on SF-36, compared to the controls, based on eleven studies with 5682 patients. Since the publication of this Cochrane review, Janus kinase (JAK) inhibitors have been licensed for use in RA treatment. Tofacitinib and baricitinib have been shown to significantly reduce fatigue in patients in clinical trials. Specifically, there were significant reductions in the FACIT-F fatigue scores, and improvements in work productivity, pain and function, compared to placebo and methotrexate, particularly in inadequate responders [68,70–73]. Data from all four of the major RA studies with baricitinib, at both 2 mg and 4 mg dosing, showed that 63–75% of participants had improvements in FACIT-F scores after 12 weeks, compared to 48–65% of patients in the control groups [71–73]. The use of JAK inhibitors and most biologics are limited to RA, highlighting an area for further research in SLE. It is also important to note that not all patients will respond to treatment in the same way, particularly with regards to a multi-facetted symptom such as fatigue.

Treatment-related fatigue should also be addressed, and is not limited to the use of corticosteroids. Fatigue has been found to be one of the main causes of methotrexate non-adherence in quantitative studies, although evidence from qualitative studies demonstrates a variation in patients' experiences of fatigue when taking this drug [74,75].

The management of associated symptoms of depression, anxiety and pain may be helpful. Given the overlap in characteristics between fatigue in RA and SLE in this regard, this is an area where similar management strategies may be effective for both (Figure 3). The empowerment of patients to undertake the self-management of their fatigue is crucial, given the disparate characteristics of this symptom between patients. The recent EULAR guidance on the self-management of symptoms in inflammatory arthritis describes the benefits of this for both patients and clinicians, including a more holistic patient-centred approach, ultimately leading to better outcomes [76]. Approaches include psychosocial and physical approaches, with appropriate input, as required from health professionals [76–78]. Group cognitive behavioural therapy has also proven effective in reducing the burden, impact and severity of fatigue in RA. Physical exercise and psychological interventions also have proven benefits for fatigue in SLE, as well as depression and quality of life, demonstrating the similarities in effective strategies between the two conditions [79]. More simple measures, such as encouraging smoking cessation and a healthy diet, are likely to potentiate the effects of such self-management strategies—particularly the former, which adversely affects disease activity and response to treatment in both RA and SLE [33,80–82].

In severe or unremitting cases of fatigue, psychological assessment should be sought, with the appropriate behavioural or psychological interventions. Unremitting and debilitating fatigue for more than six months warrants further assessment for other possible co-existing causes, such as chronic fatigue syndrome [83]. The role of antidepressants in fatigue in SLE requires particularly careful thought, due to potential interactions with hydroxychloroquine and prolonged QT [5]. They may be more suitable in the setting of RA, provided the patient is not also taking hydroxychloroquine or a similar interacting medication, although evidence on this is limited.

Ultimately, across both conditions, due to the multifactorial aetiology of fatigue and its impact on multiple facets of patients' lives, a holistic approach to its management is vital, tailored to the patient's individual needs and circumstances.

**Figure 3.** Common themes in the management of fatigue in rheumatoid arthritis and systemic lupus erythematosus [5,62].

## 7. Conclusions and Future Research Considerations

Fatigue is one of the most prevalent and debilitating symptoms in patients with rheumatic diseases. Although differing in pathology and symptomology, it is clear that there is much overlap in the nature, associated features (pain, low mood, poor sleep, por function) and proposed management of fatigue in both of these conditions. However, it is also appropriate to appreciate the differences, such as the impact of different associated organ involvement, treatments and disease activity in determining the aetiology of and, therefore, the finer ways of managing fatigue.

Multiple measures and scales have been trialled to assess fatigue in both RA and SLE. Despite a universal recognition of the importance of appropriately measuring (and addressing) fatigue, a single standardised measure, even within a single condition, is lacking. Due to the similarities in the various aspects of fatigue across both RA and SLE, a single tool of measure across such diseases may encourage clinicians to measure and place greater focus on this symptom when assessing and managing patients. Research in this area would also be greatly aided by a single tool. However, the aetiology and more detailed characteristics of fatigue in RA and SLE are sufficiently different to warrant disease-specific measures, for example, accounting for co-existing organ involvement, neurological function, disease activity measures and treatment.

Ultimately, the improved management of fatigue in patients with rheumatic diseases will lead to better overall well-being and encourage a holistic approach to patient care. Self-management strategies are crucial to relieving fatigue and related symptoms of pain, low mood and disturbed sleep, and have an overall impact on general health, function and disease activity. The assessment of fatigue in clinical practice remains infrequent. More robust and simple measures, and heightened awareness amongst clinicians and patients of the multifactorial nature of fatigue, is likely to improve its management.

**Author Contributions:** Conceptualisation, M.D., I.P. and E.N.; writing—original draft preparation, M.D.; writing—review and editing, M.D., I.P. and E.N.; supervision, E.N.; funding acquisition, M.D. and I.P. All authors have read and agreed to the published version of the manuscript.

**Funding:** M.D. is an NIHR-funded academic clinical fellow. I.P. is supported by grants from the Swedish Rheumatism Association (R-932236), King Gustaf V's 80-year Foundation (FAI-2019-0635), Professor Nanna Svartz Foundation (2019-00290), Ulla and Roland Gustafsson Foundation (2019-12), Region Stockholm and Karolinska Institutet.

**Conflicts of Interest:** The authors declare no conflict of interest.

# References


1. Sharpe, M.; Wilks, D. Fatigue. *BMJ* **2002**, *325*, 480. [CrossRef]
2. Prevalence I Background Information I Tiredness/Fatigue in Adults I CKS I NICE. Available online: https://cks.nice.org.uk/topics/tiredness-fatigue-in-adults/background-information/prevalence/ (accessed on 28 May 2021).
3. McAteer, A.; Elliott, A.M.; Hannaford, P.C. Ascertaining the size of the symptom iceberg in a UK-wide community-based survey. *Br. J. Gen. Pract.* **2011**, *61*, e1–e11. [CrossRef] [PubMed]
4. Wolfe, F.; Hawley, D.J.; Wilson, K. The prevalence and meaning of fatigue in rheumatic disease. *J. Rheumatol.* **1996**, *23*, 1407–1417.
5. Mertz, P.; Schlencker, A.; Schneider, M.; Gavand, P.E.; Martin, T.; Arnaud, L. Towards a practical management of fatigue in systemic lupus erythematosus. *Lupus Sci. Med.* **2020**, *7*, e000441. [CrossRef] [PubMed]
6. Cleanthous, S.; Tyagi, M.; Isenberg, D.A.; Newman, S.P. What do we know about self-reported fatigue in systemic lupus erythematosus? *Lupus* **2012**, *21*, 465–476. [CrossRef] [PubMed]
7. Arnaud, L.; Gavand, P.E.; Voll, R.; Schwarting, A.; Maurier, F.; Blaison, G.; Magy-Bertrand, N.; Pennaforte, J.L.; Peter, H.H.; Kieffer, P.; et al. Predictors of fatigue and severe fatigue in a large international cohort of patients with systemic lupus erythematosus and a systematic review of the literature. *Rheumatology* **2019**, *58*, 987–996. [CrossRef]
8. Penner, I.K.; Raselli, C.; Stöcklin, M.; Opwis, K.; Kappos, L.; Calabrese, P. The Fatigue Scale for Motor and Cognitive Functions (FSMC): Validation of a new instrument to assess multiple sclerosis-related fatigue. *Mult. Scler. J.* **2009**, *15*, 1509–1517. [CrossRef]
9. Druce, K.L.; Basu, N. Predictors of fatigue in rheumatoid arthritis. *Rheumatology* **2019**, *58*, v29–v34. [CrossRef]
10. Horisberger, A.; Courvoisier, D.; Ribi, C. The Fatigue Assessment Scale as a simple and reliable tool in systemic lupus erythematosus: A cross-sectional study. *Arthritis Res. Ther.* **2019**, *21*, 80. [CrossRef]
11. Katz, P. Causes and consequences of fatigue in rheumatoid arthritis. *Curr. Opin. Rheumatol.* **2017**, *29*, 269–276. [CrossRef]
12. Anders, H.J.; Vielhauer, V. Renal co-morbidity in patients with rheumatic diseases. *Arthritis Res. Ther.* **2011**, *13*, 222. [CrossRef]
13. Ziade, N.; El Khoury, B.; Zoghbi, M.; Merheb, G.; Abi Karam, G.; Mroue, K.; Missaykeh, J. Prevalence and pattern of comorbidities in chronic rheumatic and musculoskeletal diseases: The COMORD study. *Sci Rep.* **2020**, *10*, 7683. [CrossRef] [PubMed]
14. Choy, E.H.; Dures, E. Fatigue in rheumatoid arthritis. *Rheumatology* **2019**, *58*, v1–v2. [CrossRef]
15. Kirwan, J.; Heiberg, T.; Hewlett, S.; Hughes, R.; Kvien, T.; Ahlmèn, M.; Boers, M.; Minnock, P.; Saag, K.; Shea, B.; et al. Outcomes from the Patient Perspective Workshop at OMERACT 6. *J. Rheumatol.* **2003**, *30*, 868–872. [PubMed]
16. Kirwan, J.R.; Fries, J.F.; Hewlett, S.; Osborne, R.H. Patient perspective: Choosing or developing instruments. *J. Rheumatol.* **2011**, *38*, 1716–1719. [CrossRef]
17. Hewlett, S.; Cockshott, Z.; Byron, M.; Kitchen, K.; Tipler, S.; Pope, D.; Hehir, M. Patients' perceptions of fatigue in rheumatoid arthritis: Overwhelming, uncontrollable, ignored. *Arthritis Care Res.* **2005**, *53*, 697–702. [CrossRef]
18. Nikolaus, S.; Bode, C.; Taal, E.; van de Laar, M.A.F.J. New insights into the experience of fatigue among patients with rheumatoid arthritis: A qualitative study. *Ann. Rheum. Dis.* **2010**, *69*, 895–897. [CrossRef]
19. Felten, R.; Sagez, F.; Gavand, P.-E.; Martin, T.; Korganow, A.-S.; Sordet, C.; Javier, R.M.; Soulas-Sprauel, P.; Rivière, M.; Scher, F.; et al. 10 most important contemporary challenges in the management of S.L.E. *Lupus Sci. Med.* **2019**, *6*, e000303. [CrossRef]
20. Katz, P. Fatigue in Rheumatoid Arthritis. *Curr. Rheumatol. Rep.* **2017**, *19*, 25. [CrossRef] [PubMed]
21. Pazmino, S.; Lovik, A.; Boonen, A.; de Cock, D.; Stouten, V.; Joly, J.; Bertrand, D.; Van Der Elst, K.; Westhovens, R.; Verschueren, P. Does including pain, fatigue, and physical function when assessing patients with early rheumatoid arthritis provide a comprehensive picture of disease burden? *J. Rheumatol.* **2021**, *48*, 174–178. [CrossRef] [PubMed]
22. Nikiphorou, E.; Radner, H.; Chatzidionysiou, K.; Desthieux, C.; Zabalan, C.; van Eijk-Hustings, Y.; Dixon, W.G.; Hyrich, K.L.; Askling, J.; Gossec, L. Patient global assessment in measuring disease activity in rheumatoid arthritis: A review of the literature. *Arthritis Res. Ther.* **2016**, *18*, 251. [CrossRef] [PubMed]
23. Norden, D.M.; Bicer, S.; Clark, Y.; Jing, R.; Henry, C.J.; Wold, L.E.; Reiser, P.J.; Godbout, J.P.; McCarthy, D.O. Tumor growth increases neuroinflammation, fatigue and depressive-like behavior prior to alterations in muscle function. *Brain Behav. Immun.* **2015**, *43*, 76–85. [CrossRef]
24. Bluthé, R.M.; Beaudu, C.; Kelley, K.W.; Dantzer, R. Differential effects of IL-1ra on sickness behavior and weight loss induced by IL-1 in rats. *Brain Res.* **1995**, *677*, 171–176. [CrossRef]
25. Louati, K.; Berenbaum, F. Fatigue in chronic inflammation-a link to pain pathways. *Arthritis Res. Ther.* **2015**, *17*, 254. [CrossRef]

26. Bergman, M.J.; Shahouri, S.S.; Shaver, T.S.; Anderson, J.D.; Weidensaul, D.N.; Busch, R.E.; Wang, S.; Wolfe, F. Is fatigue an inflammatory variable in rheumatoid arthritis (RA)? Analyses of fatigue in ra, osteoarthritis, and fibromyalgia. *J. Rheumatol.* **2009**, *36*, 2788–2794. [CrossRef] [PubMed]

27. Groth Madsen, S.; Danneskiold-Samsøe, B.; Stockmarr, A.; Bartels, E.M. Correlations between fatigue and disease duration, disease activity, and pain in patients with rheumatoid arthritis: A systematic review. *Scand. J. Rheumatol.* **2016**, *45*, 255–261. [CrossRef] [PubMed]

28. Van Steenbergen, H.W.; Tsonaka, R.; Huizinga, T.W.J.; Boonen, A.; Van Der Helm-Van Mil, A.H.M. Fatigue in rheumatoid arthritis; A persistent problem: A large longitudinal study. *RMD Open* **2015**, *1*, e000041. [CrossRef] [PubMed]

29. Matthijssen, X.M.E.; Wouters, F.; Sidhu, N.; Van Der Helm-Van Mil, A.H.M. Value of imaging detected joint inflammation in explaining fatigue in RA at diagnosis and during the disease course: A large MRI study. *RMD Open* **2021**, *7*, e001599. [CrossRef]

30. Burgos, P.I.; Alarcón, G.S.; McGwin, G.; Crews, K.Q.; Reveille, J.D.; Vilá, L.M. Disease activity and damage are not associated with increased levels of fatigue in systemic lupus erythematosus patients from a multiethnic cohort: LXVII. *Arthritis Care Res.* **2009**, *61*, 1179–1186. [CrossRef]

31. Azizoddin, D.R.; Gandhi, N.; Weinberg, S.; Sengupta, M.; Nicassio, P.M.; Jolly, M. Fatigue in systemic lupus: The role of disease activity and its correlates. *Lupus* **2019**, *28*, 163–173. [CrossRef]

32. Strand, V.; Levy, R.A.; Cervera, R.; Petri, M.A.; Birch, H.; Freimuth, W.W.; Zhong, Z.J.; Clarke, A.E. Improvements in health-related quality of life with belimumab, a B-lymphocyte stimulator-specific inhibitor, in patients with autoantibody-positive systemic lupus erythematosus from the randomised controlled BLISS trials. *Ann. Rheum. Dis.* **2014**, *73*, 838–844. [CrossRef]

33. Parodis, I.; Sjöwall, C.; Jönsen, A.; Ramsköld, D.; Zickert, A.; Frodlund, M.; Sohrabian, A.; Arnaud, L.; Rönnelid, J.; Malmström, V.; et al. Smoking and pre-existing organ damage reduce the efficacy of belimumab in systemic lupus erythematosus. *Autoimmun. Rev.* **2017**, *16*, 343–351. [CrossRef]

34. Parodis, I.; Benavides, A.L.; Zickert, A.; Pettersson, S.; Möller, S.; Henriksson, E.W.; Voss, A.; Gunnarsson, I. The Impact of Belimumab and Rituximab on Health-Related Quality of Life in Patients With Systemic Lupus Erythematosus. *Arthritis Care Res.* **2019**, *71*, 811–821. [CrossRef]

35. Gomez, A.; Qiu, V.; Cederlund, A.; Borg, A.; Lindblom, J.; Emamikia, S.; Enman, Y.; Lampa, J.; Parodis, I. Adverse Health-Related Quality of Life Outcome Despite Adequate Clinical Response to Treatment in Systemic Lupus Erythematosus. *Front. Med.* **2021**, *8*, 651249. [CrossRef]

36. Moazzami, M.; Strand, V.; Su, J.; Touma, Z. Dual trajectories of fatigue and disease activity in an inception cohort of adults with systemic lupus erythematosus over 10 years. *Lupus* **2021**, *30*, 578–586. [CrossRef] [PubMed]

37. Lopes, F.H.A.; Freitas, M.V.C.; de Bruin, V.M.S.; de Bruin, P.F.C. Depressive symptoms are associated with impaired sleep, fatigue, and disease activity in women with rheumatoid arthritis. *Adv. Rheumatol.* **2021**, *61*, 18. [CrossRef]

38. Esbensen, B.A.; Stallknecht, S.E.; Madsen, M.E.; Hagelund, L.; Pilgaard, T. Correlations of fatigue in Danish patients with rheumatoid arthritis, psoriatic arthritis and spondyloarthritis. *PLoS ONE* **2020**, *15*, e0237117. [CrossRef] [PubMed]

39. Sturgeon, J.A.; Finan, P.H.; Zautra, A.J. Affective disturbance in rheumatoid arthritis: Psychological and disease-related pathways. *Nat. Rev. Rheumatol.* **2016**, *12*, 532–542. [CrossRef] [PubMed]

40. Dietz, B.; Katz, P.; Dall'Era, M.; Murphy, L.B.; Lanata, C.; Trupin, L.; Criswell, L.A.; Yazdany, J. Major Depression and Adverse Patient-Reported Outcomes in Systemic Lupus Erythematosus: Results From a Prospective Longitudinal Cohort. *Arthritis Care Res.* **2021**, *73*, 48–54. [CrossRef] [PubMed]

41. HQIP. *National Early Inflammatory Arthritis Audit (NEIAA) 1st Annual Report*; HQIP: London, UK, 2019.

42. Harrold, L.R.; Shan, Y.; Rebello, S.; Kramer, N.; Connolly, S.E.; Alemao, E.; Kelly, S.; Kremer, J.M.; Rosenstein, E.D. Prevalence of Sjögren's syndrome associated with rheumatoid arthritis in the USA: An observational study from the Corrona registry. *Clin. Rheumatol.* **2020**, *39*, 1899–1905. [CrossRef] [PubMed]

43. Pasoto, S.G.; de Oliveira Martins, V.A.; Bonfa, E. Sjögren's syndrome and systemic lupus erythematosus: Links and risks. *Open Access Rheumatol. Res. Rev.* **2019**, *11*, 33–45.

44. Dias, L.H.; Miyamoto, S.T.; Giovelli, R.A.; de Magalhães, C.I.M.; Valim, V. Pain and fatigue are predictors of quality of life in primary Sjögren's syndrome. *Adv. Rheumatol.* **2021**, *61*, 28. [CrossRef] [PubMed]

45. Carsons, S.E.; Vivino, F.B.; Parke, A.; Carteron, N.; Sankar, V.; Brasington, R.; Brennan, M.T.; Ehlers, W.; Fox, R.; Scofield, H.; et al. Treatment Guidelines for Rheumatologic Manifestations of Sjögren's Syndrome: Use of Biologic Agents, Management of Fatigue, and Inflammatory Musculoskeletal Pain. *Arthritis Care Res.* **2017**, *69*, 517–527. [CrossRef] [PubMed]

46. Tsokos, G.C. Autoimmunity and organ damage in systemic lupus erythematosus. *Nat. Immunol.* **2020**, *21*, 605–614. [CrossRef] [PubMed]

47. Wilkinson, T.J.; Nixon, D.G.D.; Palmer, J.; Lightfoot, C.J.; Smith, A.C. Differences in physical symptoms between those with and without kidney disease: A comparative study across disease stages in a UK population. *BMC Nephrol.* **2021**, *22*, 147. [CrossRef] [PubMed]

48. Gregg, L.P.; Jain, N.; Carmody, T.; Minhajuddin, A.T.; Rush, A.J.; Trivedi, M.H.; Hedayati, S.S. Fatigue in Nondialysis Chronic Kidney Disease: Correlates and Association with Kidney Outcomes. *Am. J. Nephrol.* **2019**, *50*, 37–47. [CrossRef]

49. Crepaldi, G.; Scirè, C.A.; Carrara, G.; Sakellariou, G.; Caporali, R.; Hmamouchi, I.; Dougados, M.; Montecucco, C. Cardiovascular comorbidities relate more than others with disease activity in rheumatoid arthritis. *PLoS ONE* **2016**, *11*, e0146991. [CrossRef]

50. Sofat, N.; Malik, O.; Higgens, C.S. Neurological involvement in patients with rheumatic disease. *QJM* **2006**, *99*, 69–79. [CrossRef]

51. Atzeni, F.; Talotta, R.; Masala, I.F.; Gerardi, M.C.; Casale, R.; Sarzi-Puttini, P. Central nervous system involvement in rheumatoid arthritis patients and the potential implications of using biological agents. *Best Pract. Res. Clin. Rheumatol.* **2018**, *32*, 500–510. [CrossRef]

52. Barraclough, M.; McKie, S.; Parker, B.; Jackson, A.; Pemberton, P.; Elliott, R.; Bruce, I.N. Altered cognitive function in systemic lupus erythematosus and associations with inflammation and functional and structural brain changes. *Ann. Rheum. Dis.* **2019**, *78*, 934–940. [CrossRef]

53. Davis, L.S.; Hutcheson, J.; Mohan, C. The role of cytokines in the pathogenesis and treatment of systemic lupus erythematosus. *J. Interf. Cytokine Res.* **2011**, *31*, 781–789. [CrossRef] [PubMed]

54. Mackay, M. Lupus brain fog: A biologic perspective on cognitive impairment, depression, and fatigue in systemic lupus erythematosus. *Immunol. Res.* **2015**, *63*, 26–37. [CrossRef]

55. Harboe, E.; Greve, O.J.; Beyer, M.; Gøransson, L.G.; Tjensvoll, A.B.; Maroni, S.; Omdal, R. Fatigue is associated with cerebral white matter hyperintensities in patients with systemic lupus erythematosus. *J. Neurol. Neurosurg. Psychiatry* **2008**, *79*, 199–201. [CrossRef] [PubMed]

56. Colombo, B.; Boneschi, F.M.; Rossi, P.; Rovaris, M.; Maderna, L.; Filippi, M.; Comi, G. MRI and motor evoked potential findings in nondisabled multiple sclerosis patients with and without symptoms of fatigue. *J. Neurol.* **2000**, *247*, 506–509. [CrossRef]

57. Carrión-Barberà, I.; Salman-Monte, T.C.; Castell, S.; Castro, F.; Ojeda, F.; Carbonell, J. Prevalence and factors associated with fatigue in female patients with systemic lupus erythematosus. *Med. Clin.* **2018**, *151*, 353–358. [CrossRef]

58. Grøn, K.L.; Ørnbjerg, L.M.; Hetland, M.L.; Aslam, F.; Khan, N.A.; Jacobs, J.W.G. The association of fatigue, comorbidity burden, disease activity, disability and gross domestic product in patients with rheumatoid arthritis. Results from 34 countries participating in the Quest-RA programme. *Clin. Exp. Rheumatol.* **2014**, *32*, 869–877. [PubMed]

59. Gomez, A.; Butrus, F.H.; Johansson, P.; Åkerström, E.; Soukka, S.; Emamikia, S.; Enman, Y.; Pettersson, S.; Parodis, I. Impact of overweight and obesity on patient-reported health-related quality of life in systemic lupus erythematosus. *Rheumatology* **2021**, *60*, 1260–1272. [CrossRef]

60. Aletaha, D.; Landewe, R.; Karonitsch, T.; Bathon, J.; Boers, M.; Bombardier, C.; Choi, H.; Combe, B.; Dougados, M.; Emery, P. Reporting disease activity in clinical trials of patients with rheumatoid arthritis: EULAR/ACR collaborative recommendations. *Arthritis Care Res.* **2008**, *59*, 1371–1377. [CrossRef]

61. Hewlett, S.; Hehir, M.; Kirwan, J.R. Measuring fatigue in rheumatoid arthritis: A systematic review of scales in use. *Arthritis Care Res.* **2007**, *57*, 429–439. [CrossRef]

62. Pope, J.E. Management of Fatigue in Rheumatoid Arthritis. *RMD Open* **2020**, *6*, e001084. [CrossRef]

63. Bingham, C.O.; Gutierrez, A.K.; Butanis, A.; Bykerk, V.P.; Curtis, J.R.; Leong, A.; Lyddiatt, A.; Nowell, W.B.; Orbai, A.M.; Bartlett, S.J. PROMIS Fatigue short forms are reliable and valid in adults with rheumatoid arthritis. *J. Patient Rep. Outcomes* **2019**, *3*, 14. [CrossRef] [PubMed]

64. Bingham, C.O.; Butanis, A.L.; Orbai, A.M.; Jones, M.; Ruffing, V.; Lyddiatt, A.; Schrandt, M.S.; Bykerk, V.P.; Cook, K.F.; Bartlett, S.J. Patients and clinicians define symptom levels and meaningful change for PROMIS pain interference and fatigue in RA using bookmarking. *Rheumatology* **2021**, keab014. [CrossRef]

65. Beaumont, J.L.; Davis, E.S.; Curtis, J.R.; Cella, D.; Yun, H. Meaningful Change Thresholds for Patient-Reported Outcomes Measurement Information System (PROMIS) Fatigue and Pain Interference Scores in Patients With Rheumatoid Arthritis. *J. Rheumatol.* **2021**, *48*, 200990. [CrossRef]

66. Hewlett, S.; Dures, E.; Almeida, C. Measures of fatigue: Bristol Rheumatoid Arthritis Fatigue Multi-Dimensional Questionnaire (BRAF MDQ), Bristol Rheumatoid Arthritis Fatigue Numerical Rating Scales (BRAF NRS) for Severity, Effect, and Coping, Chalder Fatigue Questionnaire (CFQ), Checklist Individual Strength (CIS20R and CIS8R), Fatigue Severity Scale (FSS), Functional Assessment Chronic Illness Therapy (Fatigue) (FACIT-F), Multi-Dimensional Assessment of Fatigue (MAF), Multi-Dimensional Fatigue Inventory (MFI), Pediatric Quality Of Life (PedsQL) Multi-Dimensional Fatigue Scale, Profile of Fatigue (ProF), Short Form 36 Vitality Subscale (SF-36 VT), and Visual Analog Scales (VAS). *Arthritis Care Res.* **2011**, *63* (Suppl. 11), S263–S286.

67. Druce, K.L.; Bhattacharya, Y.; Jones, G.T.; Macfarlane, G.J.; Basu, N. Most patients who reach disease remission following anti-TNF therapy continue to report fatigue: Results from the British Society for Rheumatology Biologics Register for Rheumatoid Arthritis. *Rheumatology* **2016**, *55*, 1786–1790. [CrossRef] [PubMed]

68. Choy, E.H. Effect of biologics and targeted synthetic disease-modifying anti-rheumatic drugs on fatigue in rheumatoid arthritis. *Rheumatology* **2019**, *58*, v51–v55. [CrossRef]

69. Almeida, C.; Choy, E.H.S.; Hewlett, S.; Kirwan, J.R.; Cramp, F.; Chalder, T.; Pollock, J.; Christensen, R. Biologic interventions for fatigue in rheumatoid arthritis. *Cochrane Database Syst. Rev.* **2016**, *2016*, CD008334. [CrossRef]

70. Keystone, E.C.; Taylor, P.C.; Tanaka, Y.; Gaich, C.; Delozier, A.M.; Dudek, A.; Zamora, J.V.; Cobos, J.A.C.; Rooney, T.; de Bono, S.; et al. Patient-reported outcomes from a phase 3 study of baricitinib versus placebo or adalimumab in rheumatoid arthritis: Secondary analyses from the RA-BEAM study. *Ann. Rheum. Dis.* **2017**, *76*, 1853–1861. [CrossRef] [PubMed]

71. Emery, P.; Blanco, R.; Maldonado Cocco, J.; Chen, Y.C.; Gaich, C.L.; Delozier, A.M.; de Bono, S.; Liu, J.; Rooney, T.; Chang, C.H.C.; et al. Patient-reported outcomes from a phase III study of baricitinib in patients with conventional synthetic DMARD-refractory rheumatoid arthritis. *RMD Open* **2017**, *3*, e000410. [CrossRef]

72. Smolen, J.S.; Kremer, J.M.; Gaich, C.L.; Delozier, A.M.; Schlichting, D.E.; Xie, L.; Stoykov, I.; Rooney, T.; Bird, P.; Bursón, J.M.S.; et al. Patient-reported outcomes from a randomised phase III study of baricitinib in patients with rheumatoid arthritis and an inadequate response to biological agents (RA-BEACON). *Ann. Rheum. Dis.* **2017**, *76*, 694–700. [CrossRef]

73. Schiff, M.; Takeuchi, T.; Fleischmann, R.; Gaich, C.L.; DeLozier, A.M.; Schlichting, D.; Kuo, W.L.; Won, J.E.; Carmack, T.; Rooney, T.; et al. Patient-reported outcomes of baricitinib in patients with rheumatoid arthritis and no or limited prior disease-modifying antirheumatic drug treatment. *Arthritis Res. Ther.* **2017**, *19*, 208. [CrossRef]

74. Hope, H.F.; Hyrich, K.L.; Anderson, J.; Bluett, J.; Sergeant, J.C.; Barton, A.; Cordingley, L.; Verstappen, S.M. The predictors of and reasons for non-adherence in an observational cohort of patients with rheumatoid arthritis commencing methotrexate. *Rheumatology* **2020**, *59*, 213–223. [CrossRef]

75. Hayden, C.; Neame, R.; Tarrant, C. Patients' adherence-related beliefs about methotrexate: A qualitative study of the role of written patient information. *BMJ Open* **2015**, *5*, 6918. [CrossRef] [PubMed]

76. Nikiphorou, E.; Santos, E.J.F.; Marques, A.; Böhm, P.; Bijlsma, J.W.J.; Daien, C.I.; Esbensen, B.A.; Ferreira, R.J.; Fragoulis, G.E.; Holmes, P.; et al. 2021 EULAR recommendations for the implementation of self-management strategies in patients with inflammatory arthritis. *Ann. Rheum. Dis.* **2021**. Available online: http://ard.bmj.com/ (accessed on 12 June 2021).

77. Cramp, F.; Hewlett, S.; Almeida, C.; Kirwan, J.R.; Choy, E.H.S.; Chalder, T.; Pollock, J.; Christensen, R. Non-pharmacological interventions for fatigue in rheumatoid arthritis. *Cochrane Database Syst. Rev.* **2013**, *2013*, CD008322. [CrossRef] [PubMed]

78. Hewlett, S.; Ambler, N.; Almeida, C.; Cliss, A.; Hammond, A.; Kitchen, K.; Knops, B.; Pope, D.; Spears, M.; Swinkels, A.; et al. Self-management of fatigue in rheumatoid arthritis: A randomised controlled trial of group cognitive-behavioural therapy. *Ann. Rheum. Dis.* **2011**, *70*, 1060–1067. [CrossRef] [PubMed]

79. Fangtham, M.; Kasturi, S.; Bannuru, R.R.; Nash, J.L.; Wang, C. Non-pharmacologic therapies for systemic lupus erythematosus. *Lupus* **2019**, *28*, 703–712. [CrossRef] [PubMed]

80. Arnaud, L.; Tektonidou, M.G. Long-term outcomes in systemic lupus erythematosus: Trends over time and major contributors. *Rheumatology* **2020**, *59* (Suppl. 5), v29–v38. [CrossRef] [PubMed]

81. Westhoff, G.; Rau, R.; Zink, A. Rheumatoid arthritis patients who smoke have a higher need for DMARDs and feel worse, but they do not have more joint damage than non-smokers of the same serological group. *Rheumatology* **2008**, *47*, 849–854. [CrossRef]

82. Chasset, F.; Francès, C.; Barete, S.; Amoura, Z.; Arnaud, L. Influence of smoking on the efficacy of antimalarials in cutaneous lupus: A meta-analysis of the literature. *J. Am. Acad. Dermatol.* **2015**, *72*, 634–639. [CrossRef]

83. James, N.; Baraniuk, M.D. Chronic fatigue syndrome (Myalgic encephalomyelitis)-Symptoms, diagnosis and treatment. *BMJ Best Pract.* **2021**, *5*, 121.

Journal of
*Clinical Medicine*

*Article*

# Fighting Fatigue in Systemic Lupus Erythematosus: Experience of Dehydroepiandrosterone on Clinical Parameters and Patient-Reported Outcomes

Oliver Skoglund [†], Tomas Walhelm [†], Ingrid Thyberg, Per Eriksson and Christopher Sjöwall [*]

Department of Biomedical and Clinical Sciences, Division of Inflammation and Infection/Rheumatology, Linköping University, SE-581 85 Linköping, Sweden
* Correspondence: christopher.sjowall@liu.se; Tel.: +46-10-1032416
† These authors contributed equally to this work.

**Abstract:** Manifestations related to ongoing inflammation in systemic lupus erythematosus (SLE) are often adequately managed, but patient-reported outcome measures (PROMs) support that fatigue and low quality of life (QoL) in the absence of raised disease activity remain major burdens. The adrenal hormone dehydroepiandrosterone (DHEA) has shown potential as a pharmacological agent for managing fatigue in mild SLE. We retrospectively evaluated data on dosage, disease activity, corticosteroid doses, concomitant antirheumatic drugs, and PROMs regarding pain intensity, fatigue, and well-being (visual analogue scales), QoL (EQ-5D-3L) and functional disability. A total of 15 patients with SLE were exposed to DHEA and 15 sex- and age-matched non-exposed SLE patients served as comparators. At baseline, 83% of the DHEA-exposed patients had subnormal DHEA concentration. The 15 subjects prescribed DHEA were exposed during a median time of 12 months (IQR 16.5) [range 3–81] and used a median daily dose of 50 mg of DHEA (IQR 25.0) [range 25–200]. Neither disease activity, nor damage accrual, changed significantly over time among patients using DHEA, and no severe adverse events were observed. Numerical improvements of all evaluated PROMs were seen in the DHEA-treated group, but none reached statistical significance. For DHEA-exposed patients, a non-significant trend was found regarding fatigue comparing baseline and 36 months ($p = 0.068$). In relation to SLE controls, the DHEA-exposed group initially reported significantly worse fatigue, pain, and well-being, but the differences diminished over time. In conclusion, DHEA was safe, but evidence for efficacy of DHEA supplementation in relation to PROMs were not found. Still, certain individuals with mild SLE, plagued by fatigue and absence of increased disease activity, appear to benefit from DHEA in terms of improved fatigue and QoL. Testing of DHEA concentration in blood should be performed before initiation, and investigation of other conditions, or reasons responsible for fatigue, must always be considered first.

**Keywords:** dehydroepiandrosterone; systemic lupus erythematosus; patient-reported outcomes; fatigue; SLEDAI-2K

**Citation:** Skoglund, O.; Walhelm, T.; Thyberg, I.; Eriksson, P.; Sjöwall, C. Fighting Fatigue in Systemic Lupus Erythematosus: Experience of Dehydroepiandrosterone on Clinical Parameters and Patient-Reported Outcomes. *J. Clin. Med.* **2022**, *11*, 5300. https://doi.org/10.3390/jcm11185300

Academic Editor: Maria G. Tektonidou

Received: 2 August 2022
Accepted: 5 September 2022
Published: 8 September 2022

## 1. Introduction

Systemic lupus erythematosus (SLE) is a systemic autoimmune condition with potential to affect virtually any of organ system. The female-to-male ratio of patients with SLE is approximately 9:1 and most cases are diagnosed between 15 and 44 years of age [1,2]. The pathogenesis of SLE remains to be fully uncovered, but is a product of a complex interplay between hereditary and environmental factors, such as ultraviolet light exposure, certain infections, and drugs, leading to dysfunctional disposal of cellular debris [1,2]. Periods of raised disease activity may be followed by longtime remission and the disease severity ranges from mild skin and joint manifestations to life-threatening cytopenia and central nervous system (CNS) disease [2].

While the number of treatment options for SLE steadily have increased, health related quality of life (QoL) and fatigue remain major burdens for patients in their everyday living. In a recent review article, based on data from 570 patients with SLE, 68% reported fatigue and 37% severe fatigue [3]. In an older systematic review, involving 9886 cases, it was shown that 34% (95% confidence interval [CI] 24–44%) had some form of work disability related directly or indirectly to their disease [4]. Data from Sweden confirm that the indirect costs for SLE are substantial [5].

The cause of fatigue experienced by patients with SLE is likely to be multifactorial [6,7]. An association between increased disease activity and fatigue exists, but fatigue is often present even in the absence of any detectable SLE activity [8,9]. Dehydroepiandrosterone (DHEA), derived from cholesterol, has achieved attention as a potential candidate to reduce fatigue and mild disease activity in certain patients with SLE [10]. The motive to use DHEA in SLE is strengthened by several aspects. Firstly, in animal studies, supplementation of DHEA has shown clear anti-inflammatory effects on the immune system and beneficial effects in lupus-prone mice have been observed [11–13]. Secondly, DHEA plasma concentrations are subnormal in a subset of subjects with SLE [14,15]. Thirdly, some 20 years ago, DHEA was evaluated in randomized controlled trials (RCTs) with encouraging results as a potential pharmacological agent in the treatment of mild SLE [16–19].

At our university unit, we have approximately 20 years' experience of DHEA as rescue therapy for severe fatigue in mild SLE where other pharmaceutical and non-pharmaceutical interventions were unsuccessful. Herein, we systematically evaluated our retrospective DHEA data in SLE in relation to tolerance, dosage, affected organ systems, disease activity measures, corticosteroid use, concomitant immunosuppressive therapies, and patient-reported outcome measures (PROMs). Sex- and age-matched SLE patients, unexposed to DHEA, served as controls. In addition, as all DHEA-exposed SLE patients had joint/musculoskeletal involvement, we included a second comparator group of patients with early rheumatoid arthritis (RA).

## 2. Methods

### 2.1. Data Source, Patients, and Study Design

This study was a retrospective unblinded observational study including 15 patients with SLE prescribed daily DHEA in various doses under careful follow-up. All patients with SLE were part of the research and quality register *Clinical Lupus Register in North-eastern Gothia* (Swedish acronym KLURING) at Linköping University Hospital, a tertiary referral center with a long experience of management of SLE [20]. These 15 patients represent all patients with SLE exposed to DHEA within the catchment area of Linköping healthcare district since the year 2000. As a control population, another 15 selected subjects with SLE from KLURING, living in the same geographical area but unexposed to DHEA, were included and subsequently age- and sex-matched to the group exposed to DHEA.

As an additional comparator group, 45 patients with early RA from the *2nd Timely Intervention in Early Rheumatoid Arthritis* (Swedish acronym TIRA-2) at Linköping University Hospital [21], living in the same geographical area, were included and matched 3:1 to each participant in the group of patients with SLE exposed to DHEA (Table 1).

Retrospective patient data were retrieved from March 2002 to March 2022 for the three groups based on physical visits to the Rheumatology unit, Linköping University Hospital. The patients were followed up to 36 months with (at least) annual visits. Inclusion criteria for the two SLE groups were age $\geq 18$ years and fulfillment of the 1982 American College of Rheumatology (ACR) and/or the 2012 Systemic Lupus International Collaborating Clinics (SLICC) classification criteria [22,23]. In the DHEA-exposed SLE group, 2/15 fulfilled the SLICC criteria in the absence of meeting ACR criteria; and among SLE comparators, 3/15 fulfilled the SLICC criteria in the absence of meeting ACR criteria. Data from the TIRA-2 cohort was collected 2006–2011 and inclusion criteria were symptom duration (defined as first observed joint swelling <12 months), and either fulfilment of the 1987 American Rheumatism Association criteria or suffering from morning stiffness >60 min, symmetrical arthritis, and small joint engagement [21,24]. As reflected in Table 1, the indication for DHEA supplementation was unmanageable fatigue where other pharmaceutical and non-pharmaceutical interventions had been unsuccessful. In most cases, DHEA concentration in plasma (measured by electrochemiluminescence immunoassay) before treatment initiation was available. Since reference intervals for DHEA are dependent on age and sex, the percentage of the lower reference limit for each included patient was provided.

**Table 1.** Characteristics of the included patients.

| Background Characteristics | Median Value (Range) [IQR] or % | | |
|---|---|---|---|
| | **SLE: DHEA+ (*n* = 15)** | **SLE: DHEA− (*n* = 15)** | **RA: DHEA− (*n* = 45)** |
| Females, *n* (%) | 14 (93.3) | 14 (93.3) | 42 (93.3) |
| Caucasian ethnicity, *n* (%) | 14 (93.3) | 14 (93.3) | N/A |
| Age at disease onset (years) | 42 (12–76) [20] | 43 (15–55) [21] | 49 (20–76) [19] |
| Age at baseline (years) | 51 (24–76) [19.5] | 46 (21–71) [13.5] | 50 (21–76) [19] |
| Disease duration at baseline (years) | 9 (0–31) [17] | 4 (0–19) [8.5] | 1 (0–5) [1] |
| SLEDAI-2K (score) | 0 (0–4) [1] | 2 (0–15) [4] | N/A |
| Physician's global assessment (0–4) | 0 (0–1) [0] | 0 (0–2) [0] | N/A |
| BMI at baseline (kg/m$^2$) | 27.6 (19.2–40.4) [5.2] | 24.5 (19.9–35.8) [6] | N/A |
| SDI at baseline (score) | 0 (0–2) [1] | 0 (0–4) [1] | N/A |
| ACR criteria fulfilled, *n* | 5 (3–7) | 4 (3–7) | N/A |
| **1982 ACR criteria, *n* (%)** | | | |
| Malar rash | 7 (46.7) | 8 (53.3) | N/A |
| Discoid rash | 12 (80) | 7 (46.7) | N/A |
| Photosensitivity | 1 (6.7) | 1 (6.7) | N/A |
| Oral ulcers | 3 (20) | 0 (0) | N/A |
| Arthritis | 11 (73.3) | 13 (86.7) | N/A |
| Serositis | 4 (26.7) | 5 (33.3) | N/A |
| Renal disorder | 6 (40) | 6 (40) | N/A |
| Neurological disorder | 1 (6.7) | 1 (6.7) | N/A |
| Hematological disorder | 9 (60) | 6 (40) | N/A |
| Immunological disorder | 7 (46.7) | 8 (53.3) | N/A |
| Anti-nuclear antibody | 15 (100) | 15 (100) | N/A |

ACR, American College of Rheumatology; BMI, body-mass index; DHEA, dehydroepiandrosterone; n.s., not significant; N/A, not applicable; RA, rheumatoid arthritis; SDI, Systemic Lupus International Collaborating Clinics (SLICC)/ACR damage index; SLE, systemic lupus erythematosus; SLEDAI-2K, systemic lupus erythematosus disease activity index 2000.

### 2.2. Assessments

We assessed PROMs at month 0 (baseline), 12, 24, and 36. The included PROMs were collected by questionnaires, including the Swedish version of Health Assessment Questionnaire (HAQ) to assess functional disability (0 = no disability, 3 = severe disability) [25]. The Euro-QoL 5 dimensions (EQ-5D-3L) was used to assess general health based on five different dimensions (mobility, self-care, usual activities, pain/discomfort and anxiety/depression) in order to derive a utility index which provides a score indexed at 1 (perfect health) and 0 (dead) [26] and the visual analogue scale (VAS) of pain, fatigue, and well-being (0–100; 0 = no symptoms, 100 = worst imaginable symptoms), that patients completed at every visit to the Rheumatology unit [27].

SLE disease activity was assessed by the SLE disease activity index-2000 (SLEDAI-2K) and the physician's global assessment (PGA, graded 0–4; 0 = remission, 4 = maximum disease activity), irreversible organ damage was assessed by the SLICC/ACR damage index (SDI) [28,29].

### 2.3. Laboratory Analyses

Longitudinal blood samples were evaluated to detect effects, or any side-effects related to laboratory variables. Hemoglobin concentration, blood cell counts, estimated glomerular filtration rate (eGFR) according to the MDRD 4-variable equation [30], erythrocyte sedimentation rate (ESR), C-reactive protein (CRP), creatine kinase (CK), complement protein 3 (C3) and 4 (C4) were available. Among patients with early RA, only ESR and CRP were available.

### 2.4. Statistics

The groups were compared using the Kruskal–Wallis test to determine any significance between the three groups regarding PROMs, baseline characteristics and for laboratory values where appropriate. The Mann–Whitney $U$ test was used to confirm any significance between two of the groups. Spearman's rho was applied to measure the strength of association between two variables. No adjustments of uncensored data were made. For comparison regarding number of fulfilled ACR criteria between the SLE groups, $\chi^2$ testing was used. In addition, the group exposed to DHEA was examined using the Wilcoxon signed-rank test to study changes over time in comparison with baseline values. Finally, descriptive statistics were used to display patient characteristics, PROMs, and laboratory values. Statistical analyses were performed using the SPSS software version 28.0.0.0 (SPSS Inc., Chicago, IL, USA) and Prism 9.3.1 (GraphPad Software Inc., La Jolla, CA, USA) for construction of graphs.

### 2.5. Ethics Approvals

Oral and written informed consents were obtained from all patients. The study was conducted according to the Declaration of Helsinki and approved by the Regional Ethics Boards regarding SLE (Linköping M75–08/2008) and early RA (Linköping M168–05).

## 3. Results

### 3.1. Baseline Differences between Patient Groups

The three patient groups did not significantly differ in sex, age at baseline (start of follow-up), or age at onset of rheumatic disease (Table 1). Neither were ethnicity, BMI, SLEDAI-2K scores, PGA, steroid dosage, or disease phenotypes (fulfilled ACR criteria) at baseline different between the two groups of SLE patients. However, the disease duration among patients with early RA was significantly shorter ($p < 0.001$) compared to the DHEA-exposed SLE group.

### 3.2. Disease Activity and Organ Damage

Accrual of organ damage, assessed by SDI, was not different between the two SLE groups ($p = 0.65$) at baseline, nor at the 36-month follow-up ($p = 0.46$). SDI did not change significantly over the 36 months among patients exposed to DHEA ($p = 1.0$; not shown). Global disease activity, assessed by the SLEDAI-2K, was unchanged over time ($p = 0.32$) (Figure 1). Similarly, PGA did not change significantly over time and no severe flares were observed.

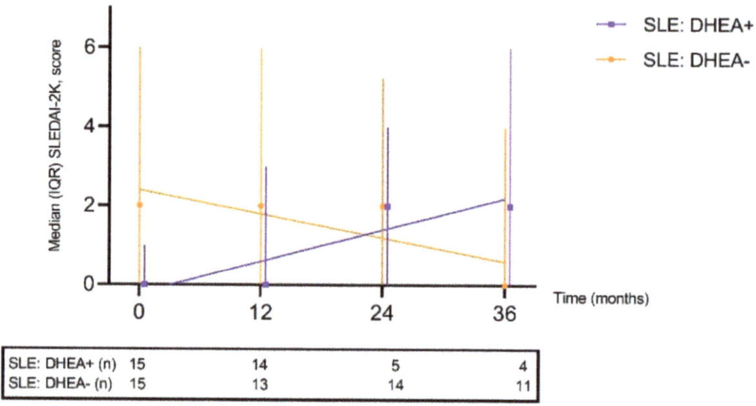


**Figure 1.** Global disease activity assessed by the systemic lupus erythematosus disease activity index-2000 (SLEDAI-2K) over time depicted for patients exposed to DHEA and for sex- and age-matched controls. No significant changes over time were observed.

### 3.3. Background Medication

At baseline, 12 of 15 patients treated with DHEA used hydroxychloroquine (HCQ) compared to 11 of 15 in the SLE group unexposed to DHEA. No other obvious differences between the groups in use of other disease-modifying antirheumatic drugs (DMARDs) were seen. The median daily dose of prednisolone at DHEA initiation was 5 mg (interquartile range [IQR] 3.75) compared to 2.5 mg (IQR 5) in the SLE controls ($p = 0.87$). As shown in Table 2, three of the DHEA-exposed patients were able to reduce their dose of prednisolone during the study period. Furthermore, among the SLE controls, 3 of 15 reduced the prednisolone dose during follow-up.

**Table 2.** Individual descriptions of the 15 patients with SLE exposed to dehydroepiandrosterone.

| Sex | Age at Start (years) | DHEA Exposure (months) | DHEA Concentration, Baseline ($\mu$mol/L) | DHEA Concentration, Percent of Lower Reference Limit (%) | Initial Daily DHEA Dose (mg) | Concomitant DMARDs | Steroid Dose at DHEA Initiation (mg) | Change in Steroid Dose at Last Follow Up (mg) | Cause of Cessation |
|---|---|---|---|---|---|---|---|---|---|
| F | 57 | 8 | 0.38 | 75 | 50 | HCQ | 0 | 0 | Treatment ongoing |
| F | 47 | 4 | N/A | N/A | 50 | MMF | 7.5 | 0 | Without specification * |
| F | 54 | 4 | 0.55 | 57 | 200 | HCQ | 0 | 0 | Lack of efficacy * |
| F | 50 | 6 | 0.38 | 40 | 50 | HCQ | 2.5 | 0 | Lack of efficacy |
| M | 43 | 81 | N/A | N/A | 50 | HCQ, MTX | 5 | +2.5 | Treatment ongoing |
| F | 56 | 3 | N/A | N/A | 50 | None | 7.5 | −2.5 | Acne, scaly hair * |
| F | 31 | 12 | 0.54 | 20 | 25 | HCQ, AZA | 2.5 | 0 | Lack of efficacy |
| F | 37 | 14 | 2.2 | 140 | 25 | HCQ | 0 | 0 | Without specification |
| F | 27 | 30 | 2.7 | 100 | 25 | HCQ, MMF | 2.5 | 0 | Treatment ongoing |
| F | 61 | 9 | 0.22 | 43 | 50 | AZA | 5 | 0 | Lack of efficacy |
| F | 47 | 69 | 0.35 | 36 | 25 | HCQ | 0 | 0 | Acne, fear of thrombosis |
| F | 58 | 17 | 0.44 | 86 | 50 | HCQ, MMF | 5 | −5 | Treatment ongoing |
| F | 76 | 36 | 0.14 | 42 | 200 | HCQ | 5 | +2.5 | Treatment ongoing |
| F | 23 | 16 | 2.7 | 68 | 50 | HCQ | 5 | −5 | Treatment ongoing |
| F | 50 | 10 | 0.43 | 45 | 25 | HCQ | 5 | 0 | Lack of efficacy |

* Early cessation ($\leq$4 months). AZA, azathioprine; DHEA, dehydroepiandrosterone; DMARDs, disease-modifying anti-rheumatic drugs; HCQ, hydroxychloroquine; MMF, mycophenolate mofetil; N/A, not applicable.

### 3.4. DHEA Exposure and Safety

Of the 15 patients exposed to DHEA, 2 (13%) had DHEA concentrations within reference intervals, 10 (67%) showed plasma levels below the lower reference limit and in 3 (20%) cases DHEA had not been analyzed at baseline. In all individuals where a second assessment of DHEA concentration was performed (i.e., after initiation of DHEA), the levels had increased to concentrations within, or even above, the age- and sex-specific reference limits.

The 15 subjects treated with DHEA were exposed for a median of 12 months (IQR 16.5) [range 3–81] and used a median daily dose of 50 mg of DHEA (IQR 25.0) [range 25–200]. As shown in Table 2, DHEA treatment with no major adverse events were observed but 9/15 ceased DHEA therapy during the 36 months. Three patients (20%) had early cessations ($\leq$4 months) due to lack of efficacy or androgenic side effects (acne). Later terminations were usually related to lack of efficacy rather than to side-effects, which mainly were of androgenic nature and deemed as mild (Table 2). Two patients remained on DHEA much longer than the 36-month follow-up and were monitored regularly as part of clinical routine, at least annually.

### 3.5. Longitudinal Effects on PROMs among DHEA-Treated Patients

PROMs at the 12-, 24- and 36-month follow-up for the exposed group were compared with respect to the baseline values (Figure 2A–E). In the DHEA-treated SLE group, numerical improvements of all evaluated PROMs were seen but none of them reached statistical significance over 36 months. A comparison of VAS fatigue between baseline and 36 months yielded a non-significant trend ($p = 0.068$). VAS fatigue at baseline did not correlate significantly with DHEA either expressed as percentage of lower reference limit (Spearman's rho = 0.078, $p = 0.82$) or as μmol/L (Spearman's rho = 0.312, $p = 0.35$). The response to DHEA was not different among patients fulfilling the ACR criteria and those who met the SLICC criteria only.

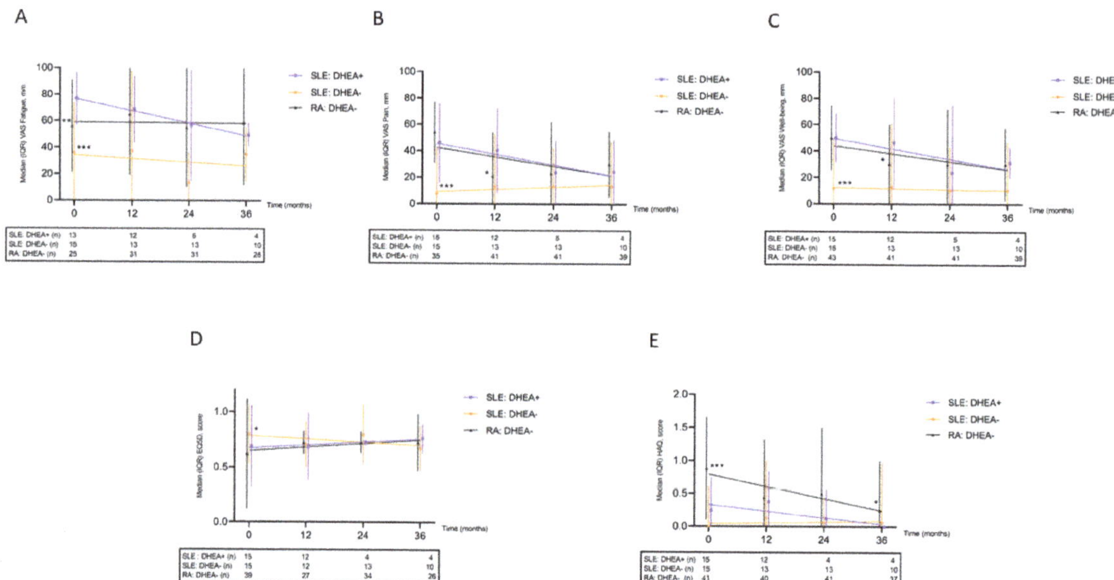

**Figure 2.** Longitudinal patient-reported outcome measures shown for patients with systemic lupus erythematosus (SLE) exposed/unexposed to DHEA and for sex- and age-matched controls with rheumatoid arthritis (RA); (**A**) visual analogue scale (VAS) fatigue, worse in DHEA-exposed SLE than in SLE/RA controls; (**B**) VAS pain, worse in DHEA-exposed SLE compared with SLE controls; (**C**) VAS well-being, worse in DHEA-exposed SLE compared with SLE controls; (**D**) EQ-5D, worse in DHEA-exposed SLE compared with SLE controls; (**E**) Health Assessment Questionnaire (HAQ), worse in RA compared with both SLE groups. * $p < 0.05$, ** $p < 0.01$, *** $p < 0.005$.

### 3.6. Effects on PROMs between the Patient Groups

Prior to DHEA supplementation (baseline), the DHEA-exposed group reported significantly worse fatigue, pain, well-being, and QoL compared to the unexposed SLE group (Figure 2A–D), but the differences diminished over time. In contrast, the functional disability was worse in the RA group compared to the other two groups (Figure 2E).

### 3.7. Effects on Laboratory Variables

Data on ESR and CRP were available for comparison in all patient groups. ESR was higher in patients with RA at baseline but, in contrast to CRP, the significance diminished over time (Figure 3A,B). C3, C4, hemoglobin concentration, and leukocyte count remained stable over time in the two SLE groups (Figure 3C–F). Although the data indicate a slight worsening of eGFR over time in both SLE groups, no significant differences in eGFR at baseline ($p = 0.12$) or at the 36-month follow-up ($p = 0.41$) were observed (Figure 3G). Further analyses of platelet, neutrophil, and lymphocyte counts as well as CK showed no significant changes over time.

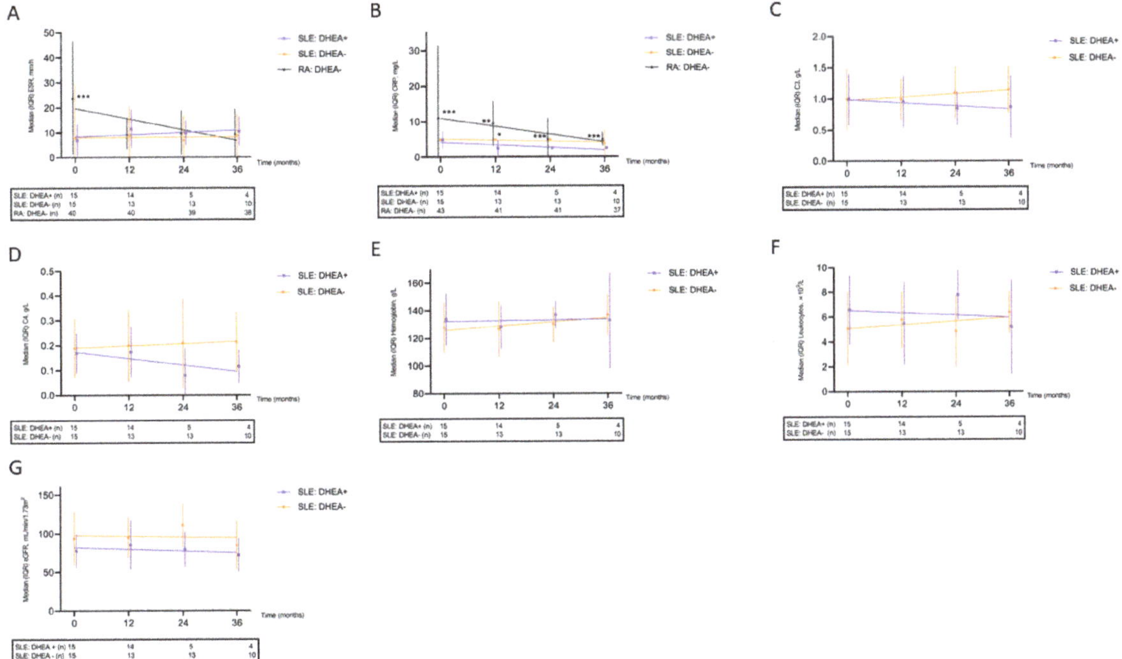

**Figure 3.** Longitudinal laboratory data demonstrated for patients with systemic lupus erythematosus (SLE) exposed/unexposed to DHEA and for sex- and age-matched controls with rheumatoid arthritis (RA); (**A**) erythrocyte sedimentation rate (ESR), higher in RA compared with both SLE groups; (**B**) C-reactive protein (CRP), higher in RA compared with DHEA-exposed SLE and higher among SLE controls than in DHEA-exposed SLE at 12 months; (**C**) Complement protein 3 (C3); (**D**) Complement protein 4 (C4); (**E**) Hemoglobin concentration; (**F**) Leukocyte count; (**G**) Estimated glomerular filtration rate (eGFR). * $p < 0.05$, ** $p < 0.01$, *** $p < 0.005$.

## 4. Discussion

Treatment options for fatigue are limited and remain an unmet need for many patients with SLE, and studies evaluating interventions for fatigue isolated from raised disease activity are rare. The scientific evidence of using DHEA for severe fatigue in mild SLE remains limited. However, albeit small, this retrospective observational unblinded study includes longitudinal follow-up data of a well-characterized DHEA-treated population in a real-life clinical setting and complements previously published RCTs [16–19].

In contrast to most studies, we herein primarily investigated improvement of PROMs. Pain, fatigue, wellbeing, QoL and functional disability are repeatedly ranked by patients as very important parameters [6]. Unfortunately, according to our data, the effects of DHEA on PROMs on a group level were mediocre or absent. This does not exclude that certain individuals could still benefit from DHEA treatment. Four of fifteen patients had been using DHEA for $\geq$30 months at the study's last follow-up. We further show that supplementation of DHEA to patients with SLE is generally safe. Mild side-effects were seen and some had an early cessation, but no severe adverse events were observed. Our patients had mild SLE with low disease activity at baseline, and no severe flares were seen during follow-up. Reassuring was that the DHEA-treated group did not accumulate more organ damage than their unexposed controls.

Originally, the idea to use DHEA in lupus arose from animal studies. In lupus-prone mice (NZB/W F1), administration of DHEA at 2 months of age significantly prolonged survival in exposed animals [12]; at 41 weeks, 71% of the DHEA-treated mice were alive compared to 22% of controls ($p = 0.04$). Moreover, DHEA injections (beginning at 2 months of age) delayed the formation of anti-double-stranded (ds) DNA antibodies in 62% of the treated animals although the antibody levels eventually rose regardless of treatment [12]. In a similar study, NZB/W mice were given DHEA and compared to unexposed controls [11]. The survival between the control and treatment group differed greatly at 12 months, with 64% survival in the DHEA-treated group versus 17% in the control group. In addition, the formation of anti-dsDNA antibodies was halted and remained comparably low in the DHEA-treated group [11]. Finally, immunofluorescence of renal tissue from controls and exposed mice at 6 months showed that DHEA-treated mice had less deposits of immunoglobulin complexes in the kidney, indicating less severe disease progression [11].

Studies evaluating DHEA in patients with SLE show mixed results [16,17]. In a multicenter double blinded RCT, 120 female patients with mild to moderate SLE were given 200 mg of DHEA or placebo over 6 months [18]. Disease activity, assessed by the Systemic Lupus Activity Measure (SLAM) and SLEDAI, [28] showed no statistical difference between the groups after 6 months, but the patient's global assessment scale in the DHEA-treated group were significantly better than placebo ($p = 0.005$). Moreover, fewer disease flares were noted over the 6-month period in the treatment group (18.3% vs. 33.9%, $p = 0.010$) [18]. In another RCT, including 381 cases with SLE over 27 centers in the United States, patients were randomized to placebo or 200 mg of DHEA for 52 weeks [19]. In patients with active SLE at baseline (defined as SLEDAI > 2), a significant improvement in SLAM ($p = 0.017$) was observed. Moreover, significantly more patients receiving placebo noted a worsening of the patient's global assessment compared to patients given DHEA (10.9% of DHEA group versus 22.6% in the placebo group, $p = 0.007$). The authors concluded that 200 mg of DHEA can improve and stabilize SLE disease activity in women with mild to moderate SLE and is generally well tolerated [19].

In a small Swedish study, lower doses of DHEA (20–30 mg daily) were investigated. The first 6 months of the study was blinded, and the latter 6 months open-label, in which all patients received DHEA [31]. DHEA was given to 20 patients and 17 received placebo. Physical and mental self-rated QoL was evaluated after 6 and 12 months of treatment. At the 6-month follow-up, the DHEA-treated group reported significant improvements in physical and emotional self-rated health in the questionnaire SF-36 compared to placebo ($p < 0.05$). Despite the small sample size, an observation was made that women with DHEA within reference limits at baseline showed similar improvements in the questionnaires as those with low DHEA. Overall, the results were less clear during the open-label phase and the authors concluded that for some patients, a lower dose of DHEA may be enough to improve well-being and QoL [31].

Our study has several limitations. It was not an RCT, which must be considered. The retrospective observational nature of the data inevitably leads to selection bias. Patients who experienced beneficial effects of DHEA were likely to continue, and thus reporting improved PROMs, compared to those who ceased and were excluded from the analysis. The included study population, with only 15 subjects exposed to DHEA, limits the statistical power and possibility of detecting significant and meaningful differences. The fact that dropouts were higher among DHEA-exposed patients compared to comparators, unequivocally leading to uncensored data, was a major limitation. In addition, all DHEA-treated patients had mild SLE without significantly raised disease activity at baseline; this makes it impossible to evaluate any effects of DHEA on SLE activity. Furthermore, functional disability assessed by HAQ may not be relevant to all patients with SLE and, although it has been used in SLE, HAQ is only validated for RA [32]. Androgenic side-effects were indeed seen in some patients, but no systematic assessment of other gonadal hormones than DHEA was performed. However, none of the patients with hypothyroidism and diabetes had to adjust their doses of levothyroxine or insulin during DHEA exposure. In contrast, major strengths of the study include the Swedish healthcare system's universal access as well as the long experience of treating patients with SLE at one tertiary referral center and longitudinal follow-up by a limited number of experienced rheumatologists. In addition, we included data from relevant SLE and RA comparators, who experienced similar clinical manifestations as the patients exposed to DHEA, living in the same geographical region of Sweden.

## 5. Conclusions

To conclude, this observational study, including longtime real-life use of DHEA in SLE, is one of very few to date. No serious adverse events were observed, but generally we did not find support for efficacy of DHEA supplementation on PROMs. Still, some individuals with mild SLE, plagued by fatigue and absence of increased disease activity, may obviously benefit from DHEA supplementation in terms of improved fatigue. Testing of DHEA concentration in blood should be performed before initiation, and investigation of other conditions, or reasons responsible for fatigue, must always be considered first.


**Author Contributions:** All authors were involved in drafting the article or revising it critically for important intellectual content, and all authors approved the final version to be published. C.S. had full access to all of the data in the study and takes responsibility for the integrity of the data and the accuracy of the data analysis. Study conception and design: O.S., T.W., I.T., P.E. and C.S. Acquisition of data: O.S., T.W., P.E. and C.S. Analysis and interpretation of data: O.S., T.W. and C.S. All authors have read and agreed to the published version of the manuscript.

**Funding:** This work was supported by grants from the Swedish Rheumatism Association [R-939149], the Region Östergötland (ALF Grants) [RÖ-932055], the Gustafsson Foundation [2021–22], the King Gustaf V's 80-year Anniversary foundation [FAI-2020-0663] and the King Gustaf V and Queen Victoria's Freemasons foundation [2022].

**Institutional Review Board Statement:** In Sweden, drugs are allowed to be used off-label. Nevertheless, informed consent was obtained from all subjects. The study was conducted according to the Declaration of Helsinki and approved by the Regional Ethics Boards regarding SLE (Linköping M75–08/2008) and early RA (Linköping M168–05).

**Informed Consent Statement:** Oral and written informed consent were obtained from all subjects involved in the study.

**Data Availability Statement:** Data available on request from the authors.

**Acknowledgments:** The authors thank Magnus Husberg for data extraction.

**Conflicts of Interest:** The authors declare that the research was conducted in the absence of any commercial or financial relationships that could be construed as a potential conflict of interest.



## References

1. Barber, M.R.W.; Drenkard, C.; Falasinnu, T.; Hoi, A.; Mak, A.; Kow, N.Y.; Svenungsson, E.; Peterson, J.; Clarke, A.E.; Ramsey-Goldman, R. Global epidemiology of systemic lupus erythematosus. *Nat. Rev. Rheumatol.* **2021**, *17*, 515–532. [CrossRef]
2. Kaul, A.; Gordon, C.; Crow, M.K.; Touma, Z.; Urowitz, M.B.; van Vollenhoven, R.; Ruiz-Irastorza, G.; Hughes, G. Systemic lupus erythematosus. *Nat. Rev. Dis. Primers* **2016**, *2*, 16039. [CrossRef]
3. Arnaud, L.; Gavand, P.E.; Voll, R.; Schwarting, A.; Maurier, F.; Blaison, G.; Magy-Bertrand, N.; Pennaforte, J.L.; Peter, H.H.; Kieffer, P.; et al. Predictors of fatigue and severe fatigue in a large international cohort of patients with systemic lupus erythematosus and a systematic review of the literature. *Rheumatology* **2019**, *58*, 987–996. [CrossRef]
4. Baker, K.; Pope, J. Employment and work disability in systemic lupus erythematosus: A systematic review. *Rheumatology* **2009**, *48*, 281–284. [CrossRef]
5. Jönsen, A.; Hjalte, F.; Willim, M.; Carlsson, K.S.; Sjöwall, C.; Svenungsson, E.; Leonard, D.; Bengtsson, C.; Rantapää-Dahlqvist, S.; Pettersson, S.; et al. Direct and indirect costs for systemic lupus erythematosus in Sweden. A nationwide health economic study based on five defined cohorts. *Semin Arthritis Rheum.* **2016**, *45*, 684–690. [CrossRef]
6. Nguyen, M.H.; Huang, F.F.; O'Neill, S.G. Patient-Reported Outcomes for Quality of Life in SLE: Essential in Clinical Trials and Ready for Routine Care. *J. Clin. Med.* **2021**, *10*, 3754. [CrossRef]
7. Dey, M.; Parodis, I.; Nikiphorou, E. Fatigue in Systemic Lupus Erythematosus and Rheumatoid Arthritis: A Comparison of Mechanisms, Measures and Management. *J. Clin. Med.* **2021**, *10*, 3566. [CrossRef]
8. Tench, C.M.; McCurdie, I.; White, P.D.; D'Cruz, D.P. The prevalence and associations of fatigue in systemic lupus erythematosus. *Rheumatology* **2000**, *39*, 1249–1254. [CrossRef]

9. Da Costa, D.; Dritsa, M.; Bernatsky, S.; Pineau, C.; Ménard, H.A.; Dasgupta, K.; Keschani, A.; Rippen, N.; Clarke, A.E. Dimensions of fatigue in systemic lupus erythematosus: Relationship to disease status and behavioral and psychosocial factors. *J. Rheumatol.* **2006**, *33*, 1282–1288.

10. Sawalha, A.H.; Kovats, S. Dehydroepiandrosterone in systemic lupus erythematosus. *Curr. Rheumatol. Rep.* **2008**, *10*, 286–291. [CrossRef]

11. Lucas, J.A.; Ahmed, S.A.; Casey, M.L.; MacDonald, P.C. Prevention of autoantibody formation and prolonged survival in New Zealand black/New Zealand white F1 mice fed dehydroisoandrosterone. *J. Clin. Investig.* **1985**, *75*, 2091–2093. [CrossRef]

12. Norton, S.D.; Harrison, L.L.; Yowell, R.; Araneo, B.A. Administration of dehydroepiandrosterone sulfate retards onset but not progression of autoimmune disease in NZB/W mice. *Autoimmunity* **1997**, *26*, 161–171. [CrossRef] [PubMed]

13. Poynter, M.E.; Daynes, R.A. Peroxisome proliferator-activated receptor alpha activation modulates cellular redox status, represses nuclear factor-kappaB signaling, and reduces inflammatory cytokine production in aging. *J. Biol. Chem.* **1998**, *273*, 32833–32841. [CrossRef] [PubMed]

14. Lahita, R.G.; Bradlow, H.L.; Ginzler, E.; Pang, S.; New, M. Low plasma androgens in women with systemic lupus erythematosus. *Arthritis Rheum.* **1987**, *30*, 241–248. [CrossRef]

15. Hedman, M.; Nilsson, E.; de la Torre, B. Low sulpho-conjugated steroid hormone levels in systemic lupus erythematosus (SLE). *Clin. Exp. Rheumatol.* **1989**, *7*, 583–588. [PubMed]

16. van Vollenhoven, R.F.; Engleman, E.G.; McGuire, J.L. Dehydroepiandrosterone in systemic lupus erythematosus. Results of a double-blind, placebo-controlled, randomized clinical trial. *Arthritis Rheum.* **1995**, *38*, 1826–1831. [CrossRef]

17. van Vollenhoven, R.F.; Park, J.L.; Genovese, M.C.; West, J.P.; McGuire, J.L. A double-blind, placebo-controlled, clinical trial of dehydroepiandrosterone in severe systemic lupus erythematosus. *Lupus* **1999**, *8*, 181–187. [CrossRef]

18. Chang, D.M.; Lan, J.L.; Lin, H.Y.; Luo, S.F. Dehydroepiandrosterone treatment of women with mild-to-moderate systemic lupus erythematosus: A multicenter randomized, double-blind, placebo-controlled trial. *Arthritis Rheum.* **2002**, *46*, 2924–2927. [CrossRef]

19. Petri, M.A.; Mease, P.J.; Merrill, J.T.; Lahita, R.G.; Iannini, M.J.; Yocum, D.E.; Ginzler, E.M.; Katz, R.S.; Gluck, O.S.; Genovese, M.C.; et al. Effects of prasterone on disease activity and symptoms in women with active systemic lupus erythematosus. *Arthritis Rheum.* **2004**, *50*, 2858–2868. [CrossRef]

20. Ziegelasch, M.; Boman, A.; Martinsson, K.; Thyberg, I.; Jacobs, C.; Nyhäll-Wåhlin, B.M.; Svärd, A.; Berglin, E.; Rantapää-Dahlqvist, S.; Skogh, T.; et al. Anti-cyclic citrullinated peptide antibodies are associated with radiographic damage but not disease activity in early rheumatoid arthritis diagnosed in 2006–2011. *Scand. J. Rheumatol.* **2020**, *49*, 434–442. [CrossRef]

21. Ziegelasch, M.; van Delft, M.A.; Wallin, P.; Skogh, T.; Magro-Checa, C.; Steup-Beekman, G.M.; Trouw, L.A.; Kastbom, A.; Sjöwall, C. Antibodies against carbamylated proteins and cyclic citrullinated peptides in systemic lupus erythematosus: Results from two well-defined European cohorts. *Arthritis Res. Ther.* **2016**, *18*, 289. [CrossRef]

22. Tan, E.M.; Cohen, A.S.; Fries, J.F.; Masi, A.T.; McShane, D.J.; Rothfield, N.F.; Schaller, J.G.; Talal, N.; Winchester, R.J. The 1982 revised criteria for the classification of systemic lupus erythematosus. *Arthritis Rheum.* **1982**, *25*, 1271–1277. [CrossRef] [PubMed]

23. Ighe, A.; Dahlström, Ö.; Skogh, T.; Sjöwall, C. Application of the 2012 Systemic Lupus International Collaborating Clinics classification criteria to patients in a regional Swedish systemic lupus erythematosus register. *Arthritis Res. Ther.* **2015**, *17*, 3. [CrossRef] [PubMed]

24. Arnett, F.C.; Edworthy, S.M.; Bloch, D.A.; McShane, D.J.; Fries, J.F.; Cooper, N.S.; Healey, L.A.; Kaplan, S.R.; Liang, M.H.; Luthra, H.S.; et al. The American Rheumatism Association 1987 revised criteria for the classification of rheumatoid arthritis. *Arthritis Rheum.* **1988**, *31*, 315–324. [CrossRef]

25. Ekdahl, C.; Eberhardt, K.; Andersson, S.I.; Svensson, B. Assessing disability in patients with rheumatoid arthritis. Use of a Swedish version of the Stanford Health Assessment Questionnaire. *Scand. J. Rheumatol.* **1988**, *17*, 263–271. [CrossRef]

26. Dolan, P. Modeling valuations for EuroQol health states. *Med. Care* **1997**, *35*, 1095–1108. [CrossRef]

27. Heijke, R.; Björk, M.; Thyberg, I.; Kastbom, A.; McDonald, L.; Sjöwall, C. Comparing longitudinal patient-reported outcome measures between Swedish patients with recent-onset systemic lupus erythematosus and early rheumatoid arthritis. *Clin. Rheumatol.* **2022**, *41*, 1561–1568. [CrossRef]

28. Griffiths, B.; Mosca, M.; Gordon, C. Assessment of patients with systemic lupus erythematosus and the use of lupus disease activity indices. *Best. Pract. Res. Clin. Rheumatol.* **2005**, *19*, 685–708. [CrossRef]

29. Gladman, D.; Ginzler, E.; Goldsmith, C.; Fortin, P.; Liang, M.; Urowitz, M.; Bacon, P.; Bombardieri, S.; Hanly, J.; Hay, E.; et al. The development and initial validation of the Systemic Lupus International Collaborating Clinics/American College of Rheumatology damage index for systemic lupus erythematosus. *Arthritis Rheum.* **1996**, *39*, 363–369. [CrossRef]

30. Levey, A.S.; Coresh, J.; Greene, T.; Stevens, L.A.; Zhang, Y.L.; Hendriksen, S.; Kusek, J.W.; Van Lente, F. Using standardized serum creatinine values in the modification of diet in renal disease study equation for estimating glomerular filtration rate. *Ann. Intern. Med.* **2006**, *145*, 247–254. [CrossRef]

31. Nordmark, G.; Bengtsson, C.; Larsson, A.; Karlsson, F.A.; Sturfelt, G.; Rönnblom, L. Effects of dehydroepiandrosterone supplement on health-related quality of life in glucocorticoid treated female patients with systemic lupus erythematosus. *Autoimmunity* **2005**, *38*, 531–540. [CrossRef]

32. Björk, M.; Dahlström, Ö.; Wetterö, J.; Sjöwall, C. Quality of life and acquired organ damage are intimately related to activity limitations in patients with systemic lupus erythematosus. *BMC Musculoskelet. Disord.* **2015**, *16*, 188. [CrossRef]

Journal of
*Clinical Medicine*

*Article*

# How Can We Enhance Adherence to Medications in Patients with Systemic Lupus Erythematosus? Results from a Qualitative Study

Sharzad Emamikia [1,*], Cidem Gentline [1], Yvonne Enman [1] and Ioannis Parodis [1,2,*]

[1] Division of Rheumatology, Department of Medicine Solna, Karolinska Institutet and Karolinska University Hospital, 17176 Stockholm, Sweden; cidem.gentline@regionstockholm.se (C.G.); yvonne.enman@ki.se (Y.E.)
[2] Department of Rheumatology, Faculty of Medicine and Health, Örebro University, 70182 Örebro, Sweden
* Correspondence: sharzad.emamikia@ki.se (S.E.); ioannis.parodis@ki.se (I.P.)

**Abstract:** Medication non-adherence is common among patients with systemic lupus erythematosus (SLE) and may lead to poor clinical outcomes. Our aim was to identify influenceable contributors to medication non-adherence and suggest interventions that could increase adherence. Patients with SLE from two Swedish tertiary referral centres ($n = 205$) participated in a survey assessing self-reported adherence to medications. Responses were used to select patients for qualitative interviews ($n = 15$). Verbatim interview transcripts were analysed by two researchers using content analysis methodology. The median age of the interviewees was 32 years, 87% were women, and their median SLE duration was nine years. Reasons for non-adherence were complex and multifaceted; we categorised them thematically into (i) patient-related (e.g., unintentional non-adherence due to forgetfulness or intentional non-adherence due to disbelief in medications); (ii) healthcare-related (e.g., untrustworthy relationship with the treating physician, authority fear, and poor information about the prescribed medications or the disease); (iii) medication-related (e.g., fear of side-effects); and (iv) disease-related reasons (e.g., lacking acceptance of a chronic illness or perceived disease quiescence). Interventions identified that healthcare could implement to improve patient adherence to medications included (i) increased communication between healthcare professionals and patients; (ii) patient education; (iii) accessible healthcare, preferably with the same personnel; (iv) well-coordinated transition from paediatric to adult care; (v) regularity in addressing adherence to medications; (vi) psychological support; and (vii) involvement of family members or people who are close to the patient.

**Keywords:** systemic lupus erythematosus; medication adherence; compliance; patient perspective; qualitative research

**Citation:** Emamikia, S.; Gentline, C.; Enman, Y.; Parodis, I. How Can We Enhance Adherence to Medications in Patients with Systemic Lupus Erythematosus? Results from a Qualitative Study. *J. Clin. Med.* **2022**, *11*, 1857. https://doi.org/10.3390/jcm11071857

Academic Editor: Chrong-Reen Wang

Received: 30 January 2022
Accepted: 22 March 2022
Published: 27 March 2022

**Publisher's Note:** MDPI stays neutral with regard to jurisdictional claims in published maps and institutional affiliations.

## 1. Introduction

Systemic lupus erythematosus (SLE) is a multisystem, autoimmune disease that most commonly affects women during their fertile years of age. While SLE can affect all organ systems or tissues, involvement of the skin, joints, central nervous system, and kidneys are among the most frequent manifestations [1]. Given that SLE is a chronic disease, the patients are generally treated with long-term regimens or even life-long immunomodulatory or immunosuppressive medications including antimalarial agents (AMA), glucocorticoids (GCs), and conventional synthetic or biologic disease-modifying anti-rheumatic drugs [1]. Adhering to medications can be a challenge for patients with chronic diseases; non-adherence has been reported to occur frequently in patients with SLE [2] and be associated with poor treatment outcomes [3,4] and a higher likelihood of developing irreversible organ damage [5]. Proportions of patients with SLE who are non-adherent to medications range from 43% to 75% in different studies, depending on how non-adherence was assessed [2]. Various methods are used to measure the degree of adherence to medications including patient-reported tools [6–9] and measures that are designed to be filled-in by others than the patients themselves (e.g., healthcare professionals [10–13]), all primarily used for research purposes.

The reasons for intentional non-adherence are less precise than those for unintentional, the latter being commonly associated with forgetfulness. Overall, intentional non-adherence include problems related to taking medications (e.g., side-effects, inability to pay for the medications, which is more prominent in developing countries, or disagreement regarding the need for pharmacological treatment) [14]. Negative beliefs regarding medications in general or about particular medications are also likely to be associated with intentional non-adherence [15]. For example, the most common reasons for discontinuation of AMA on the patient's own initiative include the perception of AMA not being an effective treatment and apprehension about the potential side-effects [11].

There are different ways to positively impact on the level of adherence to medications. Explaining to the patients how taking a specific drug will benefit them and communicating information about the potential side-effects of the drug have been shown to be important elements, as have providing in-depth information about the disease [16,17]. In this respect, well-informed patients may acquire more positive beliefs on medications and make grounded shared decisions together with their physician, which they are more inclined to follow [18–23].

Under the overall aim of identifying influenceable factors that contribute to medication non-adherence, the specific objective of the present study was to interview patients with SLE from the Karolinska and Örebro University Hospitals to identify such factors in the Swedish healthcare context, and thus propose interventions that could enhance adherence to medications.

## 2. Materials and Methods

### 2.1. Study Design

We performed structured qualitative interviews with individual SLE patients between August 2021 and November 2021. The patients were recruited from a larger online survey study that was conducted in 2021 and investigated the impact of different factors on medication adherence. The location of the interviews (digital or in person meetings) was decided by the interviewees in order to ensure a safe environment for the person being interviewed.

### 2.2. Patient-Reported Medication Adherence

Medication adherence was self-reported by the patients. The frequency of intake of GCs, AMA, and other medications was measured separately using the 6-item Medication Adherence Self-Report Inventory (MASRI) [7], a questionnaire that includes two parts (A and B) where part A contains specific questions on the amount of medication taken recently to prepare the person for the last follow-up question, which is a visual analogue scale (VAS) estimating the overall medication adherence during the past month from 0 (no medication intake) to 100% (full intake). The results from the VAS are used to determine the medication adherence level. Part B of the MASRI is focused on the exact timing of the medication intake; this part was not administered to the survey study participants. The MASRI (part A) has been shown to be reliable for measuring medication adherence in patients with SLE [24]. In addition, we used the Compliance Questionnaire Rheumatology (CQR) [9,25], a 19-item instrument that is more specific for rheumatic diseases and provides a comprehensive assessment of the adherence status. In the CQR, the respondent indicates the level of agreement with different statements made by patients with rheumatic disease. The CQR yields adherence levels from 0% to 100%.

### 2.3. Patient-Reported Disease Activity and Organ Damage

Disease activity was assessed with the Systemic Lupus Activity Questionnaire (SLAQ) [26,27], which captures symptoms and disease activity related to SLE in the preceding three months. The SLAQ assesses the presence and severity of flares and comprises 24 items that investigate SLE-related symptoms, yielding a symptom score from 0 to 24, depending on the presence or absence of symptoms. Higher scores indicate higher disease activity. The questionnaire ends with a VAS (global health score) that estimates the disease activity, where 0% reflects no disease activity and 100% represents the highest imaginable disease activity.

Patient-reported organ damage was estimated using the Self-Administered Brief Index of Lupus Damage (SA-BILD) instrument [28]. The SA-BILD includes 28 items covering ocular, neuropsychiatric, renal, pulmonary, cardiovascular, peripheral vascular, gastrointestinal, musculoskeletal, skin damage, premature gonadal failure, diabetes, and malignancy. Higher scores indicate greater levels of accumulated irreversible organ damage; the scale ranges from 0 to 30.

### 2.4. Selection of Patients for Qualitative Interviews

Patients from two Swedish tertiary referral centres (i.e., the Karolinska and Örebro University Hospitals) with an International Classification of Diseases (ICD) code indicating that they are diagnosed with SLE were asked to participate in a survey-based study addressing medication non-adherence. In this survey, patients were asked to indicate whether they were willing to be contacted for participation in a subsequent qualitative interview-based study. From 205 patients who completed the survey, 15 patients from the Karolinska University Hospital in Stockholm ($n = 14$) and Örebro University Hospital in Örebro ($n = 1$) were selected to be invited to participate in individual in-depth interviews for the purpose of the present study. To obtain in-depth information from a representative sample from the survey respondents, our selection of interviewees aimed at a varying degree of experience of challenges in adhering to medications (i.e., varying levels of adherence) and a similar age range and sex distribution to that of the population of patients who filled in the survey.

### 2.5. Interview Protocol

The English version of the interview protocol is presented in the online Supplementary Materials. The protocol was designed to address the main research question of the present investigation (i.e., what kind of implementations in healthcare could improve medication adherence in patients with SLE), as derived from the perspective of the interviewees. The main interview question was: "What do you think could make you take your medicine exactly as prescribed by your treating physician?". The questions posed to the participants were short and easy to understand. The interview was initiated with questions starting with "what" and "why", referring to a concrete event. Once those questions were answered, questions involving "how" were pursued. Issues starting with "why" and dealing with the patient's reasons for their potential lack of medication adherence were addressed towards the end of the interview. Questions of a "probing" nature were prepared with the aim of following up on certain answers with subsequent questions, gradually leading to more comprehensive information [29].

### 2.6. Interview Procedure

All interviews but one were performed in Swedish. In that interview, the questions were posed in Swedish, but the interviewee preferred responding in English. All interviews were performed by the same investigator (SE). The interviewees were informed prior to the start of the recording that the interviews would be recorded to preclude recall bias and eliminate the risk of misinterpreting the patient, and recording started upon the patient's approval. Verbatim transcription was performed by the same investigator who performed the interviews (SE), directly after each interview with the aim of implementing improvements in the subsequent interviews. Even though the interview guide was structured with predefined questions, it was inevitable that some parts of the interview were conducted ad hoc with the help of probing questions [30], and was thus dependent on the interaction between the interviewee and the interviewer. A pilot interview preceded the aforementioned 15 interviews as an aid to design and fine-tune the interview guide in compliance with qualitative research conduct methodology. Due to the ongoing severe acute respiratory syndrome coronavirus 2 (SARS-CoV-2) pandemic, most interviews were performed in a digital format using video conferencing services to virtually meet the interviewees ($n = 13$). One of the interviews was conducted at the inpatient care unit at the Karolinska University Hospital and another interview at the patient's working place.

### 2.7. Analysis of Qualitative Data

Content analysis, a flexible method to analyse text data [31–33], was performed by two investigators (SE and CG) who independently read and reread the transcripts in detail, in order to index and compare the identified subcategories. The subcategories were derived inductively (data-driven) and were centred on types of behaviours (i.e., reasons for not displaying adequate adherence to medications). The role of subcategories was to increase the granularity of data collected, ensuring a broad and inclusive process that would help comprehend the data as much as possible. Categories with substantial similarities were later unified to create key themes. Depending on how the patients described their behaviour and reasons for not adhering to medications, we stratified the reasons into intentional and unintentional based on whether the patient had actively decided to be non-adherent or not, respectively.

Data collection continued until the discussion material had reached saturation (i.e., until no additional data were identified by analysing the transcribed interviews) [34,35]. Data saturation was based on identification of new themes and categories while analysing the transcripts (i.e., inductive thematic saturation) [36]. We considered the data saturated when results from the last five interviewees did not add novel reasons for not adhering to medications as prescribed by the treating physician and when no new interventions for improving medication adherence were mentioned during the interviews. Potential discrepancies between the two investigators (SE, CG) were resolved by reaching a consensus together with a third investigator (IP), who is a rheumatologist.

### 2.8. Patient and Public Involvement

The research question of this study was developed in collaboration with a patient research partner (YE), based on shared priority to investigate adherence to medications. The patient research partner was involved from the initial stage of the research process and throughout the development and conduct of the study.

### 2.9. Ethics

The study design and conduct complied with the ethical principles of the Declaration of Helsinki. The protocol of the study was approved by the Swedish Ethical Review Authority on 1 April 2021 (reference number: 2021-00662). Written informed consent was obtained from all study participants prior to enrolment.

## 3. Results

### 3.1. Patients

All study participants were adults (age range: 20–68 years) and most of the participants were women ($n = 13$; 87%). In addition to these participants, two patients were invited to participate but declined, one who was a public figure and worried that their identity might be revealed and one without providing any specific reason. Patient characteristics including patient-reported disease activity and organ damage are presented in Table 1 ($n = 14$; data were missing in one patient). One of the participants had experienced a severe flare within three months prior to the time of filling in the initial survey. Adherence levels based on inter alia responses to the MASRI and CQR are detailed in Table 2. The patients in the original survey study ($n = 205$; age range: 20–87 years) had a median disease duration of 12 (5–22) years, and 86% among them were women.

**Table 1.** Patient characteristics. Data are presented as median and interquartile range. SLAQ and SA-BILD scores were missing for one patient.

| | |
|---|---|
| Age (years); median (IQR) | 32 (27–50) |
| Country of birth; *n* (%) <br> Sweden <br> Other | <br> 7 (46.7) <br> 8 (53.3) |
| Living alone; *n* (%) | 6 (40.0) |
| Highest education level; *n* (%) <br> University <br> High school | <br> 11 (73.3) <br> 4 (26.7) |
| Employment status; *n* (%) <br> Full time <br> Part time <br> Retired | <br> 9 (60.0) <br> 5 (33.3) <br> 1 (6.7) |
| Disease duration (years); median (IQR) | 9 (5–20) |
| SLAQ Symptom Score *; median (IQR) | 10 (6.75–15.25) |
| SLAQ Global Health Score [†]; median (IQR) | 35 (20–65) |
| SA-BILD Total Score [‡]; median (IQR) | 0 (0–1.25) |

IQR: Interquartile range; SLAQ: Systemic Lupus Activity Questionnaire; SA-BILD: Self-Administered Brief Index of Lupus Damage. * SLAQ Symptom Score ranges from 0 to 24. [†] SLAQ Global Health Sore ranges from 0 (no disease activity) to 100 (maximum disease activity). [‡] SA-BILD Total Score ranges from 0 to 30.

**Table 2.** Adherence levels in the fifteen interviewees.

| Patient | Polypharmacy (i.e., ≥Five Medications) (Y/N) | Prescribed Medications | Overall Medication Adherence According to: | | | Intentional Non-adherence (Y/N) |
|---|---|---|---|---|---|---|
| | | | MASRI (0–100%) | CQR (0–100%) | Direct Question (Y/N) * | |
| 1 | N | PRED | 96 | 61 | N | Y |
| | | HCQ | 96 | | | |
| | | AZA | 96 | | | |
| 2 | Y | PRED | 100 | 77 | Y | NA |
| | | HCQ | 100 | | | |
| | | MTX (pills) | 100 | | | |
| 3 | N | PRED | 100 | 81 | Y | NA |
| 4 | N | HCQ | 90 | 58 | N | Y |
| 5 | N | HCQ | 100 | 54 | N | Y |
| 6 | N | PRED | 70 | 65 | N | N |
| | | HCQ | 70 | | | |
| | | AZA | 69 | | | |
| 7 | N | PRED | 67 | 47 | N | Y |
| | | HCQ | 50 | | | |
| | | AZA | 20 | | | |
| 8 | N | HCQ | 88 | 72 | N | N |
| 9 | N | PRED | 95 | 74 | N | Y |
| | | HCQ | 96 | | | |
| | | MMF RTX (iv) | 99 | | | |
| 10 | N | PRED | 100 | 66 | N | N |
| | | HCQ | 100 | | | |
| | | AZA BEL (sc) | 80 | | | |

**Table 2.** *Cont.*

| Patient | Polypharmacy (i.e., ≥Five Medications) (Y/N) | Prescribed Medications | Overall Medication Adherence According to: | | | Intentional Non-adherence (Y/N) |
| | | | MASRI (0–100%) | CQR (0–100%) | Direct Question (Y/N) * | |
|---|---|---|---|---|---|---|
| 11 | Y | PRED | 96 | 54 | N | N |
| | | HCQ | 100 | | | |
| | | MMF RTX | 98 | | | |
| 12 | N | PRED | 89 | 58 | N | N |
| | | HCQ | 92 | | | |
| 13 | N | PRED | 100 | 84 | Y | NA |
| | | CYS | 100 | | | |
| 14 | N | HCQ | 100 | 74 | Y | NA |
| | | MTX (pills) | 100 | | | |
| 15 | Y | PRED | 90 | 67 | N | N |
| | | HCQ | 90 | | | |
| | | MMF | 80 | | | |

MASRI: Medications Adherence Self Report Inventory; CQR: Compliance Questionnaire Rheumatology; Y/N: yes/no; NA: not applicable; iv: intravenous; sc: subcutaneous; PRED: prednisolone; HCQ: hydroxychloroquine; AZA: azathioprine; MTX: methotrexate; MMF: mycophenolate mofetil; RTX: rituximab; BEL: belimumab; CYS: cyclosporine. * Y = adherence assent to direct question; $n$ = non-adherence assent to direct question.

### 3.2. Barriers to Medication Adherence

We identified four main themes that explained medication non-adherence in this cohort, which are demonstrated below along with selected illustrative quotations from the interviewees translated from Swedish to English.

#### 3.2.1. Patient-Related Barriers

In total, the proportion of patients who expressed patient-related barriers at least once was 93.3% ($n = 14$). Patients expressed that being diagnosed with a chronic disease at a young age was important to account for ($n = 5$; 33.3%), as were their professions ($n = 8$; 53.3%). Unintentional non-adherence was mainly caused by forgetfulness (e.g., due to interruptions in the patient's routines) ($n = 5$; 33.3%).

*I know if I'm traveling, it's easy to forget. (Patient 15)*

*And sometimes I forget why it is a good idea [to take the medications], but I think the practicalities of life sometimes and all the things you have going through in your head [ … ] get in the way … (Patient 08)*

*I'm in too much of a hurry. There's no time. It takes time to prepare a sandwich. This is the kind of thing that is done as the last… last priority. (Patient 12)*

In this respect, having to keep track of multiple pills and multiple or varying time-points for the intake were mentioned as barriers ($n = 5$; 33.3%). Unwillingness to take multiple medications ($n = 3$; 20.0%) was discussed. Factors related to the administration of the medications also seemed to cause irregular adherence (e.g., having to split the pill or having to take varying doses of a medication each day) ($n = 1$; 6.7%). The size and taste of the pills hindered some patients in taking their medications ($n = 2$; 13.3%), while some other patients experienced a resistance at a mental level when having to inject themselves with a medication ($n = 1$; 6.7%).

*I can't stand fiddling with all the pills in the morning. (Patient 12)*

*You don't want to feel so sick; you skip some. Sometimes you take a few and sometimes you don't take anything. (Patient 09)*

*I don't know if there are too many, it's just that it's too much to stand and fiddle with. I have six medicines and stand and push them out, pick them out, it sounds silly when I say it like this. I can't stand it, it's so hard. I don't know why it's so hard. (Patient 12)*

*When I was younger, I ate a lot of pills. Some days it felt like you were taking ten pills for breakfast. It's too much. I felt nauseous, some of the medications made me nauseous. And then you know you're going to do that for the rest of your life, feeling bad about medications that are supposed to make you feel good. (Patient 10)*

Moreover, various practicalities were mentioned as potential causes of non-adherence (e.g., difficulties in obtaining medications from the pharmacy) ($n = 1$; 6.7%) or because of other specific circumstances (e.g., due to the SARS-CoV-2 pandemic) ($n = 1$; 6.7%).

Several patients raised beliefs indicating that they themselves or family members were sceptical towards medications that they were prescribed, which negatively impacted on their adherence ($n = 5$; 33.3%). The mother of one of the interviewees wanted the treatment to also include herbal remedies. One of them mentioned that the scepticism was due to a previous adverse event related to an intravenous treatment, and because the rheumatologist was too focused on the present condition rather than the prospective aspects of the disease. Another patient mentioned the importance that religion had in their family.

*I grew up in a very religious [religion specified] home, and in such a home you think you don't need medicines because your body knows how to treat itself. (Patient 07)*

Scepticism could also be expressed in general terms.

*I can sometimes think, do I need all those medications? How do we know that this particular medication is helping me, or how do we know that this medication I am given now is helping me? Why am I given so much? Why haven't we sticked with some of the medications? (Patient 10)*

### 3.2.2. Healthcare-Related Barriers

The overall percentage of patients who expressed barriers related to the healthcare system at least once was 73.3% ($n = 11$). The relationship with the treating physician was central according to most interviewees ($n = 11$; 73.3%). An untrustworthy relationship with the physician could be caused by the feeling of overall inaccessibility to healthcare providers ($n = 2$; 13.3%), not being listened to ($n = 4$; 26.7%), and application of treatment strategies based on group-level evidence rather than person-centred approaches ($n = 3$; 20.0%).

*My doctor didn't really listen to me that much and it was more of a generalisation every time I saw my doctor. (Patient 05)*

*They basically prescribe the same thing to all SLE patients. (Patient 07)*

Some patients felt like they were seen as "an SLE diagnosis" rather than "an individual behind the SLE diagnosis" ($n = 2$; 13.3%).

*There is a normal person behind the civic registration number. (Patient 04)*

Additionally, the impression of not being believed as a patient when telling their physician about symptoms ($n = 2$; 13.3%) and meeting different healthcare personnel from visit to visit ($n = 2$; 13.3%) seemed to negatively impact on adherence.

*I'm not a guinea pig . . . (Patient 01)*

There was an interest in knowing the indications of the medications in more detail (i.e., whether the medication prevents symptoms or treats the disease) ($n = 2$; 13.3%). Some expressed the desire to receive more information about the potential side-effects and mechanisms of action of the drugs ($n = 4$; 26.7%).

*And then I go home and do research myself and find out what the medication does, what the side-effects are. Because they don't always tell you about the side-effects either. It comes as a shock. (Patient 09)*

Some patients had experienced that physicians considered test results more important than the patients' experience of health-related quality of life (HRQoL), which also negatively affected the trust towards healthcare and, in turn, adherence to medications ($n$ = 3; 20.0%).

The interviewees expressed concerns resulting from lack of information about whether consumption of alcohol or certain kinds of food could interact with the medication, or whether fasting had an impact, which led to not taking their medication at certain time-points ($n$ = 3; 20.0%).

Several patients ($n$ = 6; 40.0%) experienced a power imbalance between the patient and the physician. This power imbalance could cause non-adherence due to fear of authority, resulting in not finding the courage to question the physician ($n$ = 1; 6.7%), feeling embarrassed to pose questions about the medications ($n$ = 1; 6.7%), or feeling that they take control when not adhering ($n$ = 1; 6.7%).

> *It's a bit of a position of power and you still need help from that person, and it can be hard to speak out. (Patient 05)*

### 3.2.3. Medication-Related Barriers

In total, the percentage of patients who expressed medication-related barriers at least once was 66.7% ($n$ = 10). Among those, intentional non-adherence was common, with the belief that the prescribed medication might cause side-effects being the most common reason ($n$ = 10; 66.7%).

> *Cortisone, I've always been against it … (Patient 01)*

Some of the patients ($n$ = 8; 53.3%) were worried about potential side-effects that could occur in the future, and other patients ($n$ = 7; 46.7%) were worried that side-effects they had previously experienced from the prescribed medication would emerge again. Patients expressed concerns that the medications might result in a shorter lifespan ($n$ = 1; 6.7%), weight gain ($n$ = 1; 6.7%), nausea ($n$ = 1; 6.7%), and skin dyspigmentation ($n$ = 2; 13.3%). Some interviewees also raised apprehensions about potential negative effects on the kidneys ($n$ = 1; 6.7%) and eyes ($n$ = 2; 13.3%), and that the prescribed medication might cause osteoporosis ($n$ = 1; 6.7%) and/or diabetes ($n$ = 1; 6.7%).

For some interviewees, intentional non-adherence could be a consequence of poor information about their prescribed medications ($n$ = 8; 53.3%).

> *It's often you feel, what am I putting in me? (Patient 10)*

The perspective of time seemed to be important (e.g., information about how long the medication is planned to last), especially when a substantial difference in symptoms was not experienced after commencement of a new medication ($n$ = 2; 13.3%). In such cases, non-adherence was the result of testing the efficacy of the medication on the patient's own initiative (i.e., to see whether there was a difference in symptoms when not taking the medication).

> *I don't notice any difference if I skip the medication for a day or two. (Patient 04)*

### 3.2.4. Disease-Related Barriers

The overall percentage of patients expressing disease-related barriers at least once was 53.3% ($n$ = 8). Some patients perceived the severity of their disease as mild, which negatively affected their adherence to medications ($n$ = 6; 40.0%).

> *I don't have very serious symptoms. It doesn't matter much if I don't take the medication. (Patient 05)*

Some patients believed that it was difficult to talk to other people about their disease since the disease was not always tangible, which could result in neglect ($n$ = 1; 6.7%). Along the same lines, patients were less motivated to take their medications when feeling well because no symptoms reminded them of the need to take their medications ($n$ = 6; 40.0%).

> *I have stopped when I have felt good. I felt like I didn't need them anymore. (Patient 07)*

*Usually when I'm okay. I forget that I am sick and need to take the medication. (Patient 06)*

With regard to disease manifestations and/or comorbid conditions, depression was mentioned as a reason for non-adherence ($n = 1$; 6.7%) that occurred previously in the patient's life but was not manifest at the time of the interview. Finally, acceptance of a chronic illness was difficult for some patients ($n = 5$; 33.3%), partly because in several cases the disease is not visible to others and therefore difficult to talk about, or because taking multiple medications would make them feel that they are sick.

*I guess it is that you don't feel you are sick. You have side-effects. I don't want to feel sick; I want to be a healthy person. (Patient 12)*

*I hate feeling sick . . . (Patient 01)*

*I was still very sick at that time . . . now that I'm healthy. Or well, healthy . . . (Patient 07)*

### 3.3. Positive Impact on Medication Adherence

Patients explained that they also had several reasons for taking their medications as prescribed. Many experienced that they felt balanced regarding their well-being when they took their medications ($n = 7$; 46.7%), they avoided flares ($n = 6$; 40.0%), they had less pain and more energy ($n = 5$; 33.3%), and they could be physically active and able to work ($n = 3$; 20.0%). Several patients ($n = 6$; 40.0%) described the use of pill organisers as an important tool for maintaining a high adherence level, despite some resistance in accepting them.

*I feel so old. Only old ladies have such pill organisers. (Patient 12)*

### 3.4. The Impact of Shared Decision-Making

In the interviews, the patients talked about their view on shared decision-making. Notably, several patients in our study ($n = 6$; 40.0%) had the impression that their involvement in therapeutic decision-making would not affect their adherence to medications.

*I have no medical knowledge of these medications. I have to trust their professionalism and knowledge. I can't do more than that. My doctor is a specialist in the field. (Patient 12)*

*How can you be involved in something you don't know anything about? [ . . . ] I don't know anything about these medications. It's the doctor who's a specialist, I listen and test. If it goes well, I'll be fine. (Patient 03)*

*I'm not medically trained. [ . . . ] You accept what you get. (Patient 14)*

*I probably leave it to them . . . they know what is best for me. (Patient 08)*

One patient felt that she was not given the opportunity to be involved, even though she did not believe that shared decision-making would impact on her adherence ($n = 1$; 6.7%).

*God no [reaction on whether the patient felt involved in the decision-making for the prescribed medications]. I don't think it would make any big difference quite honestly since they [the rheumatologists] know what they are prescribing, and they basically prescribe the same thing to all SLE patients. The names of the drugs may differ slightly, but they are still immunosuppressants and cortisone. So, I don't think it would make any big difference. (Patient 07)*

One patient even felt that more responsibility was put on them through shared decision-making.

*Rather, I feel that I perhaps have had a little too much participation and I have been able to decide, but I did not really receive any recommendations or advice, which I would have liked. (Patient 05)*

### 3.5. Facilitators for Improving Medication Adherence

Based on the thematic analysis and direct suggestions by the interviewees, the following distinct interventions were identified as potential strategies for enhancing medication adherence.

### 3.5.1. Increased Communication

Patients called for comprehensive information from their treating physicians about the prescribed medications (e.g., pharmacodynamics and pharmacovigilance). It was a desire that this information is person-centred and accounts for the individual situation and circumstances of the patient. Such information should ideally be provided on a regular basis.

*Maybe it's because you don't know what's in that medicine that is supposed to make you feel better. (Patient 04)*

*Or, above all, tell us about late complications if you don't take them [the medications]. Not as a threat. If you don't do this, this will happen. Take the time to tell us what can happen. Or what the risk is if you do not treat [the disease]. (Patient 12)*

Prolonged visits and interactive encounters with the treating physicians might reduce the stressful impression during a visit when patients sometimes feel that they cannot ask questions or are unable to absorb the information that is provided to them, especially when abbreviations are used in the communication, as pointed out by one interviewee.

*There is such a short time for these appointments with the doctors that… it doesn't feel like you're getting an answer to all questions. (Patient 10)*

*There are a lot of things you miss when you go to your visit because you are nervous, and you need to take notes. (Patient 02)*

### 3.5.2. Patient Education

Personalised information could be given in different formats (e.g., in writing). Additionally, congresses specifically targeted to patients could be organised, aiming for patient education about the disease of SLE and common medications used to treat it.

*They told me "You should be happy if you survive until you are an adult" and "Don't expect to be able to have a job, you will be retired early and you will not have any children". (Patient 04)*

Personalised adjustment of medications accounting for the patient's lifestyle was suggested (e.g., if a patient has problems taking their medication at certain timepoints, rearrangement of the treatment regimen could be attempted). Adjustment of the communication forms accounting for the patient's lifestyle also emerged as a desirable action (e.g., if the patient is an athlete or exercises in a regular and systematic manner, a focus on physical dimensions of how the disease afflicts the body might be useful).

### 3.5.3. Accessible Healthcare

Easily accessible healthcare services for questions also emerged as a desire. Digital options could be provided to eliminate frustration due to difficulties in reaching the healthcare providers within a reasonable time. Another suggestion from a patient was to develop an application where patients monitor their symptoms and are able to see how medication non-adherence is associated with flares, similar to one created by and for patients with rheumatoid arthritis, where one of the multiple objectives was to increase the understanding of the impact of adherence [37]. Furthermore, meeting the same treating physician and nurse at all visits (i.e., personnel that becomes familiar to the patient and knows the patient's medical history) appeared to be valuable for the patients.

*All these years when I've been able to see the same nurses in the same departments. You can tell it gives a lot. The greetings. Several nurses have seen me and asked "how are you?" Then back in action. They don't make any big deal of it. Let's go, let's go. You'll be here for 2–3 days and you'll be back on track. It gives a feeling of safety. (Patient 01)*

### 3.5.4. Structured Transition from Paediatric to Adult Care

In our study population, five of the interviewees had childhood-onset SLE. Special support may be of particular importance in certain phases of the disease course, or at certain arrangements; a smooth and structured transition from paediatric to adult care was referred to as an example. Several patients experienced challenges to accept the changes in their encounter with the healthcare services posed by this transition. This was described to have inevitable consequences regarding trust issues and, in turn, adherence to therapy.

*When I got SLE, it was my parents who read through it [leaflet] because I probably didn't understand much. [ … ] My parents used to remind me and gave me the medication every morning. (Patient 13)*

*When I went to the "Children's' Hospital", back then I felt like I got a lot out of it. It was also when I was sickest. I received a lot of help and had a close contact with my doctor. [ … ] It was easy to get in touch with them if there was anything. But then… Now [ … ] I don't get the help I feel I need. It's quite difficult to contact the doctor and [ … ] book an appointment when something is wrong. (Patient 05)*

*I've been taking the medications for so long that I don't feel like I've had an adult life as a free person, so I've never had any other life before. (Patient 14)*

### 3.5.5. Medication Reconciliation

A straightforward communication was also suggested for the process of medication reconciliation, which occasionally may be complicated. Medication reconciliation could be completed in different formats (e.g., by directly asking the patient whether they take the prescribed medications, or by asking indirectly via electronic questionnaires at the outpatient clinic before the visit, or through the Swedish online healthcare guide service).

*This would mean to me that they follow up on my illness and I would be happy to receive such a question. (Patient 02)*

*At every encounter with medical staff, because then it also becomes a reminder [ … ] it would mean to me that they follow up the patient and keep track. (Patient 07)*

### 3.5.6. Psychosocial Support

Psychosocial support aiming for acceptance of the diagnosis emerged as an important need. The chronic nature of the disease appeared to necessitate supportive or therapeutic counselling, especially during the early stages after the diagnosis is made.

*I think that you need a contact with a social counsellor … people who are sick with diseases that are difficult to live with need more mental support than what is offered today. (Patient 03)*

*I was never offered to talk to anyone. Maybe you should offer all patients to talk to someone, then if this person is a psychologist or a nurse, it doesn't matter to me. (Patient 07)*

Patients call for a focus on aspects that impact on physical and mental HRQoL.

*[I want to] build up some muscle, and I feel like nothing is happening. I don't get any support from the healthcare system… The important thing for them is that I'm "healthy" and the blood tests look good. They don't mind about my quality of life. (Patient 11)*

*They are very happy to check test results. They can say "your blood tests look great! I'm glad to see these results!" Ok, but what are we going to do about those days when I didn't feel so good? Then nothing happens. (Patient 10)*

### 3.5.7. Engaging Family Members

Some patients mentioned how people who were close to them could have an impact on them. One example was sceptical family members who weakened the belief that medications can treat diseases.

*So that you kind of find out what the family's religion and perceptions of medicine are so that you don't just throw the patient into something that the patient won't follow later. (Patient 07)*

*Relatives know best. You're a stubborn man or woman, so they [the rheumatologists] might have to approach you from somewhere else [reach out to relatives]. (Patient 01)*

*I have a mother who is very sceptical and puts pressure on me all the time. (Patient 11)*

Thus, it emerged as an action of particular importance to involve people who are close to the patient in the information about the suggested or the planned treatment regimen, as well as involve them in broader aspects of patient education.

## 4. Discussion

Through in-depth interviews, we aimed at understanding the perspectives of SLE patients on factors that contribute to medication non-adherence and identifying suitable interventions or actions that could improve adherence if implemented in healthcare. Patient education at multiple levels emerged as an imperative need; patients desired more in-depth information by the healthcare personnel compared with what they currently receive, including information about disease features, blood, and urine test results, and ongoing or suggested treatment regimens. Prolonged visits with the treating physician to ensure sufficient time to address all aspects of the disease and the current condition was requested by most patients. The importance of educating the patient and communicating information in a manner that ensures that the patient feels confident has been discussed in previous research [38,39]. Several patients in our study were reluctant towards involvement in therapeutic decision-making, while many patients desired more information from their treating physician about their disease and their medications. As a matter of course, unwillingness to participate in decisions and a feeling of insufficient knowledge about the suggested medications may be related to each other in a causal manner, with the former emerging as a consequence of the latter [40]. Regular discussions about adherence to medications also emerged as a suitable action for the identification of non-adherence patterns, and when unintentional non-adherence is detected, practical advice on how to manage to effectuate regularity in taking the medications should be provided. In some cases, psychological support may be needed; it is important to establish strategies for identifying such needs and take action to provide this resource.

Overall, our findings concurred with suggestions for interventions against medication non-adherence derived from previous studies involving patients with SLE [2,41,42] and other chronic diseases [43]. Several previous studies have described methods for enhancing adherence to medications in patients with SLE [16,17], some of which were qualitative studies (e.g., from Jamaica, Portugal, the United Kingdom (UK), and the United States (US)) [18–23], while others explored and evaluated specific interventions in patient groups [44–47]. Improved communication between the healthcare and patients along with increased information about the rationale for taking the medication as well as less complicated medication regimens improved adherence in an American study [16]. Another study from New Zealand [17] provided implications that addressing patients' concerns about side-effects may improve the relationship between the treating physician and the patient and thereby improve adherence. As in our study, results from other qualitative interview studies [18–23] suggested that improvements in communication between physicians and patients, making medications affordable and available, increasing the patients' knowledge about the disease and therapeutic options, challenging the patients' beliefs regarding medication effectiveness with well-documented evidence as well as facilitating access to healthcare, all are crucial factors that can be expected to contribute to increasing medication adherence.

The applicability and effectiveness of various interventions should also be seen from the perspective of the cultural background of the respective patient population. Targeted nursing including detailed patient-specific solutions that involved the patient's daily life has been shown to be effective in improving medication adherence in Chinese patients [44]. In that study, the patients were followed up for 20 months and received support in various aspects related to understanding the disease and its prognosis, accounting for the patient's socioeconomic background and psychological health. The patients received information about the importance of appropriate treatment, diet, prevention of infections, and other aspects [44]. However, not all studies have shown that increased communication improves adherence. A study from the US [45] examined the usefulness of cellular text messaging for improving the adherence to hydroxychloroquine (HCQ) in adolescent SLE patients, but despite this type of individualised communication, the repetitive messages did not influence the patients' degree of adherence. In this respect, it is reasonable to postulate that the means of communication may matter. Another US study showed that routine testing of the blood levels of HCQ improved adherence over time, from 56% to 80% adherent patients [46]. A study from India evaluated the role of a clinical pharmacist in terms of individualised counselling regarding the disease, medications, and lifestyle modifications using pre-developed written educational material [47]; counselling performed monthly and three times in total resulted in improved adherence to medication.

Interestingly, several patients in our study referred to their professions while being interviewed. Some patients worked within healthcare themselves, others were athletes; irrespective of their profession, patients related several of their responses to their occupation, which raises awareness of the importance of taking the patient's profession and background into account in order to provide individualised information. Another aspect that pointed to the importance of person-centred approaches was that the patient's age was important for the patients, both the current age and the age when the diagnosis was made, with a younger age in both cases being coupled with a lower level of acceptance of illness and a lower degree of motivation to adhere to regularity in taking medications. Therefore, these patient subgroups may need more attentive contact with their physicians and/or other health professionals involved in their care, and a higher level of support. The transition from paediatric to adult care appeared to be challenging, in conformity with what has been shown in other studies [48,49], especially when the patient had a strong bond to the paediatrician. All the above set a strong motive for the healthcare and treating physicians to tailor the communication, the information about the disease and therapies, and the surveillance strategies to the patient's individual background.

Some patients mentioned that multiple medications and complicated schedules of varying daily doses constituted a reason for impaired adherence. Polypharmacy, defined as five or more concurrent medications [50], has been documented as a factor contributing to non-adherence in previous studies [13,51], collectively suggesting that regular and thorough medication reconciliation and, when possible, attempts to decrease the total amount of daily administrations may help improve adherence to medications.

Suffering from depression has been shown in previous research to cause poor adherence to medications in patients with SLE [4,52], which was also detected in our interviews. Along with symptomatically treating depression, more attentive follow-up and support directed to improve adherence in this patient subgroup might prove helpful. Depression can influence forgetfulness [4] and has been described as an independent risk factor for non-adherence [13]. Considering that a substantial proportion of patients with SLE develop anxiety or depression [1], which is often resistant to therapy for SLE [53], this patient subgroup may be of particular importance to consider in the allocation of healthcare resources to actions against medication non-adherence.

In patients with SLE, the use of AMA has been associated with a wide variety of beneficial effects [54,55], including favourable associations with biological disease properties such as lower levels of B cell activating factor (BAFF) [56] and favourable effects on HRQoL [57]; this drug class constitutes the cornerstone of SLE therapy. During our interviews, it appeared cumbersome to retrieve the correct daily dose of HCQ when a different number of pills was to be taken on different days of the week according to the prescription. One patient called for more flexible dosage schedules or an easy way to divide the pill instead of varying dosages each day. This issue has also been addressed in a large study from the US comprising 3127 patients with SLE, which demonstrated that a substantially higher proportion of patients taking different doses every day reported that they forgot or mistook the correct dose (32%) compared with patients who were on the same dose every day [58]. Thus, prescriptions that are easy to follow may ascertain that the patient will follow them to a larger extent. While this notion may appear intuitive, complicated dosage schedules are often used to adjust for the patient's weight, raising actionable awareness of the need for a reduction of the degree of complexity when designing therapeutic regimens.

Financial costs of medications have been identified as potential barriers to medication adherence in patients with SLE in a study from the US [21], as well as in documentations from developing countries [19]. Along these lines, one interviewee mentioned that there had been times when the patient did not pick up a prescription for SLE from the pharmacy due to the price of the medication to prioritise other medications prescribed for other diseases. This affirmation by the patient is of interest considering the Swedish context, where costs for prescribed medications exceeding a yearly amount of 2350 SEK (equivalent to USD 262, based on the current exchange rate), underlying that therapeutic decision-making and patient education should also account for comorbidities and raising actionable awareness of the need for the identification of impecuniousness in certain cases and allocation of resources to ensure access to the prescribed medications.

Support from people close to the patient was mentioned by some study participants to be vital. This suggests that information and education should not only be provided to the patients, but also to family members or trusted people from the patient's environment. In this regard, a previous study showed that a clinical pharmacist could have an important role in counselling family members and providing them with information about the prescribed medications [47]. Last, but not least, not only the socioeconomic background, but also values important to the patient should be accounted for in therapeutic decisions and in patient education to ensure a holistic person-centred strategy against medication non-adherence. Such values may include the religious beliefs of the patient or family members, which was described by one of the interviewees to be of particular importance. Religious and spiritual beliefs have also been described to impact on medication adherence in previous research [19], especially when such beliefs specifically address or interfere with the use of medications or the time of the day when medication intake is scheduled. The same study [19] identified perceived mild severity of the disease as one of the reasons for medication non-adherence; some patients tended to take their medications only when experiencing symptoms, which is in compliance with the findings in the present study. However, high disease activity and disease flares may also result in difficulties with adhering to medications, as shown in another study [4]. While these discrepancies across studies may give the impression of being conflicting, a reasonable explanation might be that the level of disease activity impacts on how adherence is affected (e.g., with low-grade activity acting as a "reminder" and high-grade activity being a negative contributor), as do cultural and socioeconomic facets.

### 4.1. Limitations and Strengths

The small study population may be considered a limitation. However, this study was designed to be a qualitative thematic analysis of interview content and data collection from individual interviews that continued until the material had reached saturation (i.e., until no additional content or patterns were identified during the analysis). Thus, a larger number of interviewees was deemed excessive and even inappropriate, as it might result in a cumbersome dataset to analyse while not adding informative data. The study was not designed to prospectively assess the impact of the identified potential interventions on SLE patients' adherence to medications. However, the hypotheses generated herein warrant prospective investigation in future studies. Moreover, statistical determination of associations between hypotheses and data was beyond the scope of this study; thus, the analysis of the interviews was not assisted by statistics software and the presentation and discussion of our findings was mainly descriptive.

Disease duration was not considered when patients were asked to participate in the interviews. Our study participants' median disease duration of nine years was shorter than that of the respondents to the initial survey-based study (12 years). Shorter disease duration has been associated with adherence difficulties in a study comprising 834 patients with SLE [4]. Thus, since we aimed to mainly interview patients with adherence difficulties, our selection strategy may intrinsically have resulted in a shorter median disease duration for the interviewees compared with the survey respondents.

A major strength of our study was that the interviews were performed by a researcher who did not belong to the clinic personnel and with whom the study participants therefore had no prior acquaintance (SE). An additional advantage of this was that the interviewer had no interference in the therapeutic decision-making, thus ensuring objectivity during the interviews. This made the interviewees comfortable with sharing in-depth thoughts and opinions. In a few instances, the interviewees requested assurance that their treating rheumatologist would not be made aware of their responses. However, two researchers with different backgrounds (master in biomedicine, SE; resident in rheumatology, CG) were involved in the analysis of data to ensure that the analysis and subsequent interpretation of results were not dependent on one individual investigator or biased by the view of a specific profession.

### 4.2. Clinical Relevance

Medication non-adherence may have detrimental effects on patient safety, especially when poor adherence is not identified, resulting in mistaken treatment evaluations. This may in turn lead to unnecessary dose increases or changes in the therapeutic regimens, potentially resulting in additional morbidity and impairments of the patients' HRQoL. Towards the goal of improving SLE patients' adherence to medications, the first and most important step is to systematically seek modifiable factors that negatively impact on adherence at the level of the individual patient, and thereafter, together with the patient, couple these factors with person-centred interventions against non-adherence.

### 5. Conclusions

We identified several potential interventions for improving medication adherence in patients with SLE. Our findings suggest that patients should be educated by obtaining complete written and oral information about the medications that they are recommended or prescribed and about the disease that they are diagnosed with, in order to increase their knowledge and deepen their understanding of the rationale for taking the medications. Healthcare professionals should account for the patients' HRQoL, religious beliefs, and professions, and include family members in the care process. Professional mental support should be offered to help the patients accept the diagnosis of a chronic disease and the need for treatment. Direct and empathic medication reconciliation on multiple occasions emerged as a desire from the patients. Finally, digital options for providing information and for communication between healthcare providers and patients also emerged as a potential facilitator. It is warranted to apply these interventions prospectively to an SLE patient population to examine the effectiveness that these strategies may have on improving medication adherence.

**Supplementary Materials:** The following supporting information can be downloaded at: https://www.mdpi.com/article/10.3390/jcm11071857/s1, Supplementary materials: Interview schedule.

**Author Contributions:** Conceptualization, S.E. and I.P.; Methodology, S.E., Y.E. and I.P.; Interviews and verbatim transcription, S.E.; Formal analysis, S.E., C.G. and I.P.; Writing—original draft preparation, S.E.; Writing—review and editing, All authors. All authors have read and agreed to the published version of the manuscript.

**Funding:** This research was funded by the Swedish Rheumatism Association, grant number R-941095; King Gustaf V's 80-year Foundation, grant number FAI-2020-0741; Professor Nanna Svartz Foundation, grant number 2020-00368; Ulla and Roland Gustafsson Foundation, grant number 2021-26; Region Stockholm, grant number FoUI-955483, and the Karolinska Institutet.

**Institutional Review Board Statement:** The study was conducted in accordance with the Declaration of Helsinki, and the study protocol was approved by the Swedish Ethical Review Authority (reference number: 2021-00662; date of approval: 1 April 2021).

**Informed Consent Statement:** Informed consent was obtained from all subjects involved in the study.

**Data Availability Statement:** The data presented in this study are available on request from the corresponding author. The data are not publicly available due to being interview transcripts.

**Acknowledgments:** The authors would like to express gratitude to the individuals who participated in the interviews and shared their thoughts, views, and opinions.

**Conflicts of Interest:** IP has received research funding and/or honoraria from Amgen, AstraZeneca, Aurinia Pharmaceuticals, Elli Lilly and Company, Gilead Sciences, GlaxoSmithKline, Janssen Pharmaceuticals, Novartis, and F. Hoffmann-La Roche AG. The other authors declare that they have no conflict of interest. The funders had no role in the design of the study, collection, analyses, or interpretation of data, writing of the manuscript, or in the decision to publish the results.

## References

1. Kaul, A.; Gordon, C.; Crow, M.K.; Touma, Z.; Urowitz, M.B.; van Vollenhoven, R.; Ruiz-Irastorza, G.; Hughes, G. Systemic lupus erythematosus. *Nat. Rev. Dis. Primers* **2016**, *2*, 16039. [CrossRef] [PubMed]
2. Mehat, P.; Atiquzzaman, M.; Esdaile, J.M.; AviNa-Zubieta, A.; De Vera, M.A. Medication Nonadherence in Systemic Lupus Erythematosus: A Systematic Review. *Arthritis Care Res.* **2017**, *69*, 1706–1713. [CrossRef] [PubMed]
3. Feldman, C.H.; Yazdany, J.; Guan, H.; Solomon, D.H.; Costenbader, K.H. Medication Nonadherence Is Associated With Increased Subsequent Acute Care Utilization Among Medicaid Beneficiaries With Systemic Lupus Erythematosus. *Arthritis Care Res.* **2015**, *67*, 1712–1721. [CrossRef] [PubMed]
4. Julian, L.J.; Yelin, E.; Yazdany, J.; Panopalis, P.; Trupin, L.; Criswell, L.A.; Katz, P. Depression, medication adherence, and service utilization in systemic lupus erythematosus. *Arthritis Rheum.* **2009**, *61*, 240–246. [CrossRef] [PubMed]
5. Sun, K.; Eudy, A.M.; Criscione-Schreiber, L.G.; Sadun, R.E.; Rogers, J.L.; Doss, J.; Corneli, A.L.; Bosworth, H.B.; Clowse, M.E.B. Racial Disparities in Medication Adherence between African American and Caucasian Patients With Systemic Lupus Erythematosus and Their Associated Factors. *ACR Open Rheumatol.* **2020**, *2*, 430–437. [CrossRef]
6. Morisky, D.E.; Green, L.W.; Levine, D.M. Concurrent and predictive validity of a self-reported measure of medication adherence. *Med. Care* **1986**, *24*, 67–74. [CrossRef]
7. Koneru, S.; Shishov, M.; Ware, A.; Farhey, Y.; Mongey, A.B.; Graham, T.B.; Passo, M.H.; Houk, J.L.; Higgins, G.C.; Brunner, H.I. Effectively measuring adherence to medications for systemic lupus erythematosus in a clinical setting. *Arthritis Rheum.* **2007**, *57*, 1000–1006. [CrossRef]
8. Walsh, J.C.; Dalton, M.; Gazzard, B.G. Adherence to combination antiretroviral therapy assessed by anonymous patient self-report. *Aids* **1998**, *12*, 2361–2363.
9. de Klerk, E.; van der Heijde, D.; van der Tempel, H.; van der Linden, S. Development of a questionnaire to investigate patient compliance with antirheumatic drug therapy. *J. Rheumatol.* **1999**, *26*, 2635–2641.
10. Steiner, J.F.; Prochazka, A.V. The assessment of refill compliance using pharmacy records: Methods, validity, and applications. *J. Clin. Epidemiol.* **1997**, *50*, 105–116. [CrossRef]
11. Costedoat-Chalumeau, N.; Amoura, Z.; Hulot, J.S.; Aymard, G.; Leroux, G.; Marra, D.; Lechat, P.; Piette, J.C. Very low blood hydroxychloroquine concentration as an objective marker of poor adherence to treatment of systemic lupus erythematosus. *Ann. Rheum. Dis.* **2007**, *66*, 821–824. [CrossRef]
12. Cramer, J.A.; Mattson, R.H.; Prevey, M.L.; Scheyer, R.D.; Ouellette, V.L. How often is medication taken as prescribed? A novel assessment technique. *JAMA* **1989**, *261*, 3273–3277. [CrossRef]
13. Marengo, M.F.; Waimann, C.A.; de Achaval, S.; Zhang, H.; Garcia-Gonzalez, A.; Richardson, M.N.; Reveille, J.D.; Suarez-Almazor, M.E. Measuring therapeutic adherence in systemic lupus erythematosus with electronic monitoring. *Lupus* **2012**, *21*, 1158–1165. [CrossRef]
14. Lehane, E.; McCarthy, G. Intentional and unintentional medication non-adherence: A comprehensive framework for clinical research and practice? A discussion paper. *Int. J. Nurs. Stud.* **2007**, *44*, 1468–1477. [CrossRef]
15. Horne, R.; Weinman, J.; Hankins, M. The beliefs about medicines questionnaire: The development and evaluation of a new method for assessing the cognitive representation of medication. *Psychol. Health* **1999**, *14*, 1–24. [CrossRef]

16. Koneru, S.; Kocharla, L.; Higgins, G.C.; Ware, A.; Passo, M.H.; Farhey, Y.D.; Mongey, A.B.; Graham, T.B.; Houk, J.L.; Brunner, H.I. Adherence to medications in systemic lupus erythematosus. *J. Clin. Rheumatol. Pract. Rep. Rheum. Musculoskelet. Dis.* **2008**, *14*, 195–201. [CrossRef]

17. Daleboudt, G.M.; Broadbent, E.; McQueen, F.; Kaptein, A.A. Intentional and unintentional treatment nonadherence in patients with systemic lupus erythematosus. *Arthritis Care Res.* **2011**, *63*, 342–350. [CrossRef]

18. Chambers, S.A.; Raine, R.; Rahman, A.; Isenberg, D. Why do patients with systemic lupus erythematosus take or fail to take their prescribed medications? A qualitative study in a UK cohort. *Rheumatology* **2009**, *48*, 266–271. [CrossRef]

19. Chambers, S.; Raine, R.; Rahman, A.; Hagley, K.; De Ceulaer, K.; Isenberg, D. Factors influencing adherence to medications in a group of patients with systemic lupus erythematosus in Jamaica. *Lupus* **2008**, *17*, 761–769. [CrossRef]

20. Farinha, F.; Freitas, F.; Agueda, A.; Cunha, I.; Barcelos, A. Concerns of patients with systemic lupus erythematosus and adherence to therapy—A qualitative study. *Patient Prefer. Adherence* **2017**, *11*, 1213–1219. [CrossRef]

21. Garcia Popa-Lisseanu, M.G.; Greisinger, A.; Richardson, M.; O'Malley, K.J.; Janssen, N.M.; Marcus, D.M.; Tagore, J.; Suarez-Almazor, M.E. Determinants of treatment adherence in ethnically diverse, economically disadvantaged patients with rheumatic disease. *J. Rheumatol.* **2005**, *32*, 913–919.

22. Leung, J.; Baker, E.A.; Kim, A.H.J. Exploring intentional medication non-adherence in patients with systemic lupus erythematosus: The role of physician-patient interactions. *Rheumatol. Adv. Pract.* **2021**, *5*, rkaa078. [CrossRef]

23. Sun, K.; Corneli, A.L.; Dombeck, C.; Swezey, T.; Rogers, J.L.; Criscione-Schreiber, L.G.; Sadun, R.E.; Eudy, A.M.; Doss, J.; Bosworth, H.B.; et al. Barriers to taking medications for systemic lupus erythematosus: A qualitative study of racial minority patients, lupus providers, and clinic staff. *Arthritis Care Res.* **2021**, *4*, 4. [CrossRef]

24. Shishov, M.; Koneru, S.; Graham, T.B.; Houk, L.J.; Mongey, A.B.; Passo, M.; Rennebohm, R.; Ware, A.; Higgins, G.; Brunner, H.I. The medication adherence self-report inventory (MASRI) can accurately estimate adherence with medications in systemic lupus erythematosus (SLE). *Arthritis Rheum.* **2005**, *52*, S188.

25. de Klerk, E.; van der Heijde, D.; Landewe, R.; van der Tempel, H.; van der Linden, S. The compliance-questionnaire-rheumatology compared with electronic medication event monitoring: A validation study. *J. Rheumatol.* **2003**, *30*, 2469–2475.

26. Karlson, E.W.; Daltroy, L.H.; Rivest, C.; Ramsey-Goldman, R.; Wright, E.A.; Partridge, A.J.; Liang, M.H.; Fortin, P.R. Validation of a Systemic Lupus Activity Questionnaire (SLAQ) for population studies. *Lupus* **2003**, *12*, 280–286. [CrossRef]

27. Pettersson, S.; Svenungsson, E.; Gustafsson, J.; Moller, S.; Gunnarsson, I.; Welin Henriksson, E. A comparison of patients' and physicians' assessments of disease activity using the Swedish version of the Systemic Lupus Activity Questionnaire. *Scand. J. Rheumatol.* **2017**, *46*, 474–483. [CrossRef]

28. Drenkard, C.; Yazdany, J.; Trupin, L.; Katz, P.P.; Dunlop-Thomas, C.; Bao, G.; Lim, S.S. Validity of a Self-Administered Version of the Brief Index of Lupus Damage in a Predominantly African American Systemic Lupus Erythematosus Cohort. *Arthritis Care Res.* **2014**, *66*, 888–896. [CrossRef]

29. Brinkmann, S.; Kvale, S. *InterViews: Learning the Craft of Qualitative Research Interviewing*; SAGE: Thousand Oaks, CA, USA, 2018.

30. Price, B. Laddered questions and qualitative data research interviews. *J. Adv. Nurs.* **2002**, *37*, 273–281. [CrossRef]

31. Pope, C.; Mays, N. *Qualitative Research in Health Hare*, 3rd ed.; Wiley: Hoboken, NJ, USA, 2008.

32. Hsieh, H.F.; Shannon, S.E. Three approaches to qualitative content analysis. *Qual. Health Res.* **2005**, *15*, 1277–1288. [CrossRef]

33. Mayring, P. Qualitative Content Analysis. *Forum. Qual. Soc. Res.* **2000**, *1*.

34. Cutcliffe, J.R.; McKenna, H.P. When do we know that we know? Considering the truth of research findings and the craft of qualitative research. *Int. J. Nurs. Stud.* **2002**, *39*, 611–618. [CrossRef]

35. Guest, G.; Bunce, A.; Johnson, L. How Many Interviews Are Enough? An Experiment with Data Saturation and Variability. *Field Methods* **2006**, *18*, 59–82. [CrossRef]

36. Saunders, B.; Sim, J.; Kingstone, T.; Baker, S.; Waterfield, J.; Bartlam, B.; Burroughs, H.; Jinks, C. Saturation in qualitative research: Exploring its conceptualization and operationalization. *Qual. Quant.* **2018**, *52*, 1893–1907. [CrossRef] [PubMed]

37. Elsa, S. Elsa Science Website. Available online: https://www.elsa.science/en/for-healthcare/ (accessed on 15 January 2022).

38. Wohlfahrt, A.; Campos, A.; Iversen, M.D.; Gagne, J.J.; Massarotti, E.; Solomon, D.H.; Feldman, C.H. Use of rheumatology-specific patient navigators to understand and reduce barriers to medication adherence: Analysis of qualitative findings. *PLoS ONE* **2018**, *13*, 16. [CrossRef]

39. van Vollenhoven, R.F.; Mosca, M.; Bertsias, G.; Isenberg, D.; Kuhn, A.; Lerstrøm, K.; Aringer, M.; Bootsma, H.; Boumpas, D.; Bruce, I.N.; et al. Treat-to-target in systemic lupus erythematosus: Recommendations from an international task force. *Ann. Rheum. Dis.* **2014**, *73*, 958–967. [CrossRef]

40. Parodis, I.; Studenic, P. Patient-Reported Outcomes in Systemic Lupus Erythematosus. Can Lupus Patients Take the Driver's Seat in Their Disease Monitoring? *J. Clin. Med.* **2022**, *11*, 340. [CrossRef]

41. Lavielle, M.; Puyraimond-Zemmour, D.; Romand, X.; Gossec, L.; Senbel, E.; Pouplin, S.; Beauvais, C.; Gutermann, L.; Mezieres, M.; Dougados, M.; et al. Methods to improve medication adherence in patients with chronic inflammatory rheumatic diseases: A systematic literature review. *RMD Open* **2018**, *4*, e000684. [CrossRef]

42. Ritschl, V.; Stamm, T.A.; Aletaha, D.; Bijlsma, J.W.J.; Böhm, P.; Dragoi, R.; Dures, E.; Estévez-López, F.; Gossec, L.; Iagnocco, A.; et al. Prevention, screening, assessing and managing of non-adherent behaviour in people with rheumatic and musculoskeletal diseases: Systematic reviews informing the 2020 EULAR points to consider. *RMD Open* **2020**, *6*, e001432. [CrossRef]

43. Kvarnström, K.; Westerholm, A.; Airaksinen, M.; Liira, H. Factors Contributing to Medication Adherence in Patients with a Chronic Condition: A Scoping Review of Qualitative Research. *Pharmaceutics* **2021**, *13*, 1100. [CrossRef]

44. Zhang, X.; Tian, Y.; Li, J.; Zhao, X. Effect of targeted nursing applied to SLE patients. *Exp. Ther. Med.* **2016**, *11*, 2209–2212. [CrossRef] [PubMed]

45. Ting, T.V.; Kudalkar, D.; Nelson, S.; Cortina, S.; Pendl, J.; Budhani, S.; Neville, J.; Taylor, J.; Huggins, J.; Drotar, D.; et al. Usefulness of cellular text messaging for improving adherence among adolescents and young adults with systemic lupus erythematosus. *J. Rheumatol.* **2012**, *39*, 174–179. [CrossRef] [PubMed]

46. Durcan, L.; Clarke, W.A.; Magder, L.S.; Petri, M. Hydroxychloroquine Blood Levels in Systemic Lupus Erythematosus: Clarifying Dosing Controversies and Improving Adherence. *J. Rheumatol.* **2015**, *42*, 2092–2097. [CrossRef]

47. Ganachari, M.S.; Almas, S.A. Evaluation of clinical pharmacist mediated education and counselling of systemic lupus erythematosus patients in tertiary care hospital. *Indian J. Rheumatol.* **2012**, *7*, 7–12. [CrossRef]

48. Case, S.; Sinnette, C.; Phillip, C.; Grosgogeat, C.; Costenbader, K.H.; Leatherwood, C.; Feldman, C.H.; Son, M.B. Patient experiences and strategies for coping with SLE: A qualitative study. *Lupus* **2021**, *30*, 1405–1414. [CrossRef]

49. Ardoin, S.P. Transitions in Rheumatic Disease: Pediatric to Adult Care. *Pediatr. Clin. N. Am.* **2018**, *65*, 867–883. [CrossRef]

50. World Health Organization. *Medication Safety in Polypharmacy*; (WHO/UHC/SDS/2019.11). Licence: CC BY-NC-SA 3.0 IGO; World Health Organization: Geneva, Switzerland, 2019.

51. Abdul-Sattar, A.B.; Abou El Magd, S.A. Determinants of medication non-adherence in Egyptian patients with systemic lupus erythematosus: Sharkia Governorate. *Rheumatol. Int.* **2015**, *35*, 1045–1051. [CrossRef]

52. Heiman, E.; Lim, S.S.; Bao, G.; Drenkard, C. Depressive Symptoms Are Associated With Low Treatment Adherence in African American Individuals With Systemic Lupus Erythematosus. *J. Clin. Rheumatol. Pract. Rep. Rheum. Musculoskelet. Dis.* **2018**, *24*, 368–374. [CrossRef]

53. Gomez, A.; Qiu, V.; Cederlund, A.; Borg, A.; Lindblom, J.; Emamikia, S.; Enman, Y.; Lampa, J.; Parodis, I. Adverse Health-Related Quality of Life Outcome Despite Adequate Clinical Response to Treatment in Systemic Lupus Erythematosus. *Front. Med.* **2021**, *8*, 651249. [CrossRef]

54. Ruiz-Irastorza, G.; Khamashta, M.A. Hydroxychloroquine: The cornerstone of lupus therapy. *Lupus* **2008**, *17*, 271–273. [CrossRef]

55. Ruiz-Irastorza, G.; Ramos-Casals, M.; Brito-Zeron, P.; Khamashta, M.A. Clinical efficacy and side effects of antimalarials in systemic lupus erythematosus: A systematic review. *Ann. Rheum. Dis.* **2010**, *69*, 20–28. [CrossRef] [PubMed]

56. Hernández-Breijo, B.; Gomez, A.; Soukka, S.; Johansson, P.; Parodis, I. Antimalarial agents diminish while methotrexate, azathioprine and mycophenolic acid increase BAFF levels in systemic lupus erythematosus. *Autoimmun. Rev.* **2019**, *18*, 102372. [CrossRef] [PubMed]

57. Gomez, A.; Soukka, S.; Johansson, P.; Akerstrom, E.; Emamikia, S.; Enman, Y.; Chatzidionysiou, K.; Parodis, I. Use of Antimalarial Agents is Associated with Favourable Physical Functioning in Patients with Systemic Lupus Erythematosus. *J. Clin. Med.* **2020**, *9*, 1813. [CrossRef] [PubMed]

58. Wallace, D.J.; Tse, K.; Hanrahan, L.; Davies, R.; Petri, M.A. Hydroxychloroquine usage in US patients, their experiences of tolerability and adherence, and implications for treatment: Survey results from 3127 patients with SLE conducted by the Lupus Foundation of America. *Lupus Sci. Med.* **2019**, *6*, e000317. [CrossRef] [PubMed]

Journal of
*Clinical Medicine*

*Article*

# Persistence of Depression and Anxiety despite Short-Term Disease Activity Improvement in Patients with Systemic Lupus Erythematosus: A Single-Centre, Prospective Study

Myrto Nikoloudaki [1], Argyro Repa [1], Sofia Pitsigavdaki [1], Ainour Molla Ismail Sali [1], Prodromos Sidiropoulos [1,2], Christos Lionis [3] and George Bertsias [1,2,*]

1   Department of Rheumatology and Clinical Immunology, University Hospital of Heraklion and Medical School, University of Crete, 71110 Heraklion, Greece; myrtw.95@hotmail.gr (M.N.); arrepa2002@yahoo.gr (A.R.); sofiamed@windowslive.com (S.P.); aynursalih@msn.com (A.M.I.S.); sidiropp@uoc.gr (P.S.)
2   Institute of Molecular Biology and Biotechnology—FORTH, 71110 Heraklion, Greece
3   Clinic of Social and Family Medicine, University of Crete Medical School, 71110 Heraklion, Greece; lionis@galinos.med.uoc.gr
*   Correspondence: gbertsias@uoc.gr; Tel.: +30-2810-394635

**Citation:** Nikoloudaki, M.; Repa, A.; Pitsigavdaki, S.; Molla Ismail Sali, A.; Sidiropoulos, P.; Lionis, C.; Bertsias, G. Persistence of Depression and Anxiety despite Short-Term Disease Activity Improvement in Patients with Systemic Lupus Erythematosus: A Single-Centre, Prospective Study. *J. Clin. Med.* **2022**, *11*, 4316. https://doi.org/10.3390/jcm11154316

Academic Editors: Christopher Sjöwall and Ioannis Parodis

Received: 8 June 2022
Accepted: 20 July 2022
Published: 25 July 2022
Corrected: 28 December 2022



**Abstract:** Mental disorders such as anxiety and depression are prevalent in systemic lupus erythematosus (SLE) patients, yet their association with the underlying disease activity remains uncertain and has been mostly evaluated at a cross-sectional level. To examine longitudinal trends in anxiety, depression, and lupus activity, a prospective observational study was performed on 40 adult SLE outpatients with active disease (SLE Disease Activity Index [SLEDAI]-2K $\geq$ 3 [excluding serology]) who received standard-of-care. Anxiety and depression were determined at baseline and 6 months by the Hospital Anxiety and Depression Scale. Treatment adherence was assessed with a self-reported patient survey. Increased anxiety (median [interquartile range] HADS-A: 11.0 [7.8]) and depression (HADS-D: 8.0 [4.8]) were found at inclusion, which remained stable and non-improving during follow-up (difference: 0.0 [4.8] and $-0.5$ [4.0], respectively) despite reduced SLEDAI-2K by 2.0 (4.0) ($p < 0.001$). Among possible baseline predictors, paid employment—but not disease activity—correlated with reduced HADS-A and HADS-D with corresponding standardized beta-coefficients of $-0.35$ ($p = 0.017$) and $-0.27$ ($p = 0.093$). Higher anxiety and depression correlated with lower treatment adherence ($p = 0.041$ and $p = 0.088$, respectively). These results indicate a high-mental disease burden in active SLE that persists despite disease control and emphasize the need to consider socioeconomic factors as part of comprehensive patient assessment.

**Keywords:** comorbidities; mood disorders; low-disease activity; compliance; patient outcome

## 1. Introduction

Patients with Systemic Lupus Erythematosus (SLE) tend to suffer from a variety of physical and mental comorbidities [1,2]. The latter comprise predominantly depression and anxiety disorders with point prevalence rates of 35.0% (95% confidence interval [CI] 29.9–40.3%) and 25.8% (95% CI 19.2–32.9%), respectively [3], although estimations vary according to the metrics and definitions used [3,4]. In the University of California San Francisco Lupus Outcomes Study, depression (defined by the Center for Epidemiologic Studies depression scale) incidence rate was 8.8 per 100 person-years [5] and in another multi-ethnic and racial cohort from the same region, 16% of SLE patients developed depression (based on the Patient Health Questionnaire-8) over an average observation period of 26 months [6]. Anxiety disorder has been less extensively evaluated, nevertheless a small case-control study found increased prevalence in SLE patients when compared to counterparts with rheumatoid arthritis and healthy individuals [7].

Accruing evidence suggests that mental disorders, especially depression, in patients with SLE are associated with multiple adverse outcomes such as fatigue [8], cognitive difficulties [9,10], subclinical atherosclerosis [11], work [12] or functional [13] disability, and reduced health-related quality of life [14,15]. Indeed, severe forms of these disorders can have detrimental effects on daily-life activities and social roles. In a cross-sectional analysis of 80 SLE patients, Nowicka-Sauer et al. [16] found that anxiety and depression collectively explained 43% of illness perception variance. Accordingly, identifying factors contributing to these comorbidities can advance our understanding of their etiology and also rationalize their possible modification towards the improvement of patient well-being.

In this regard, controversy exists over the relationship between anxiety and depression and SLE disease activity. Thus, active lupus (quantified for example, with the SLE Disease Activity Index [SLEDAI]), especially from the mucocutaneous and musculoskeletal domains, has been correlated with increased depression and anxiety symptoms in some [14,17–19]—but not all [15,20–23]—studies. Likewise, a connection between inflammatory mediators such as lupus autoantibodies and mood disorders has not been consistently shown (reviewed in [24]). Of potential relevance is the association between depression and lower treatment adherence [25–28], a known driver for lupus flare and activity. This finding, however, lacks extensive confirmation or may be influenced by other factors such as ethnicity [29–32]. Therefore, evaluation of the frequency and determinants of mental disorders in different regions and clinical settings is important. Importantly, the majority of aforementioned studies had a cross-sectional design or included patients with no prespecified activity level at entry.

To this end, we carried out a prospective observational study in active and flaring SLE patients who were treated according to standard-of-care, in order to monitor longitudinal changes in depression and anxiety in relation to disease activity. The main hypothesis we sought to test was whether treatment-induced amelioration of the disease would result in the improvement of the aforementioned mental disorders. Taking advantage of our study context of active SLE, we also examined for the possible relationship between anxiety and depressive symptoms with reduced adherence to treatment.

## 2. Materials and Methods

### 2.1. Study Population

A prospective observational (non-interventional) study was performed at the outpatient clinics of the Department of Rheumatology and Clinical Immunology, University Hospital of Heraklion (Crete, Greece), covering from primary to tertiary care [33,34]. Patients were enrolled by consecutive sampling techniques between May 2021 and September 2021. Inclusion criteria were: (a) SLE diagnosis according to physician assessment and ascertained by the 2019 European Alliance of Associations for Rheumatology (EULAR) and American College of Rheumatology (ACR) classification criteria [35]; (b) age 18–65 years; (c) active disease defined by a clinical (excluding serology) SLEDAI-2K $\geq$ 3 [36] not present in the previous visit; (d) permanent residence in Crete; and (e) comprehension of Greek language. Patients with other coexisting rheumatic diseases, active neuropsychiatric lupus (diagnosed according to multidisciplinary approach as described elsewhere [37]), dementia, malignancy (past or present), and ongoing pregnancy were excluded. A total of 117 patients visited the outpatient clinics during the enrolment period, 50 of whom met the inclusion criteria. Ten participants did not attend their scheduled follow-up visit, thus data from 40 participants were analyzed.

### 2.2. Monitoring Protocol, Disease Evaluation, and Data Collection

Patients were monitored at two to four-month intervals over a period of six months as part of routine clinical practice and according to disease severity (based on physician judgment). Disease assessment at baseline and during follow-up included: (a) laboratory (complete blood count, liver and renal function, urinalysis) and immunological [serum anti-dsDNA, C3, C4, antiphospholipid antibodies] tests, (b) disease activity (quantified by the SLEDAI-2K [38] and the Safety of Estrogens in Lupus Erythematosus, National Assessment (SELENA)-SLEDAI Physician Global Assessment [PGA] [39]), (c) organ damage (quantified by the Systemic Lupus International Collaborating Clinics (SLICC) and ACR damage index [SDI] [40]), (d) comorbid diseases (ascertained by medical history, chart review and electronic prescription data) and, (e) use of medications, including the route of administration and dosage of glucocorticoids. Data on sociodemographic factors (age, disease duration, education level, marital status) were retrieved from medical charts and verified by patient interviews. Working status was assessed as described elsewhere [41] and included past (never or ever had paid employment) and current working status (having paid employment or not). Data were entered into a secure electronic database installed on the Department of Rheumatology and Clinical Immunology (University Hospital of Heraklion) protected server and network. The operation and maintenance of the database were strictly supervised by the scientifically accountable protocol and access was granted only to authorized users and researchers. All principles of anonymity, confidentiality, and non-traceability of data were adhered to.

### 2.3. Assessment of Anxiety, Depression, and Treatment Adherence

Anxiety and depression levels were determined at baseline and during follow-up by the Hospital Anxiety and Depression Scale (HADS), a self-rating psychometric instrument widely used in SLE [4,24,42] and validated in Greek patients [43] (including patients with chronic rheumatic diseases [44]). Briefly, HADS includes seven questions for each disorder (anxiety, depression), with a score ranging from 0–21. Scores $\leq 7$ correspond to normal levels of anxiety or depression, 8–10 to borderline pathological levels, and 11–21 to pathological levels. Patients with a diagnosis of anxiety disorder or depression were identified by reviewing the medical history, formal psychiatric evaluations, use, and indications for anxiolytic or antidepressant treatments (i.e., prescribed for underlying mental disorder as opposed to other conditions such as fibromyalgia). Treatment adherence was estimated with a methodology based on self-reported patient survey (modified from [45]). The scale is calculated by assigning one point for each positive answer, thus ranging from 0 (highest adherence) to 4 (lowest adherence).

### 2.4. Statistical Analysis

Categorical data are presented as numbers with percentages and continuous data as mean with standard deviation (continuous variables) or median with interquartile range (ordinal variables). Linear regression was used to identify factors associated with anxiety and depression. Possible predictors were first assessed by univariate analysis and variables associated with $p$-Value < 0.100 were considered for multivariate-adjusted analysis (stepwise backward selection method). To determine longitudinal changes (follow-up vs. baseline) in disease activity (SLEDAI-2K), anxiety (HADS-A), and depression (HADS-D), we applied the Wilcoxon Signed Rank test. In addition, absolute differences ($\Delta(delta)$ = follow-up *minus* baseline scores) were calculated for SLEDAI-2K, HADS-A, and HADS-D. Patients were grouped as having stable or worsening, or improving anxiety and depression ($\Delta$HADS-A/D $\geq 0$ vs. < 0, respectively) and independent samples Mann–Whitney test was used to examine for between-group differences in $\Delta$SLEDAI-2K. We also used the Spearman correlation test for the correlation of longitudinal changes in anxiety and depression. The association between treatment adherence and baseline anxiety or depression levels was evaluated by a chi-squared test. Statistical significance was indicated as a two-tailed $p$-Value < 0.05. All statistical analyses were performed using SPSS V25.0.

### 2.5. Ethical Aspects

The study was approved by the Research Ethics Committee of the University of Crete and by the Ethics Committee of the University General Hospital of Heraklion, Crete. Written informed consent was obtained from all patients. All conditions for the protection of personal data and medical confidentiality were met.

## 3. Results

### 3.1. Patients with Active SLE Manifest Increased Anxiety and Depression Levels That Persist over Time

We evaluated 40 SLE patients (39 women) with an average (SD) age and disease duration of 50.5 (10.3) and 10.3 (7.0) years, respectively (Table 1 and Supplementary Table S1).

**Table 1.** Demographic and clinical characteristics of SLE patients (*n* = 40).

|  | No. (%) or Mean (SD) [1] |
| --- | --- |
| Gender (female) | 39 (97.5%) |
| Race (white) | 40 (100.0%) |
| Age (years) | 50.5 (10.3) |
| Disease duration | 10.3 (7.0) |
| Education level | |
|    Basic or primary | 6 (15.0%) |
|    Secondary | 19 (47.5%) |
|    High or tertiary | 14 (35.0%) |
| Employment status (working) | 21 (52.5%) |
| Comorbidities | |
|    Hypertension | 7 (17.5%) |
|    Dyslipidemia | 11 (27.5%) |
|    Osteoporosis | 9 (22.5%) |
|    Thyroiditis | 7 (17.5%) |
|    Hypothyroidism | 5 (12.5%) |
|    COPD [2] or bronchial asthma | 2 (5.0%) |
|    Diabetes mellitus | 2 (5.0%) |
|    Fibromyalgia | 15 (37.5%) |
|    Mental disorder | 16 (40.0%) |
|      Depression | 13 (32.5%) |
|      Anxiety disorder | 5 (12.5%) |
| Organ damage (SDI) [3] | 18 (45.0%) |

[1] SD, standard deviation; [2] COPD, chronic obstructive pulmonary disease; [3] SDI, SLICC/ACR damage index.

Fourteen patients (35.0%) had high- or tertiary-level education and the majority (52.5%) were engaged in paid employment. A variety of comorbid conditions were present in our study sample, including mental disorders previously diagnosed by a specialist (depression in n = 13 patients). Organ damage (defined as SDI > 0) had accrued in 18 (45.0%) patients (Table 1).

At inclusion, all patients had active disease with a median (IQR) SLEDAI-2K of 6.0 (4.0) (Table 2). Assessment of mental status by the HADS index indicated an increased burden of both anxiety (HADS-A) and depression (HADS-D) with corresponding median (IQR) scores of 11.0 (7.8) and 8.0 (4.8). Accordingly, anxiety and depression of even a mild degree were detected in 70.0% and 52.5% of our patient cohort, respectively.

According to physician judgment and in line with standard clinical practice, patients were offered with treatment modifications due to active disease including initiation or dosage increase of hydroxychloroquine (n = 1), methotrexate (n = 8), azathioprine (n = 3), mycophenolate (n = 3), cyclophosphamide (n = 1), biological agent (n = 6), and glucocorticoids (n = 14). At the follow-up assessment, a significant reduction was noted in SLEDAI-2K, which reached a median of 4.0 (2.0) (Table 2). Conversely, neither anxiety (HADS-A) nor depression (HADS-D) showed significant trends. Thus, average changes in anxiety and depression scores were minimal (median [IQR]: 0.0 [4.8] and −0.5 [4.0], respectively). These results indicate that despite a short-term lowering of disease activity, the burden of mental disorders tends to remain stable and non-improving in patients with SLE.

**Table 2.** Disease activity, anxiety, and depression levels in SLE patients at inclusion and follow-up visits.

|  | Baseline [1] | Follow-Up | *p*-Value [2] |
|---|---|---|---|
| **SLEDAI-2K** [3] | 6.0 (4.0) | 4.0 (2.0) | <0.001 |
| 0 | 0 (0.0%) | 5 (12.5%) | |
| 1–4 | 14 (35.0%) | 22 (55.0%) | |
| 5–8 | 22 (55.0%) | 13 (32.5%) | |
| ≥9 | 4 (10.0%) | 9 (0.0%) | |
| **HADS-Anxiety** | 11.0 (7.8) | 11.0 (5.5) | 0.964 |
| Normal (≤7) | 12 (30.0%) | 8 (20.0%) | |
| Mild (8–10) | 7 (17.5%) | 9 (22.5%) | |
| Moderate (11–14) | 11 (27.5%) | 13 (32.5%) | |
| Severe (≥15) | 10 (25.0%) | 10 (25.0%) | |
| **HADS-Depression** | 8.0 (4.8) | 8.0 (6.8) | 0.463 |
| Normal (≤7) | 19 (47.5%) | 19 (47.5%) | |
| Mild (8–10) | 12 (30.0%) | 13 (32.5%) | |
| Moderate (11–14) | 6 (15.0%) | 7 (17.5%) | |
| Severe (≥15) | 3 (7.5%) | 1 (2.5%) | |

[1] Data are presented as median (interquartile range) or no. (%). [2] Wilcoxon Signed Rank Test. [3] SLE Disease Activity Index-2K.

### 3.2. Lack of Correlation between Longitudinal Changes in Disease Activity and Mental Disorders in SLE Patients

We sought to gain additional insights into the relationship between anxiety, depression, and disease activity in patients with SLE. Further to examining average trends, we grouped our study sample according to whether SLEDAI-2K was improved (by at least one unit; n = 24) or not (n = 16) during follow-up. We then compared the longitudinal changes in anxiety and depression levels (HADS-A and -D at follow-up minus HADS-A and -D at inclusion visit) between the two aforementioned patient subsets. Comparable changes in HADS-A and HADS-D scores were noted in SLE patients with improved vs. non-improved disease activity (Figure 1A).

In addition, we identified patients who attained a low-disease activity state according to the definitions proposed by Franklyn et al. [46] and Polachek et al. [36]. Again, HADS-A and HADS-D temporal trends did not differ significantly in patients who achieved or did not achieve low-disease activity (Supplementary Table S2). To address any confounding effects of administered treatments, the previous analyses were repeated separately in patients who were started on or received an increased dose of glucocorticoids due to active disease. In an ancillary analysis, we classified patients according to whether their level of anxiety or depression (as defined in Table 2) improved (for instance, from "severe" to "moderate"), remain stable, or worsened (for instance, from "mild" to "moderate"). By comparing the three aforementioned groups for corresponding changes in SLEDAI-2K, we found no significant trends (Supplementary Table S3).

Notwithstanding the small sample size, results were similar to the whole patient cohort (data not shown). Notably, longitudinal changes in anxiety showed a strong correlation (rho = 0.457, *p* = 0.003) with corresponding changes in depression levels (Figure 1B). Altogether, these data reiterate that SLE patients whose disease improved and even reached a low-activity state, are still burdened with mental disorders.

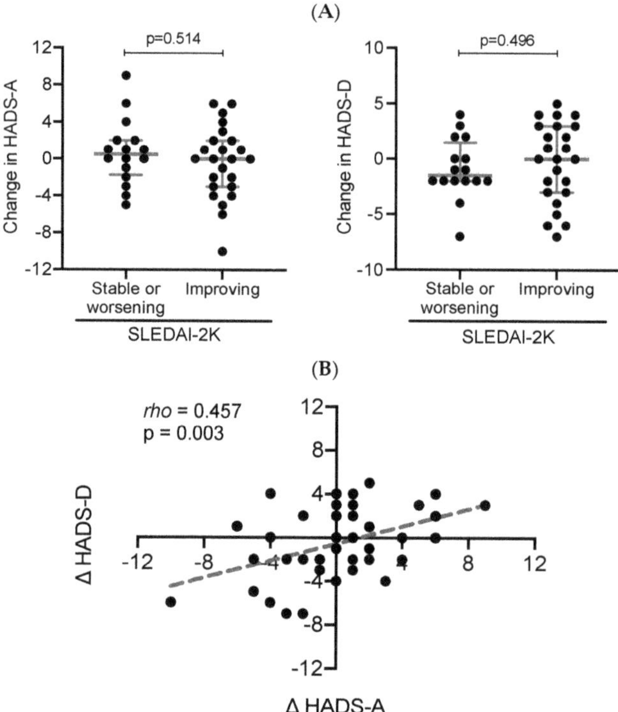



**Figure 1.** Longitudinal changes in anxiety and depression in association with improvement or not in disease activity. (**A**) Dot plots demonstrate changes (follow-up *minus* baseline) in HADS-A (left panel) and HADS-D (right panel) in SLE patients with improving vs. stable or worsening disease activity (SLEDAI-2K). Independent samples Mann–Whitney test was performed between the two patient groups. Blue lines represent medians (interquartile range). (**B**) Correlation of longitudinal changes (follow-up minus baseline) in HADS-A (Δ HADS-A) and HADS-D (Δ HADS-D) values in the SLE sample (each patient is represented by a separate black circles). The Spearman's correlation coefficient *rho* = 0.457 (*p*-Value = 0.003).

*3.3. Association of Mental Disorders with Sociodemographic Characteristics in SLE Patients*

The previous findings prompted us to search for other possible predictors of mental disorders in our study sample. To this end, we examined the baseline (i.e., registered at inclusion visit) scores of HADS-A and HADS-D in relationship with standard sociodemographic and clinical parameters. Using previously recommended cut-offs, we found no significant differences in average age, SLE duration, disease activity or severity (SLEDAI-2K), organ damage, presence of comorbidities, and education level in patients with high anxiety (HADS-A $\geq$ 11) or depression (HADS-D $\geq$ 8) levels as compared to their counterparts with lower scores (Supplementary Table S4). Conversely, patients with lower levels of anxiety reported paid employment at a significantly higher frequency than those with high anxiety (73.7% vs. 33.3%, respectively, $p = 0.011$). A similar trend in active employment status was observed in SLE patients with low when compared to high-depressive symptoms (68.2% vs. 33.3%, respectively, $p = 0.028$) (Supplementary Table S4). Next, the same parameters were analyzed by linear regression resulting in comparable findings although the association between employment status and HADS-D scores did not reach statistical significance (Table 3).

**Table 3.** Anxiety and depression in association with sociodemographic and clinical characteristics of SLE patients.

| | Anxiety Level (HADS-A) | | Depression Level (HADS-D) | |
|---|---|---|---|---|
| **Univariate Analysis** | **Standardized $\beta$ Coefficient; $p$-Value [1]** | | | |
| Age (years) | 0.05 | 0.771 | 0.04 | 0.812 |
| Education [2] | −0.14 | 0.389 | −0.22 | 0.172 |
| SLE duration (years) | 0.10 | 0.528 | 0.06 | 0.721 |
| Employment [3] | −0.42 | 0.007 | −0.27 | 0.093 |
| Comorbidities (no.) | 0.12 | 0.466 | 0.22 | 0.167 |
| SLEDAI-2K | −0.04 | 0.786 | −0.15 | 0.353 |
| Organ damage (SDI) | 0.08 | 0.632 | −0.22 | 0.174 |
| SLE treatment | | | | |
| HCQ [4,5] | −0.07 | 0.632 | 0.11 | 0.489 |
| Glucocorticoids [4] | −0.06 | 0.372 | −0.22 | 0.180 |
| Immunosuppressives [4] | −0.45 | 0.003 | −0.17 | 0.284 |
| Biologics [4] | −0.05 | 0.765 | 0.07 | 0.686 |
| **Multivariable-adjusted [5]** | | | | |
| Employment (working) [3] | −0.35 | 0.017 | −0.27 | 0.093 |
| Immunosuppressives [4] | −0.39 | 0.008 | – | – |

[1] Linear regression analysis. 95% CI (95% confidence interval); [2] Treated as ordinal variable (0 = primary level; 1 = secondary levels; 3 = tertiary level); [3] Treated as dummy variable (1 = paid employment; 0 = not paid employment); [4] Treated as dummy variable (1 = use; 0 = no use); [5] Backwards elimination model (variables with univariate $p$-Value < 0.100 were entered); HCQ, hydroxychloroquine.

Coupled with our aforementioned results, this analysis suggests that socioeconomic factors (employment)—rather than disease activity—may be linked to the excessive burden of mental disorders in patients with SLE.

### 3.4. Increased Anxiety and Depression Levels Are Associated with Lower Adherence to Treatment

Our study focused on trends of anxiety and depression in the context of active lupus. Notably, previous studies have associated mental disorders with poor treatment compliance in patients with SLE [25–28]. Using a self-reported measure, we found that 19 out of 40 patients (47.5%) had low or very low adherence to treatment. We then investigated whether the severity of mental disorders (assessed at the inclusion visit) correlated with treatment adherence. Within patients with low-anxiety levels (HADS-A < 11), the majority (68.4%) had high compliance; in contrast, among patients with high anxiety (HADS-A $\geq$ 11), only 38.1% had high compliance and 23.8% exhibited very low or no adherence to treatment ($p$ = 0.041; Table 4).

**Table 4.** Association of anxiety and depression with treatment adherence in SLE patients.

| | Treatment Adherence (Self-Reported): Highest to Lowest | | | |
|---|---|---|---|---|
| | **0–1** | **2** | **3–4** | **$p$-Value [1]** |
| **Anxiety level** | | | | |
| No or low | 13 (68.4%) | 6 (31.6%) | 0 (0.0%) | |
| Moderate or severe [2] | 8 (38.1%) | 8 (38.1%) | 5 (23.8%) | 0.041 |
| **Depression level** | | | | |
| No or low | 18 (58.1%) | 11 (35.5%) | 2 (6.5%) | 0.088 |
| Moderate or severe [3] | 3 (33.3%) | 3 (33.3%) | 3 (33.3%) | |

[1] Chi-squared test. [2] HADS-A $\geq$ 11. [3] HADS-D $\geq$ 8.

This relationship was confirmed by a statistically significant positive correlation between HADS-A and adherence scores treated as continuous variables (Spearman's rho = 0.324, $p$ = 0.041) (data not shown). Likewise, SLE patients with lower severity of depressive symptoms (HADS-D < 8) had better treatment compliance (58.1% with high compliance) when compared to those with HADS-D $\geq$ 8 (33.3%), however, this association was not statistically significant ($p$ = 0.088) probably due to the small sample size. Altogether, active SLE patients with a high burden of mental disorders are less likely to adhere to treatment of their disease.

## 4. Discussion

Mental comorbidities such as anxiety and depression are common in patients with SLE, however their association with underlying activity and likewise, their responsiveness to disease improvement remains inconclusive [24]. Our longitudinal analysis of 40 active lupus patients who received standard-of-care treatment to control their disease, demonstrated a high burden of anxiety and depression that remains unchanged at least over a short-term follow-up period and may be determined by socioeconomic factors such as employment status rather than by clinical parameters. Notably, increased levels of anxiety and depression tended to correlate with lower treatment adherence, an established determinant for disease flares [47,48], thus further emphasizing the importance of assessing mental disorders and associated risk factors as part of a comprehensive management plan in patients with SLE.

In our sample comprising of active SLE patients with an average age and disease duration of 50.5 and 10.3 years, respectively, significant anxiety (HADS-A $\geq$ 11) and depression (HADS-D $\geq$ 8) was each noted in 52.5%. This is in line with the results from previous cross-sectional observational studies [7,11,18,23,49–52] and meta-analyses of published data [3,4], although reported rates may vary according to the study design, population characteristics, and diagnostic instruments used. In the same context, a large Danish cohort study found that compared with the general population, the adjusted hazard ratio of depression was 2.22 (95% CI 1.77–2.77) for SLE patients [53]. Intriguingly, Roberts et al. [54] analyzed data from 194,483 women and found that a history of depression was linked to increased risk (adjusted hazard ratio 2.45; 95% CI 1.74–3.45) for subsequent development of SLE, irrespective of the effect of other confounding factors, thus suggesting a possible cross-interaction between the two conditions.

Although it is plausible to consider inflammation as a determining factor for mental disorders in SLE [55], there are conflicting reports regarding the relationship of disease activity with anxiety and depression [14,15,17–23]. To overcome the cross-sectional design limitations of most aforementioned studies, we enrolled active SLE individuals according to predetermined criteria and monitored them at two consecutive time points, i.e., at inclusion and six months post-treatment modification. Contrary to SLEDAI which was significantly improved over time, HADS-A and -D scores remained unchanged. Additionally, we found no reduction in mental disorders within patients who attained a state of low-lupus activity. Subgroup analysis according to intake or not of glucocorticoids yielded similar findings, thus reducing the possibility for a treatment confounding effect [56]. Our results are in agreement with a longitudinal study of 139 SLE patients which revealed four distinct anxiety trajectories that remained stable and not affected by disease activity over an average period of 30.9 months [57]. A similar analysis focusing on depression also showed persistence over time and a lack of association with temporal trends in SLEDAI-2K (average follow-up of 30.2 months) [58]. Collectively, and in line with a previous cohort study indicating that depression might be a long-term outcome of SLE [53], these data suggest that fluctuations of disease activity might not be major drivers of anxiety and depression, especially in the context of long-standing disease, although it has been argued that prolonged remission (i.e., lasting at least 5 years) might have a positive impact on depression [59].

Our previous finding coupled with the lack of association between other clinical characteristics and mental disorders prompted us to explore the possible role of sociodemographic factors. We found paid employment status to be protective against both anxiety and depression with corresponding odds ratios of 0.18 and 0.23, independent of SLE severity measures such as SLEDAI and organ damage. This is in agreement with other studies that have identified socioeconomic factors, in particular unemployment, financial strain, or low-social support, as significant correlates of depression in SLE [5,14,19,23,60]. Indeed, mediation modeling has suggested that low-socioeconomic status may impact negatively on the psychosocial resilience [60] and perceived stress [13] of lupus patients, thus contributing to higher anxiety, depression, and subsequent disability. It might be also that some SLE individuals are unable to (find) work due to the severity of the underlying disease or the concomitant anxiety or depressive symptoms [19]. These data underscore the importance of considering relevant socioeconomic factors when assessing the mental status of patients with SLE.

To our knowledge, our study is the first to evaluate medication adherence in Greek individuals with SLE. Using a self-reported survey, we found that 47.5% of patients with active lupus had low or very low compliance to treatment, a percentage that falls within the range (typically, 43–75%) of previously reported adherence rates [61]. Notably, increased levels of mental disorders tended to correlate with non-adherence, an association that has been previously shown especially for depression in several observational studies [25–28,62,63]. In this regard, anxiety and depression have been recognized as major determinants of the resilience [29] and illness perception [16] of lupus patients, which can both impact on compliance. Considering the prognostic implications of treatment adherence in terms of flares prevention and improved patient outcomes [64], these findings underline the importance of identifying and managing mental disorders in patients with SLE.

Several study limitations should be discussed such as that our results were derived from patients with distinct ethnic, demographic, and clinical characteristics, thus may not be generalizable to the whole SLE spectrum. Nevertheless, we applied specific inclusion criteria for active disease evaluated before and after treatment modifications, which facilitates the homogeneity of our data. Although the sample size can be considered relatively small to detect modest effect sizes, our prospective design enabled the generation of robust data regarding intra-individual temporal changes in SLE activity and mental disorders. Because our cohort was followed for six months, we were not able to examine the possible effect of sustained disease control on anxiety and depression. Additionally, the levels of mood disorders prior to study enrolment and how this might have affected the study findings was not available. Finally, the association between mental disorders and employment might be confounded by other parameters not captured in our analysis, still, the validity of our findings has been confirmed by other studies [5,14,19,23,60].

## 5. Conclusions

Active SLE patients exhibit a significant burden of anxiety and depressive symptoms, which remain unchanged despite treatment-induced short-term improvement in disease activity. This concurs with the fact that socioeconomic factors such as employment status, rather than clinical parameters, are significant predictors of the mental status of these patients. Despite the lack of association with disease activity, higher levels of anxiety and depression tend to coincide with lower treatment adherence, which is an established driver of adverse disease outcomes and flares. Together, our findings reiterate the importance of a comprehensive risk assessment for mental disorders in patients with SLE towards the improvement of their overall health status and prognosis.

**Supplementary Materials:** The following supporting information can be downloaded at: https://www.mdpi.com/article/10.3390/jcm11154316/s1, Table S1: Treatment of SLE patients included in the study; Table S2. Longitudinal changes in anxiety and depression in SLE patients who achieved or did not achieve a state of low-disease activity; Table S3: Reclassification of the anxiety and depression level in association with longitudinal change in disease activity in SLE patients; Table S4: Anxiety and depression in association with sociodemographic and clinical characteristics of SLE patients.

**Author Contributions:** Conceptualization, G.B. and C.L.; methodology, M.N.; formal analysis, G.B. and M.N.; investigation, A.R., S.P., A.M.I.S. and P.S.; data curation, M.N. and G.B.; writing—original draft preparation, M.N.; writing—review and editing, G.B. and P.S.; supervision, G.B. All authors have read and agreed to the published version of the manuscript.

**Funding:** This research was funded by the Special Account for Research Funds (ELKE) of the University of Crete (grant number KA10210).

**Institutional Review Board Statement:** The study was conducted in accordance with the Declaration of Helsinki and approved by the Ethics Committee of the University Hospital of Heraklion (protocol code 38/14-11-2018).

**Informed Consent Statement:** Informed consent was obtained from all subjects involved in the study.

**Data Availability Statement:** Data are available upon request.

**Conflicts of Interest:** The authors declare no conflict of interest.

# References

1. Gergianaki, I.; Garantziotis, P.; Adamichou, C.; Saridakis, I.; Spyrou, G.; Sidiropoulos, P.; Bertsias, G. High Comorbidity Burden in Patients with SLE: Data from the Community-Based Lupus Registry of Crete. *J. Clin. Med.* **2021**, *10*, 998. [CrossRef] [PubMed]
2. Rees, F.; Doherty, M.; Grainge, M.; Lanyon, P.; Davenport, G.; Zhang, W. Burden of Comorbidity in Systemic Lupus Erythematosus in the UK, 1999–2012. *Arthritis Care Res.* **2016**, *68*, 819–827. [CrossRef] [PubMed]
3. Moustafa, A.T.; Moazzami, M.; Engel, L.; Bangert, E.; Hassanein, M.; Marzouk, S.; Kravtsenyuk, M.; Fung, W.; Eder, L.; Su, J.; et al. Prevalence and metric of depression and anxiety in systemic lupus erythematosus: A systematic review and meta-analysis. *Semin. Arthritis Rheum.* **2020**, *50*, 84–94. [CrossRef] [PubMed]
4. Zhang, L.; Fu, T.; Yin, R.; Zhang, Q.; Shen, B. Prevalence of depression and anxiety in systemic lupus erythematosus: A systematic review and meta-analysis. *BMC Psychiatry* **2017**, *17*, 70. [CrossRef]
5. McCormick, N.; Trupin, L.; Yelin, E.H.; Katz, P.P. Socioeconomic Predictors of Incident Depression in Systemic Lupus Erythematosus. *Arthritis Care Res.* **2018**, *70*, 104–113. [CrossRef]
6. Patterson, S.L.; Trupin, L.; Yazdany, J.; Dall'Era, M.; Lanata, C.; Dequattro, K.; Hartogensis, W.; Katz, P. Physical Inactivity Independently Predicts Incident Depression in a Multi-Racial/Ethnic Systemic Lupus Cohort. *Arthritis Care Res.* **2021**, *74*, 1098–1104. [CrossRef]
7. Figueiredo-Braga, M.; Cornaby, C.; Cortez, A.; Bernardes, M.; Terroso, G.; Figueiredo, M.; Mesquita, C.D.S.; Costa, L.; Poole, B.D. Depression and anxiety in systemic lupus erythematosus: The crosstalk between immunological, clinical, and psychosocial factors. *Medicine* **2018**, *97*, e11376. [CrossRef]
8. Monahan, R.C.; Beaart-van de Voorde, L.J.; Eikenboom, J.; Fronczek, R.; Kloppenburg, M.; Middelkoop, H.A.; Terwindt, G.M.; van der Wee, N.J.; Huizinga, T.W.; Steup-Beekman, G.M. Fatigue in patients with systemic lupus erythematosus and neuropsychiatric symptoms is associated with anxiety and depression rather than inflammatory disease activity. *Lupus* **2021**, *30*, 1124–1132. [CrossRef]
9. Lillis, T.A.; Tirone, V.; Gandhi, N.; Weinberg, S.; Nika, A.; Sequeira, W.; Hobfoll, S.E.; Block, J.A.; Jolly, M. Sleep Disturbance and Depression Symptoms Mediate Relationship Between Pain and Cognitive Dysfunction in Lupus. *Arthritis Care Res.* **2019**, *71*, 406–412. [CrossRef]
10. Bingham, K.S.; DiazMartinez, J.; Green, R.; Tartaglia, M.C.; Ruttan, L.; Su, J.; Wither, J.E.; Kakvan, M.; Anderson, N.; Bonilla, D.; et al. Longitudinal relationships between cognitive domains and depression and anxiety symptoms in systemic lupus erythematosus. *Semin. Arthritis Rheum.* **2021**, *51*, 1186–1192. [CrossRef]
11. Jorge, A.; Lertratanakul, A.; Lee, J.; Pearce, W.; McPherson, D.; Thompson, T.; Barinas-Mitchell, E.; Ramsey-Goldman, R. Depression and Progression of Subclinical Cardiovascular Disease in Systemic Lupus Erythematosus. *Arthritis Care Res.* **2017**, *69*, 5–11. [CrossRef] [PubMed]
12. Utset, T.O.; Baskaran, A.; Segal, B.M.; Trupin, L.; Ogale, S.; Herberich, E.; Kalunian, K. Work disability, lost productivity and associated risk factors in patients diagnosed with systemic lupus erythematosus. *Lupus Sci. Med.* **2015**, *2*, e000058. [CrossRef] [PubMed]
13. Sumner, L.A.; Olmstead, R.; Azizoddin, D.R.; Ormseth, S.R.; Draper, T.L.; Ayeroff, J.R.; Zamora-Racaza, G.; Weisman, M.H.; Nicassio, P.M. The contributions of socioeconomic status, perceived stress, and depression to disability in adults with systemic lupus erythematosus. *Disabil. Rehabil.* **2020**, *42*, 1264–1269. [CrossRef] [PubMed]
14. Parperis, K.; Psarelis, S.; Chatzittofis, A.; Michaelides, M.; Nikiforou, D.; Antoniade, E.; Bhattarai, B. Association of clinical characteristics, disease activity and health-related quality of life in SLE patients with major depressive disorder. *Rheumatology* **2021**, *60*, 5369–5378. [CrossRef]
15. Dietz, B.; Katz, P.; Dall'Era, M.; Murphy, L.B.; Lanata, C.; Trupin, L.; Criswell, L.A.; Yazdany, J. Major Depression and Adverse Patient-Reported Outcomes in Systemic Lupus Erythematosus: Results From a Prospective Longitudinal Cohort. *Arthritis Care Res.* **2021**, *73*, 48–54. [CrossRef] [PubMed]

16. Nowicka-Sauer, K.; Hajduk, A.; Kujawska-Danecka, H.; Banaszkiewicz, D.; Smolenska, Z.; Czuszynska, Z.; Siebert, J. Illness perception is significantly determined by depression and anxiety in systemic lupus erythematosus. *Lupus* **2018**, *27*, 454–460. [CrossRef]

17. Tay, S.H.; Cheung, P.P.; Mak, A. Active disease is independently associated with more severe anxiety rather than depressive symptoms in patients with systemic lupus erythematosus. *Lupus* **2015**, *24*, 1392–1399. [CrossRef]

18. Abd-Alrasool, Z.A.; Gorial, F.I.; Hashim, M.T. Prevalence and severity of depression among Iraqi patients with systemic lupus erythematosus: A descriptive study. *Mediterr. J. Rheumatol.* **2017**, *28*, 142–146. [CrossRef]

19. Eldeiry, D.; Zandy, M.; Tayer-Shifman, O.E.; Kwan, A.; Marzouk, S.; Su, J.; Bingham, K.; Touma, Z. Association between depression and anxiety with skin and musculoskeletal clinical phenotypes in systemic lupus erythematosus. *Rheumatology* **2020**, *59*, 3211–3220. [CrossRef]

20. Shortall, E.; Isenberg, D.; Newman, S.P. Factors associated with mood and mood disorders in SLE. *Lupus* **1995**, *4*, 272–279. [CrossRef]

21. Segui, J.; Ramos-Casals, M.; Garcia-Carrasco, M.; de Flores, T.; Cervera, R.; Valdes, M.; Font, J.; Ingelmo, M. Psychiatric and psychosocial disorders in patients with systemic lupus erythematosus: A longitudinal study of active and inactive stages of the disease. *Lupus* **2000**, *9*, 584–588. [CrossRef] [PubMed]

22. Jarpa, E.; Babul, M.; Calderon, J.; Gonzalez, M.; Martinez, M.E.; Bravo-Zehnder, M.; Henriquez, C.; Jacobelli, S.; Gonzalez, A.; Massardo, L. Common mental disorders and psychological distress in systemic lupus erythematosus are not associated with disease activity. *Lupus* **2011**, *20*, 58–66. [CrossRef] [PubMed]

23. Narupan, N.; Seeherunwong, A.; Pumpuang, W. Prevalence and biopsychosocial factors associated with depressive symptoms among patients living with systemic lupus erythematosus in clinical settings in urban Thailand. *BMC Psychiatry* **2022**, *22*, 103. [CrossRef] [PubMed]

24. Tisseverasinghe, A.; Peschken, C.; Hitchon, C. Anxiety and Mood Disorders in Systemic Lupus Erythematosus: Current Insights and Future Directions. *Curr. Rheumatol. Rep.* **2018**, *20*, 85. [CrossRef]

25. Alsowaida, N.; Alrasheed, M.; Mayet, A.; Alsuwaida, A.; Omair, M.A. Medication adherence, depression and disease activity among patients with systemic lupus erythematosus. *Lupus* **2018**, *27*, 327–332. [CrossRef]

26. Julian, L.J.; Yelin, E.; Yazdany, J.; Panopalis, P.; Trupin, L.; Criswell, L.A.; Katz, P. Depression, medication adherence, and service utilization in systemic lupus erythematosus. *Arthritis Rheum.* **2009**, *61*, 240–246. [CrossRef]

27. Du, X.; Chen, H.; Zhuang, Y.; Zhao, Q.; Shen, B. Medication Adherence in Chinese Patients With Systemic Lupus Erythematosus. *J. Clin. Rheumatol.* **2020**, *26*, 94–98. [CrossRef]

28. Davis, A.M.; Graham, T.B.; Zhu, Y.; McPheeters, M.L. Depression and medication nonadherence in childhood-onset systemic lupus erythematosus. *Lupus* **2018**, *27*, 1532–1541. [CrossRef]

29. Mendoza-Pinto, C.; Garcia-Carrasco, M.; Campos-Rivera, S.; Munguia-Realpozo, P.; Etchegaray-Morales, I.; Ayon-Aguilar, J.; Alonso-Garcia, N.E.; Mendez-Martinez, S. Medication adherence is influenced by resilience in patients with systemic lupus erythematosus. *Lupus* **2021**, *30*, 1051–1057. [CrossRef]

30. Chang, J.C.; Davis, A.M.; Klein-Gitelman, M.S.; Cidav, Z.; Mandell, D.S.; Knight, A.M. Impact of Psychiatric Diagnosis and Treatment on Medication Adherence in Youth With Systemic Lupus Erythematosus. *Arthritis Care Res.* **2021**, *73*, 30–38. [CrossRef]

31. Geraldino-Pardilla, L.; Perel-Winkler, A.; Miceli, J.; Neville, K.; Danias, G.; Nguyen, S.; Dervieux, T.; Kapoor, T.; Giles, J.; Askanase, A. Association between hydroxychloroquine levels and disease activity in a predominantly Hispanic systemic lupus erythematosus cohort. *Lupus* **2019**, *28*, 862–867. [CrossRef] [PubMed]

32. Mosley-Williams, A.; Lumley, M.A.; Gillis, M.; Leisen, J.; Guice, D. Barriers to treatment adherence among African American and white women with systemic lupus erythematosus. *Arthritis Rheum.* **2002**, *47*, 630–638. [CrossRef] [PubMed]

33. Adamichou, C.; Nikolopoulos, D.; Genitsaridi, I.; Bortoluzzi, A.; Fanouriakis, A.; Papastefanakis, E.; Kalogiannaki, E.; Gergianaki, I.; Sidiropoulos, P.; Boumpas, D.T.; et al. In an early SLE cohort the ACR-1997, SLICC-2012 and EULAR/ACR-2019 criteria classify non-overlapping groups of patients: Use of all three criteria ensures optimal capture for clinical studies while their modification earlier classification and treatment. *Ann. Rheum. Dis.* **2020**, *79*, 232–241. [CrossRef]

34. Gergianaki, I.; Fanouriakis, A.; Repa, A.; Tzanakakis, M.; Adamichou, C.; Pompieri, A.; Spirou, G.; Bertsias, A.; Kabouraki, E.; Tzanakis, I.; et al. Epidemiology and burden of systemic lupus erythematosus in a Southern European population: Data from the community-based lupus registry of Crete, Greece. *Ann. Rheum. Dis.* **2017**, *76*, 1992–2000. [CrossRef] [PubMed]

35. Aringer, M.; Costenbader, K.; Daikh, D.; Brinks, R.; Mosca, M.; Ramsey-Goldman, R.; Smolen, J.S.; Wofsy, D.; Boumpas, D.T.; Kamen, D.L.; et al. 2019 European League Against Rheumatism/American College of Rheumatology classification criteria for systemic lupus erythematosus. *Ann. Rheum. Dis.* **2019**, *78*, 1151–1159. [CrossRef]

36. Polachek, A.; Gladman, D.D.; Su, J.; Urowitz, M.B. Defining Low Disease Activity in Systemic Lupus Erythematosus. *Arthritis Care Res.* **2017**, *69*, 997–1003. [CrossRef]

37. Fanouriakis, A.; Pamfil, C.; Rednic, S.; Sidiropoulos, P.; Bertsias, G.; Boumpas, D.T. Is it primary neuropsychiatric systemic lupus erythematosus? Performance of existing attribution models using physician judgment as the gold standard. *Clin. Exp. Rheumatol.* **2016**, *34*, 910–917.

38. Gladman, D.D.; Ibanez, D.; Urowitz, M.B. Systemic lupus erythematosus disease activity index 2000. *J. Rheumatol.* **2002**, *29*, 288–291.

39. Petri, M.; Kim, M.Y.; Kalunian, K.C.; Grossman, J.; Hahn, B.H.; Sammaritano, L.R.; Lockshin, M.; Merrill, J.T.; Belmont, H.M.; Askanase, A.D.; et al. Combined oral contraceptives in women with systemic lupus erythematosus. *N. Engl. J. Med.* **2005**, *353*, 2550–2558. [CrossRef]

40. Gladman, D.; Ginzler, E.; Goldsmith, C.; Fortin, P.; Liang, M.; Urowitz, M.; Bacon, P.; Bombardieri, S.; Hanly, J.; Hay, E.; et al. The development and initial validation of the Systemic Lupus International Collaborating Clinics/American College of Rheumatology damage index for systemic lupus erythematosus. *Arthritis Rheum.* **1996**, *39*, 363–369. [CrossRef]

41. Bultink, I.E.; Turkstra, F.; Dijkmans, B.A.; Voskuyl, A.E. High prevalence of unemployment in patients with systemic lupus erythematosus: Association with organ damage and health-related quality of life. *J. Rheumatol.* **2008**, *35*, 1053–1057. [PubMed]

42. de Almeida Macedo, E.; Appenzeller, S.; Lavras Costallat, L.T. Assessment of the Hospital Anxiety and Depression Scale (HADS) performance for the diagnosis of anxiety in patients with systemic lupus erythematosus. *Rheumatol. Int.* **2017**, *37*, 1999–2004. [CrossRef] [PubMed]

43. Michopoulos, I.; Douzenis, A.; Kalkavoura, C.; Christodoulou, C.; Michalopoulou, P.; Kalemi, G.; Fineti, K.; Patapis, P.; Protopapas, K.; Lykouras, L. Hospital Anxiety and Depression Scale (HADS): Validation in a Greek general hospital sample. *Ann. Gen. Psychiatry* **2008**, *7*, 4. [CrossRef] [PubMed]

44. Chatzitheodorou, D.; Kabitsis, C.; Papadopoulos, N.G.; Galanopoulou, V. Assessing disability in patients with rheumatic diseases: Translation, reliability and validity testing of a Greek version of the Stanford Health Assessment Questionnaire (HAQ). *Rheumatol. Int.* **2008**, *28*, 1091–1097. [CrossRef] [PubMed]

45. Stavropoulou, C. Perceived information needs and non-adherence: Evidence from Greek patients with hypertension. *Health Expect.* **2012**, *15*, 187–196. [CrossRef]

46. Franklyn, K.; Lau, C.S.; Navarra, S.V.; Louthrenoo, W.; Lateef, A.; Hamijoyo, L.; Wahono, C.S.; Chen, S.L.; Jin, O.; Morton, S.; et al. Definition and initial validation of a Lupus Low Disease Activity State (LLDAS). *Ann. Rheum. Dis.* **2016**, *75*, 1615–1621. [CrossRef]

47. Adamichou, C.; Bertsias, G. Flares in systemic lupus erythematosus: Diagnosis, risk factors and preventive strategies. *Mediterr. J. Rheumatol.* **2017**, *28*, 4–12. [CrossRef]

48. Costedoat-Chalumeau, N.; Pouchot, J.; Guettrot-Imbert, G.; Le Guern, V.; Leroux, G.; Marra, D.; Morel, N.; Piette, J.C. Adherence to treatment in systemic lupus erythematosus patients. *Best Pract. Res. Clin. Rheumatol.* **2013**, *27*, 329–340. [CrossRef]

49. Kwan, A.; Marzouk, S.; Ghanean, H.; Kishwar, A.; Anderson, N.; Bonilla, D.; Vitti, M.; Su, J.; Touma, Z. Assessment of the psychometric properties of patient-reported outcomes of depression and anxiety in systemic lupus erythematosus. *Semin. Arthritis Rheum.* **2019**, *49*, 260–266. [CrossRef]

50. Xie, X.; Wu, D.; Chen, H. Prevalence and risk factors of anxiety and depression in patients with systemic lupus erythematosus in Southwest China. *Rheumatol. Int.* **2016**, *36*, 1705–1710. [CrossRef] [PubMed]

51. Fernandez, H.; Cevallos, A.; Jimbo Sotomayor, R.; Naranjo-Saltos, F.; Mera Orces, D.; Basantes, E. Mental disorders in systemic lupus erythematosus: A cohort study. *Rheumatol. Int.* **2019**, *39*, 1689–1695. [CrossRef]

52. Gholizadeh, S.; Azizoddin, D.R.; Mills, S.D.; Zamora, G.; Potemra, H.M.K.; Hirz, A.E.; Wallace, D.J.; Weisman, M.H.; Nicassio, P.M. Body image mediates the impact of pain on depressive symptoms in patients with systemic lupus erythematosus. *Lupus* **2019**, *28*, 1148–1153. [CrossRef] [PubMed]

53. Hesselvig, J.H.; Egeberg, A.; Kofoed, K.; Gislason, G.; Dreyer, L. Increased risk of depression in patients with cutaneous lupus erythematosus and systemic lupus erythematosus: A Danish nationwide cohort study. *Br. J. Dermatol.* **2018**, *179*, 1095–1101. [CrossRef] [PubMed]

54. Roberts, A.L.; Kubzansky, L.D.; Malspeis, S.; Feldman, C.H.; Costenbader, K.H. Association of Depression With Risk of Incident Systemic Lupus Erythematosus in Women Assessed Across 2 Decades. *JAMA Psychiatry* **2018**, *75*, 1225–1233. [CrossRef] [PubMed]

55. Papadaki, E.; Kavroulakis, E.; Bertsias, G.; Fanouriakis, A.; Karageorgou, D.; Sidiropoulos, P.; Papastefanakis, E.; Boumpas, D.T.; Simos, P. Regional cerebral perfusion correlates with anxiety in neuropsychiatric SLE: Evidence for a mechanism distinct from depression. *Lupus* **2019**, *28*, 1678–1689. [CrossRef]

56. Miyawaki, Y.; Shimizu, S.; Ogawa, Y.; Sada, K.E.; Katayama, Y.; Asano, Y.; Hayashi, K.; Yamamura, Y.; Hiramatsu-Asano, S.; Ohashi, K.; et al. Association of glucocorticoid doses and emotional health in lupus low disease activity state (LLDAS): A cross-sectional study. *Arthritis Res. Ther.* **2021**, *23*, 79. [CrossRef]

57. Lew, D.; Huang, X.; Kellahan, S.R.; Xian, H.; Eisen, S.; Kim, A.H.J. Anxiety Symptoms Among Patients With Systemic Lupus Erythematosus Persist Over Time and Are Independent of SLE Disease Activity. *ACR Open Rheumatol.* **2022**, *4*, 432–440. [CrossRef]

58. Kellahan, S.R.; Huang, X.; Lew, D.; Xian, H.; Eisen, S.; Kim, A.H.J. Depressed Symptomatology Persists Over Time in Systemic Lupus Erythematosus Patients. *Arthritis Care Res.* **2021**. [CrossRef]

59. Margiotta, D.P.E.; Fasano, S.; Basta, F.; Pierro, L.; Riccardi, A.; Navarini, L.; Valentini, G.; Afeltra, A. The association between duration of remission, fatigue, depression and health-related quality of life in Italian patients with systemic lupus erythematosus. *Lupus* **2019**, *28*, 1705–1711. [CrossRef]

60. Azizoddin, D.R.; Zamora-Racaza, G.; Ormseth, S.R.; Sumner, L.A.; Cost, C.; Ayeroff, J.R.; Weisman, M.H.; Nicassio, P.M. Psychological Factors that Link Socioeconomic Status to Depression/Anxiety in Patients with Systemic Lupus Erythematosus. *J. Clin. Psychol. Med. Settings* **2017**, *24*, 302–315. [CrossRef] [PubMed]

61. Mehat, P.; Atiquzzaman, M.; Esdaile, J.M.; AviNa-Zubieta, A.; De Vera, M.A. Medication Nonadherence in Systemic Lupus Erythematosus: A Systematic Review. *Arthritis Care Res.* **2017**, *69*, 1706–1713. [CrossRef] [PubMed]

62. Heiman, E.; Lim, S.S.; Bao, G.; Drenkard, C. Depressive Symptoms Are Associated With Low Treatment Adherence in African American Individuals With Systemic Lupus Erythematosus. *J. Clin. Rheumatol.* **2018**, *24*, 368–374. [CrossRef] [PubMed]
63. Shenavandeh, S.; Mani, A.; Eazadnegahdar, M.; Nekooeian, A. Medication Adherence of Patients with Systemic Lupus Erythematosus and Rheumatoid Arthritis Considering the Psychosocial Factors, Health Literacy and Current Life Concerns of Patients. *Curr. Rheumatol. Rev.* **2021**, *17*, 412–420. [CrossRef]
64. Gomez, A.; Soukka, S.; Johansson, P.; Akerstrom, E.; Emamikia, S.; Enman, Y.; Chatzidionysiou, K.; Parodis, I. Use of Antimalarial Agents is Associated with Favourable Physical Functioning in Patients with Systemic Lupus Erythematosus. *J. Clin. Med.* **2020**, *9*, 1813. [CrossRef]

Journal of
*Clinical Medicine*

*Article*

# Long-Term Clinical Outcome in Systemic Lupus Erythematosus Patients Followed for More Than 20 Years: The Milan Systemic Lupus Erythematosus Consortium (SMiLE) Cohort


Maria Gerosa [1,2,*], Lorenzo Beretta [3], Giuseppe Alvise Ramirez [4], Enrica Bozzolo [4], Martina Cornalba [2], Chiara Bellocchi [3], Lorenza Maria Argolini [2], Luca Moroni [4], Nicola Farina [4], Giulia Segatto [3], Lorenzo Dagna [4] and Roberto Caporali [1,2]

[1] Research Centre for Adult and Pediatric Rheumatic Diseases, Department of Clinical Sciences and Community Health, University of Milan, 20129 Milan, Italy; roberto.caporali@unimi.it
[2] ASST Pini CTO, Lupus Clinic, Division of Clinical Rheumatology, 20122 Milan, Italy; martinacornalba2@gmail.com (M.C.); lorenza.argolini@hotmail.it (L.M.A.)
[3] Fondazione IRCCS Ca' Granda Ospedale Maggiore Policlinico di Milano, Referral Centre for Systemic Autoimmune Diseases, 20122 Milan, Italy; lorberimm@hotmail.com (L.B.); chiara.bellocchi@unimi.it (C.B.); giulia.segatto@unimi.it (G.S.)
[4] IRCCS Ospedale San Raffaele, Unit of Immunology, Rheumatology, Allergy and Rare Diseases, 20132 Milan, Italy; ramirez.giuseppealvise@hsr.it (G.A.R.); bozzolo.enrica@hsr.it (E.B.); moroni.luca@hsr.it (L.M.); farina.nicola@hsr.it (N.F.); dagna.lorenzo@hsr.it (L.D.)
* Correspondence: maria.gerosa@unimi.it; Tel.: +39-0258296272



**Citation:** Gerosa, M.; Beretta, L.; Ramirez, G.A.; Bozzolo, E.; Cornalba, M.; Bellocchi, C.; Argolini, L.M.; Moroni, L.; Farina, N.; Segatto, G.; et al. Long-Term Clinical Outcome in Systemic Lupus Erythematosus Patients Followed for More Than 20 Years: The Milan Systemic Lupus Erythematosus Consortium (SMiLE) Cohort. *J. Clin. Med.* **2022**, *11*, 3587. https://doi.org/10.3390/jcm11133587

Academic Editors: Christopher Sjöwall and Ioannis Parodis

Received: 25 April 2022
Accepted: 16 June 2022
Published: 22 June 2022

**Publisher's Note:** MDPI stays neutral with regard to jurisdictional claims in published maps and institutional affiliations.





**Abstract:** Tackling active disease to prevent damage accrual constitutes a major goal in the management of patients with systemic lupus erythematosus (SLE). Patients with early onset disease or in the early phase of the disease course are at increased risk of developing severe manifestations and subsequent damage accrual, while less is known about the course of the disease in the long term. To address this issue, we performed a multicentre retrospective observational study focused on patients living with SLE for at least 20 years and determined their disease status at 15 and 20 years after onset and at their last clinical evaluation. Disease activity was measured through the British Isles Lupus Assessment Group (BILAG) tool and late flares were defined as worsening in one or more BILAG domains after 20 years of disease. Remission was classified according to attainment of lupus low-disease-activity state (LLDAS) criteria or the Definitions Of Remission In SLE (DORIS) parameters. Damage was quantitated through the Systemic Lupus Erythematosus International Collaborating Clinics/American College of Rheumatology damage index (SLICC/ACR-DI). LLAS/DORIS remission prevalence steadily increased over time. In total, 84 patients had a late flare and 88 had late damage accrual. Lack of LLDAS/DORIS remission status at the 20 year timepoint ($p = 0.0026$ and $p = 0.0337$, respectively), prednisone dose $\geq 7.5$ mg ($p = 9.17 \times 10^{-5}$) or active serology (either dsDNA binding, low complement or both; $p = 0.001$) were all associated with increased late flare risk. Late flares, in turn, heralded the development of late damage ($p = 2.7 \times 10^{-5}$). These data suggest that patients with longstanding SLE are frequently in remission but still at risk of disease flares and eventual damage accrual, suggesting the need for tailored monitoring and therapeutic approaches aiming at effective immunomodulation besides immunosuppression, at least by means of steroids.

**Keywords:** lupus; flare; damage; long disease duration; trajectories; remission; low disease activity


## 1. Introduction

Systemic lupus erythematosus (SLE) is a chronic autoimmune disease of unknown aetiology, characterized by a broad spectrum of clinical presentations, and linked to the production of autoantibodies leading to inflammation with multi-organ involvement. Over time, SLE morbidity may be affected not only by exacerbations, but also by progressive organ damage, either related to SLE, treatments or comorbidities [1–4]. The treat-to-target (T2T) therapeutic approach, aiming at "remission of systemic symptoms and organ manifestations" is founded on this notion, with the perspective of minimising patient disability and improving long-term survival [5–7]. In the past, different definitions of remission were proposed [1,8], but none were universally accepted, hindering the selection of an appropriate outcome measure for any T2T strategy. Starting from 2016, the Definitions Of Remission In SLE (DORIS) Initiative provided a framework for defining remission in SLE [2,9]. Then, the task force performed a thorough revision of accumulating evidence, information and data to reach the final recommendations for a definition of remission in SLE in 2021 [10].

While remission is an achievable outcome, it may seldom be reached or sustained with the current therapies and thus the development of a more attainable target associated with reduced damage accrual protection has been advocated. The Asia Pacific Lupus Collaboration (APLC) performed a series of studies to validate the so-called Lupus Low Disease Activity State (LLDAS). The LLDAS is a composite score evaluating five different aspects of the disease, and has proven to be a valuable alternative for the implementation of treat-to-target therapeutic strategies. It has been demonstrated that patients in persistent LLDAS show a significantly lower frequency of disease flare-ups and lower accumulation of organ damage [11,12].

With the introduction of new therapies, there has been considerable improvement in the survival of individuals with SLE. In contrast, patients living longer with the disease may present chronic organ damage and disability as a result of persistent disease activity and/or treatment side effects [13–15].

Few studies are available on patients affected by long-standing SLE. According to these, skin and joint involvement are associated with a lower likelihood of achieving LLDAS or remission, probably because of the still suboptimal control of these symptoms based on current therapies [16,17].

Based on the previous considerations, the objectives of this study are: the evaluation of the proportion of patients achieving remission according to DORIS definitions or low disease activity, LLDAS, in a multicentre cohort of SLE patients with disease duration of more than 20 years; the identification of valid prognostic markers to support such remission or LLDAS, and prevent possible disease flare-ups; the estimation of the effect of prolonged remission or LLDAS on damage accumulation in patients with long-standing disease.

## 2. Materials and Methods

### 2.1. Patients

The Milan Systemic Lupus Erythematosus Consortium (SMiLE) cohort is a longitudinal observational cohort of SLE patients regularly followed at three rheumatology tertiary centres in Milan: Lupus Clinic of the Clinical Rheumatology Unit of ASST Pini-CTO, the Referral Center for Systemic Autoimmune Diseases of Fondazione Ca' Granda IRCCS Policlinico and the Lupus Clinic of the Unit of Immunology, Rheumatology, Allergy and Rare Diseases at IRCCS Ospedale San Raffaele. The cohort was created with the aim to better characterise the biological and clinical features of SLE patients and includes all patients fulfilling the 1997 American College of Rheumatology revised classification criteria or the 2019 SLICC/ACR revised criteria for the diagnosis of SLE. The protocol was approved by the IRCCS San Raffaele Hospital Ethics Committee and the Comitato Etico Milano Area 2 (approval no. 0002450/2020) (for both the "Pini" and "Policlinico" hospitals). When entering the cohort, all patients signed an informed consent form.

Within this population, we selected all the patients with a disease duration $\geq$20 years. Medical records of patients were retrospectively evaluated and clinical and laboratory information were collected at baseline, at 15 and 20 years from the first visit and at the time of the last observation.

*2.2. Outcome Measures*

Disease activity, severity of organ involvement and flares were assessed by the British Isles Lupus Assessment Group 2004 (BILAG-2004) index [18–20]. In addition, the Systemic lupus erythematosus disease activity index 2000 (SLEDAI-2K) [21] and the Physician Global Assessment (PGA) 0–3 scale [22] were also scored. Remission was defined according to the DORIS final recommendations [10] and namely as a clinical SLEDAI (cSLEDAI) = 0, a PGA < 0.5, irrespective of serology; patient may be on antimalarials, low-dose glucocorticoids (prednisone or equivalents ≤ 5 mg/day), and/or stable immunosuppressives including biologics. Low-disease-activity state (LLDAS) was defined as recently reported by Franklyn et al. [12], that is, an SLEDAI-2K ≤ 4, no disease activity in major organ systems (renal, central nervous system [CNS], cardiopulmonary, vascular, fever), no occurrence of haemolytic anaemia or gastrointestinal activity, no new disease activity in relation to previous evaluations, a PGA ≤ 1; patient may be on low-dose glucocorticoids (prednisone or equivalents) ≤ 7.5 mg/day), standard maintenance dose of well-tolerated, approved, immunosuppressive drug and biologic therapies, excluding the investigational drugs. Estimated damage accumulation was calculated using the Systemic Lupus International Collaborating Clinics/American College of Rheumatology Damage Index (SLICC/ACR-DI) [23].

Low complement was defined as C3 and/or C4 levels below the reference value. Anti-dsDNA positivity was considered when values where above twice the threshold value, while and anti-PL positivity was confirmed when at least one among aCL, anti-b2GPI and LAC was positive in 2 determinations 12 weeks apart.

*2.3. Statistical Analysis*

For descriptive statistics, continuous variables were summarized as mean ± standard deviation, except for skewed data that were described as median (interquartile range). The chi-squared test was used to compare categorical variables from $2 \times 2$ contingency tables.

Survival analysis was conducted to estimate the time to the first flare occurring after the 20th year of disease. The analysis was restricted to the 20–30th year interval to ensure that at least 20% of cases were available at the end of observation and to provide reliable survival estimates. Time-to-event analysis was conducted via Cox regression with the estimation of hazard ratios (HR) and relative 95% confidence intervals (CI95).

## 3. Results

*3.1. Clinical Features at Enrolment and at Study End*

Long-term data were available for 221 patients with a mean follow-up of 28.5 ± 6.6 years from diagnosis (Table 1). Musculoskeletal and mucocutaneous involvements were the most prevalent manifestations in patient history, and nearly half of the patients ($n = 106$) experienced lupus nephritis. Renal involvement was the presenting manifestation in 32.2% of cases and usually occurred early in the medical history (81.3% of times within 10 years, 91.6% of times within 15 years). The vast majority of patients ($n = 202, 91.4\%$) were treated with hydroxychloroquine and 172 subjects (77.8%) were exposed to conventional immunosuppressants. At the end of the observation (28.5 ± 6.6 years from diagnosis), 129 patients were both in LLDAS and DORIS remission and 41 patients were neither in LLDAS or DORIS remission. The two remission classifications were significantly associated ($\chi^2 = 63.940; p < 0.0001$). A total of 172 patients (77.8%) had accumulated one or more SLICC/ACR-DI items with a mean of 1.7 ± 1.73 items/patient. Cataract was the most prevalent damage item ($n = 33, 14.9\%$), followed by erosive/deforming arthritis ($n = 26, 11.8\%$) and osteoporosis with fractures ($n = 25, 11.3\%$), as detailed in Supplemental Table S1.

**Table 1.** Clinical, laboratory and demographic characteristics.

| Variable | *n* = 221 |
| --- | --- |
| Female, *n* (%) | 198 (89.6%) |
| Age at diagnosis, years | 25.6 ± 10.6 |
| Follow-up, years | 28.5 ± 6.6 |
| Serology, *n* (%) | |
| Anti-dsDNA | 177 (80.1%) |
| Anti-Sm | 33 (14.9%) |
| aPL | 98 (44.3%) |
| Low complement | 181 (81.9%) |
| Clinical features ever, *n* (%) | |
| Musculoskeletal involvement | 189 (85.5%) |
| Mucocutaneous involvement | 180 (81.4%) |
| Renal involvement | 106 (48.8%) |
| Neuropsychiatric SLE | 49 (22.2%) |
| Cardiopulmonary involvement | 75 (33.9%) |
| Haematological involvement | 138 (62.4%) |
| Constitutional symptoms | 168 (76%) |
| Gastrointestinal involvement | 11 (5%) |
| Ophthalmic involvement | 19 (8.6%) |
| Treatment ever, *n* (%) | |
| Hydroxychloroquine | 202 (91.4%) |
| Prednisone ≥5 mg | 221 (100%) |
| Methotrexate | 51 (23.1%) |
| Mycophenolate mofetil/mycophenolic acid | 72 (32.6%) |
| Azathioprine | 113 (51.1%) |
| Cyclosporine | 49 (22.2%) |
| Cyclophosphamide | 72 (32.6%) |
| High-dose intravenous steroids | 117 (52.9%) |
| Rituximab | 14 (6.3%) |
| Belimumab | 25 (11.3%) |

ds-DNA: double strain-DNA; anti-Sm: anti-Smith; aPL: anti-phosholipid antibodies; SLE: Systemic Lupus Erythematosus.

## 3.2. Changes in Disease Activity and Damage Accrual over Time

The chance of being in LLDAS or in remission, according to DORIS definition, increased over time. Nonetheless, chronic damage according to the SLICC/ACR-DI, was also higher when patients were observed at later timepoints (Table 2). BILAG scores for each disease domain were collected at each timepoint and are reported in Supplemental Figure S1. Disease flare rates after 20 years of follow up were calculated based on worsening BILAG scores in one or more domains. Eighty-four subjects (38.9%) had one or more flares, yielding a 10-year flare risk of nearly 50% (Figure 1). Most flares were experienced in the musculoskeletal domain, as shown in Table 3. In 10 cases (11.9%), a BILAG A flare was preceded by a BILAG B score, in 19 cases (22.6%) a BILAG A or B flare was preceded by a BILAG C score, by a D score in 41 (48.8%) and by an E score in 14 (16.7%). Pre-flare serological status was altered in 46 (55.4%) patients: 40 had (47.6%) low complement, 26 (30.9%) had increased dsDNA binding, and 18 (21.4%) had a negative serological status. Regarding therapy before flare, 21 subjects (25%) were treated with prednisone >5 mg/day, 45 (53.7%) with hydroxychloroquine and 39 (46.4%) with immunosuppressants. In total, 88 patients of 216 with available data (40.7%) accrued additional damage items after the 20-year timepoint.

**Table 2.** LLDAS, remission ad SLICC/ACR-DI indexes at long-term endpoints.

| Status | 15 Years (*n* = 199) | 20 Years (*n* = 205) | Last Observation (*n* = 221) |
|---|---|---|---|
| LLDAS | 137 (68.8%) | 155 (75.6%) | 177 (80%) |
| DORIS remission | 107 (53.8%) | 115 (58%) * | 132 (59.7%) |
| SLICC/ACR-DI | 0.57 ± 1.04 ** | 0.89 ± 1.42 *** | 1.57 ± 1.9 |

Full dataset available at the last observation; exploratory analysis on available data at 15 and 20 years from diagnosis. Data from: * 183 patients; ** 198 patients; *** 216 patients. LLDAS: Lupus low disease activity state; DORIS: Definitions Of Remission In SLE; SLICC/ACR-DI: Systemic Lupus Erythematosus International Collaborating Clinics/American College of Rheumatology damage index.

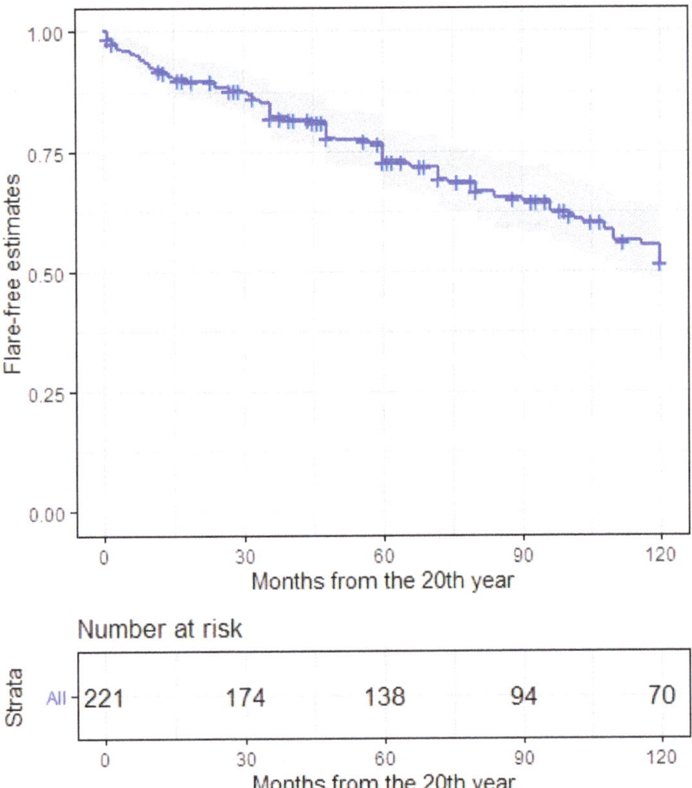

**Figure 1.** Flare-free estimates in long-term SLE patients up to 10 years from the 20th year of disease.

**Table 3.** Clinical features of patients with flares after 20 years of follow up by BILAG domain.

| Organ/Apparatus | Flares (*n* = 84) |
|---|---|
| Musculoskeletal | 38 (45.2%) |
| Mucocutaneous | 15 (17.9%) |
| Renal | 9 (10.7%) |
| Neuropsychiatric SLE | 6 (7.1%) |
| Cardiopulmonary | 3 (3.6%) |
| Haematological | 6 (7.1%) |
| Constitutional | 2 (2.4%) |
| Gastrointestinal | 3 (3.6%) |

SLE: Systemic Lupus Erythematosus.

### 3.3. Factors Associated with Late Flares and Damage Accrual

Patients in LLDAS at the 20-year timepoint had nearly half the risk of a flare within the following ten years compared to patients who were not in LLDAS (HR = 0.487, CI95 = 0.305–0.778, $p$ = 0.0026; Figure 2). Similar results were observed considering the attainment of DORIS remission at 20 years of disease (HR = 0.611, CI95 = 0.338–0.963, $p$ = 0.0337). Patients with low complement and increased dsDNA binding at pre-flare evaluation had a higher 10-year risk of flare compared to patients with a fully negative serology (HR = 2.645, CI95 = 1.478–4.73, $p$ = 0.001), while the occurrence of either serological alteration was not associated with any risk (Figure 3). No other clinical features at the 20-year timepoint were associated with eventual flares. Patients taking prednisone or equivalent at a dosage $\geq$ 7.5 mg/day before flare had a higher flare risk in the subsequent follow up compared to patients with a lower-dose or off steroids (HR = 2.684, CI95 = 1.637–4.403, $p$ = 9.17 $\times$ $10^{-5}$).

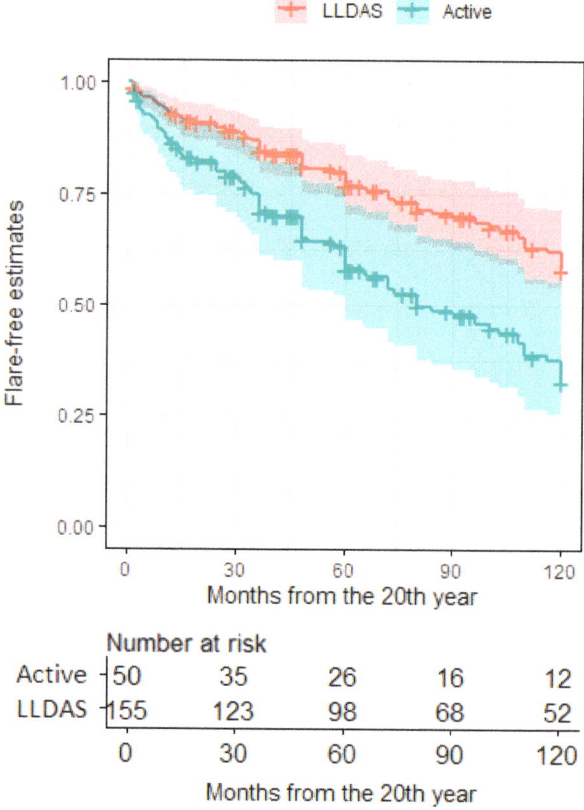


**Figure 2.** Ten-year flare-free estimates in long-term SLE patients according to the LLDAS status at 20 years. Flare-free survival estimates after the 20th year of disease according to the activity state at the 20th year. LLDAS, lupus low-disease-activity state; Active, absence of LLDAS (active disease).

Late flares (that is, occurring after 20 years of follow up, N = 52/88, 59%) were more frequent among patients with progressing SLICC/ACR-DI (N = 52/88, 59.0%) than among patients who did not accrue additional damage after 20 years of follow up (N = 41/128, 32.0%; OR = 3.411, CI95 = 1.90–6.12, $p$ = 2.7 $\times$ $10^{-5}$).

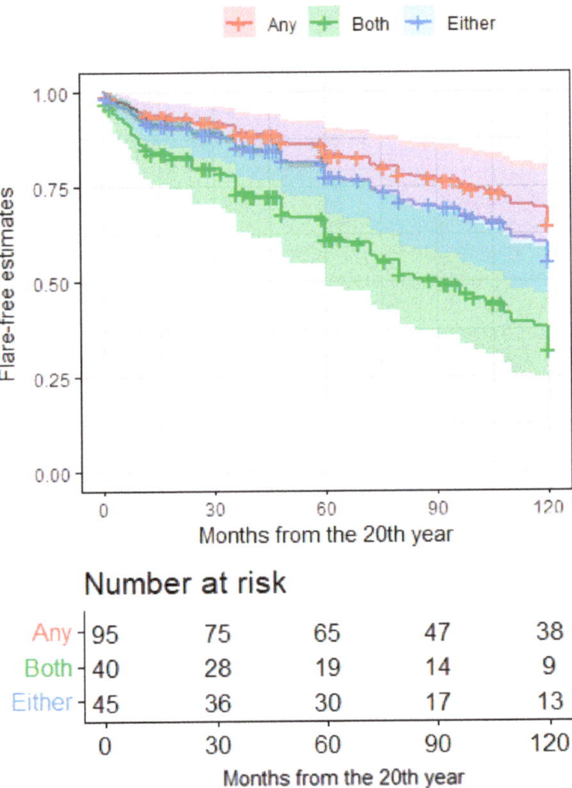


**Figure 3.** Ten-year flare-free estimates in long-term SLE patients according to serology at 20th years. Flare-free survival estimates after the 20th year of disease according to the serological status at the 20th year. Both—presence of both serological alterations (low dsDNA AND hypocomplementemia); Either—presence of either serological alterations (low dsDNA OR hypocomplementemia); Any—absence of any serological alteration.

## 4. Discussion

In this multicentre study, we analysed the trajectories of a relatively large cohort of patients with SLE with long disease duration. We found that most patients have low to no disease activity when observed from the 15th year of disease on, and that the proportion of patients in remission increases over time. However, late damage accrual was also experienced by 40% of patients, possibly with a linear trend. Patients who were not in remission or LLDAS at the 20-year timepoint had a higher risk of late flares, which in turn were associated with damage accrual. Consistently, patients needing higher prednisone doses at pre-flare evaluation had higher chances of developing a flare than patients who had been able to de-escalate or discontinue their treatment with prednisone. Increased dsDNA binding and low complement were also associated with the development of late flares.

Tackling disease progression from persisting activity to damage and disability constitutes a major goal for patients with SLE [3]. Damage accrual has in fact also been associated with higher mortality rates, besides its impact on quality of life [7,24–27]. Patients with early-onset disease and patients in the early phase of their disease course have consistently been reported to experience more severe manifestations and accrue damage more rapidly [28–30]. Damage accrual is predicted to grow linearly at least during the first two decades of disease duration, while less is currently known for later timepoints [27,29,31]. Our data suggest that this linear trend might progress even in very late phases of the disease, which prompts the identification of factors that may modulate this unfavourable course. Consistent with previous studies, we observed that the achievement of a stable remission might protect from late flares and eventually damage [15,17,28,32,33]. Notably, we also confirmed that LLDAS largely overlaps with clinical remission [15] and might constitute a feasible treatment endpoint both in clinical trials and routine rheumatology practice [26,34,35]. In line with other reports, we also found that the musculoskeletal system was most frequently involved in late flares and more prone to be involved in damage accrual [27,29]. In fact, deforming/erosive arthritis and osteoporotic fractures were among the most frequent SLICC/ACR-DI items in our cohort. The relatively lower rate of avascular necrosis compared to other studies might be attributed to ethnic or geographic factors affecting vitamin D metabolism [29,36]. Indeed, high rates of cataract development were also observed in our cohort along with osteoporosis fractures, cardiovascular and retinal complications, possibly suggesting a role of corticosteroid-related mechanisms in contributing to the accrual of damage [37,38]. Taking these data together with the association of high-dose corticosteroid treatments with late flares and that of late flares with damage suggests that distinct treatment approaches might apply to patients at distinct stages of the disease. Immunomodulation through antimalarials and belimumab has been associated with slower damage progression [37,39,40] and might be favoured in older patients with longstanding disease over maintenance immunosuppression. Nonetheless, data from the literature also suggest that immunomodulation is most effective before the onset of initial damage [41], damage itself being a risk factor for further SLICC/ACR-DI progression [7]. The role of immunomodulatory treatment in minimising the effects of deranged B-cell responses in SLE might also be consistent with the predictive role of active combined increased dsDNA binding and low complement towards late flares.

The results of this study suggest that late flares herald damage accrual in patients with longstanding disease and might correlate with higher doses of corticosteroids. This evidence should however be put into the context of some study limitations. First, data regarding the preceding disease history were incomplete, as digital clinical records were only recently introduced into our clinical setting and hard copies of older documents were not available for all subjects. Second, data collection was planned at discrete timepoints rather than visit-by-visit, preventing comprehensive tracking of disease fluctuations over time. Third, treatment features were not homogeneous among subjects, possibly reflecting the evolution of lupus care during recent decades but introducing potential confounders in terms of deflection of disease activity and damage accrual trajectories. Fourth, we focused on a relatively limited number of clinical variables without complementary assessment of potential biomarkers, preventing the development of further patient stratification by pathophysiological or phenotype features.

## 5. Conclusions

Notwithstanding these limitations, our data provide novel hints regarding disease- and treatment-related morbidity in patients with longstanding SLE. Even patients with longer distance from disease onset, especially those with persistent active serology, are at risk of developing disease flares, which in turn constitute a risk factor for late damage accrual. Taken together, our data support the need for closer monitoring of these patients, even in the long term.

**Supplementary Materials:** The following supporting information can be downloaded at: https://www.mdpi.com/article/10.3390/jcm11133587/s1, Table S1: SLICC/ACR-DI items at the last time observation; Figure S1: BILAG scores at relevant time-points. Figure S2: Risk of flares according to use of steroids at the 20th year or thereafter.

**Author Contributions:** Conceptualization, M.G., G.A.R. and L.M.A.; methodology, M.G., G.A.R. and L.B.; software, G.A.R. and L.B.; validation, E.B.; formal analysis, G.A.R. and L.B.; investigation, L.M.A., L.M., G.S. and M.C.; data curation, N.F., G.S. and C.B.; writing—original draft preparation, M.G., M.C., G.A.R. and L.B.; writing—review and editing, M.G., G.A.R., L.B., E.B., L.D., G.S. and R.C.; supervision, E.B., L.D. and R.C.; project administration, M.G.; funding acquisition, L.B. All authors have read and agreed to the published version of the manuscript.

**Funding:** This study was partially funded by the Italian Ministry of Health—current research IRCCS.

**Institutional Review Board Statement:** The study was conducted in accordance with the Declaration of Helsinki, and approved by the IRCCS San Raffaele Hospital Ethics Committee and the Comitato Etico Milano Area 2 (approval no. 0002450/2020).

**Informed Consent Statement:** Informed consent was obtained from all subjects involved in the study.

**Data Availability Statement:** Data are available upon request from the corresponding author. The data are not publicly available due to privacy reasons.

**Acknowledgments:** The authors acknowledge Gabriele Gallina (IRCCS Ospedale San Raffaele, Milan, Italy).

**Conflicts of Interest:** The authors declare no conflict of interest.

## References


1. Tsokos, G.C. Systemic lupus erythematosus. *N. Engl. J. Med.* **2011**, *365*, 2110–2121. [CrossRef] [PubMed]
2. Tsokos, G.C.; Lo, M.S.; Costa Reis, P.; Sullivan, K.E. New insights into the immunopathogenesis of systemic lupus erythematosus. *Nat. Rev. Rheumatol.* **2016**, *12*, 716–730. [CrossRef] [PubMed]
3. Fanouriakis, A.; Kostopoulou, M.; Alunno, A.; Aringer, M.; Bajema, I.; Boletis, J.N.; Cervera, R.; Doria, A.; Gordon, C.; Govoni, M.; et al. 2019 update of the EULAR recommendations for the management of systemic lupus erythematosus. *Ann. Rheum. Dis.* **2019**, *78*, 736–745. [CrossRef] [PubMed]
4. Murphy, G.; Isenberg, D.A. New therapies for systemic lupus erythematosus—Past imperfect, future tense. *Nat. Rev. Rheumatol.* **2019**, *15*, 403–412. [CrossRef]
5. Rios-Garces, R.; Espinosa, G.; van Vollenhoven, R.; Cervera, R. Treat-to-target in systemic lupus erythematosus: Where are we? *Eur. J. Intern. Med.* **2020**, *74*, 29–34. [CrossRef]
6. Nuttall, A.; Isenberg, D.A. Assessment of disease activity, damage and quality of life in systemic lupus erythematosus: New aspects. *Best Pract. Res. Clin. Rheumatol.* **2013**, *27*, 309–318. [CrossRef]
7. Bruce, I.N.; O'Keeffe, A.G.; Farewell, V.; Hanly, J.G.; Manzi, S.; Su, L.; Gladman, D.D.; Bae, S.C.; Sanchez-Guerrero, J.; Romero-Diaz, J.; et al. Factors associated with damage accrual in patients with systemic lupus erythematosus: Results from the Systemic Lupus International Collaborating Clinics (SLICC) Inception Cohort. *Ann. Rheum. Dis.* **2015**, *74*, 1706–1713. [CrossRef]
8. Golder, V.; Tsang, A.S.M.W.P. Treatment targets in SLE: Remission and low disease activity state. *Rheumatology* **2020**, *59* (Suppl. 5), v19–v28. [CrossRef]
9. van Vollenhoven, R.; Voskuyl, A.; Bertsias, G.; Aranow, C.; Aringer, M.; Arnaud, L.; Askanase, A.; Balážová, P.; Bonfa, E.; Bootsma, H.; et al. A framework for remission in SLE: Consensus findings from a large international task force on definitions of remission in SLE (DORIS). *Ann. Rheum. Dis.* **2017**, *76*, 554–561. [CrossRef]
10. van Vollenhoven, R.F.; Bertsias, G.; Doria, A.; Isenberg, D.; Morand, E.; Petri, M.A.; Pons-Estel, B.A.; Rahman, A.; Ugarte-Gil, M.F.; Voskuyl, A.; et al. 2021 DORIS definition of remission in SLE: Final recommendations from an international task force. *Lupus Sci. Med.* **2021**, *8*, e000538. [CrossRef]
11. Polachek, A.; Gladman, D.D.; Su, J.; Urowitz, M.B. Defining Low Disease Activity in Systemic Lupus Erythematosus. *Arthritis Care Res.* **2017**, *69*, 997–1003. [CrossRef]
12. Franklyn, K.; Lau, C.S.; Navarra, S.V.; Louthrenoo, W.; Lateef, A.; Hamijoyo, L.; Wahono, C.S.; Le Chen, S.; Jin, O.; Morton, S.; et al. Definition and initial validation of a Lupus Low Disease Activity State (LLDAS). *Ann. Rheum. Dis.* **2016**, *75*, 1615–1621. [CrossRef] [PubMed]
13. Tani, C.; Vagelli, R.; Stagnaro, C.; Carli, L.; Mosca, M. Remission and low disease activity in systemic lupus erythematosus: An achievable goal even with fewer steroids? Real-life data from a monocentric cohort. *Lupus Sci. Med.* **2018**, *5*, e000234. [CrossRef] [PubMed]
14. Saccon, F.; Zen, M.; Gatto, M.; Margiotta, D.P.E.; Afeltra, A.; Ceccarelli, F.; Conti, F.; Bortoluzzi, A.; Govoni, M.; Frontini, G.; et al. Remission in systemic lupus erythematosus: Testing different definitions in a large multicentre cohort. *Ann. Rheum. Dis.* **2020**, *79*, 943–950. [CrossRef] [PubMed]
15. Zen, M.; Iaccarino, L.; Gatto, M.; Saccon, F.; Larosa, M.; Ghirardello, A.; Punzi, L.; Doria, A. Lupus low disease activity state is associated with a decrease in damage progression in Caucasian patients with SLE, but overlaps with remission. *Ann. Rheum. Dis.* **2018**, *77*, 104–110. [CrossRef]

16. Golder, V.; Kandane-Rathnayake, R.; Hoi, A.Y.; Huq, M.; Louthrenoo, W.; An, Y.; Li, Z.G.; Luo, S.F.; Sockalingam, S.; Lau, C.S.; et al. Frequency and predictors of the lupus low disease activity state in a multi-national and multi-ethnic cohort. *Arthritis Res. Ther.* **2016**, *18*, 260. [CrossRef]

17. Tsang, A.S.M.W.; Bultink, I.E.; Heslinga, M.; Voskuyl, A.E. Both prolonged remission and Lupus Low Disease Activity State are associated with reduced damage accrual in systemic lupus erythematosus. *Rheumatology* **2017**, *56*, 121–128. [CrossRef]

18. Isenberg, D.A.; Rahman, A.; Allen, E.; Farewell, V.; Akil, M.; Bruce, I.N.; D'cruz, D.; Griffiths, B.; Khamashta, M.; Maddison, P.; et al. BILAG 2004. Development and initial validation of an updated version of the British Isles Lupus Assessment Group's disease activity index for patients with systemic lupus erythematosus. *Rheumatology* **2005**, *44*, 902–906. [CrossRef]

19. Yee, C.S.; Cresswell, L.; Farewell, V.; Rahman, A.; Teh, L.S.; Griffiths, B.; Bruce, I.N.; Ahmad, Y.; Prabu, A.; Akil, M.; et al. Numerical scoring for the BILAG-2004 index. *Rheumatology* **2010**, *49*, 1665–1669. [CrossRef]

20. Yee, C.S.; Farewell, V.; Isenberg, D.A.; Griffiths, B.; Teh, L.S.; Bruce, I.N.; Ahmad, Y.; Rahman, A.; Prabu, A.; Akil, M.; et al. The BILAG-2004 index is sensitive to change for assessment of SLE disease activity. *Rheumatology* **2009**, *48*, 691–695. [CrossRef]

21. Gladman, D.D.; Ibanez, D.; Urowitz, M.B. Systemic lupus erythematosus disease activity index 2000. *J. Rheumatol.* **2002**, *29*, 288–291. [PubMed]

22. Luijten, K.M.; Tekstra, J.; Bijlsma, J.W.; Bijl, M. The Systemic Lupus Erythematosus Responder Index (SRI); a new SLE disease activity assessment. *Autoimmun. Rev.* **2012**, *11*, 326–329. [CrossRef]

23. Gladman, D.D.; Goldsmith, C.H.; Urowitz, M.B.; Bacon, P.; Fortin, P.; Ginzler, E.; Gordon, C.; Hanly, J.G.; Isenberg, D.A.; Petri, M.; et al. The Systemic Lupus International Collaborating Clinics/American College of Rheumatology (SLICC/ACR) Damage Index for Systemic Lupus Erythematosus International Comparison. *J. Rheumatol.* **2000**, *27*, 373–376. [PubMed]

24. Chambers, S.A.; Allen, E.; Rahman, A.; Isenberg, D. Damage and mortality in a group of British patients with systemic lupus erythematosus followed up for over 10 years. *Rheumatology* **2009**, *48*, 673–675. [CrossRef] [PubMed]

25. Rahman, P.; Gladman, D.D.; Urowitz, M.B.; Hallett, D.; Tam, L.S. Early damage as measured by the SLICC/ACR damage index is a predictor of mortality in systemic lupus erythematosus. *Lupus* **2001**, *10*, 93–96. [CrossRef]

26. Connelly, K.; Golder, V.; Kandane-Rathnayake, R.; Morand, E.F. Clinician-reported outcome measures in lupus trials: A problem worth solving. *Lancet Rheumatol.* **2021**, *3*, E595–E603. [CrossRef]

27. Taraborelli, M.; Cavazzana, I.; Martinazzi, N.; Lazzaroni, M.G.; Fredi, M.; Andreoli, L.; Franceschini, F.; Tincani, A. Organ damage accrual and distribution in systemic lupus erythematosus patients followed-up for more than 10 years. *Lupus* **2017**, *26*, 1197–1204. [CrossRef] [PubMed]

28. Piga, M.; Floris, A.; Cappellazzo, G.; Chessa, E.; Congia, M.; Mathieu, A.; Cauli, A. Failure to achieve lupus low disease activity state (LLDAS) six months after diagnosis is associated with early damage accrual in Caucasian patients with systemic lupus erythematosus. *Arthritis Res. Ther.* **2017**, *19*, 247. [CrossRef] [PubMed]

29. Lim, L.S.H.; Pullenayegum, E.; Feldman, B.M.; Lim, L.; Gladman, D.D.; Silverman, E.D. From Childhood to Adulthood: Disease Activity Trajectories in Childhood-Onset Systemic Lupus Erythematosus. *Arthritis Care Res.* **2018**, *70*, 750–757. [CrossRef]

30. Ramirez Gomez, L.A.; Uribe Uribe, O.; Osio Uribe, O.; Grisales Romero, H.; Cardiel, M.H.; Wojdyla, D.; Pons-Estel, B.A.; Grupo Latinoamericano de Estudio del Lupus (GLADEL). Childhood systemic lupus erythematosus in Latin America. The GLADEL experience in 230 children. *Lupus* **2008**, *17*, 596–604. [CrossRef]

31. Gladman, D.D.; Urowitz, M.B.; Rahman, P.; Ibanez, D.; Tam, L.S. Accrual of organ damage over time in patients with systemic lupus erythematosus. *J. Rheumatol.* **2003**, *30*, 1955–1959. [PubMed]

32. Zen, M.; Bassi, N.; Nalotto, L.; Canova, M.; Bettio, S.; Gatto, M.; Ghirardello, A.; Iaccarino, L.; Punzi, L.; Doria, A. Disease activity patterns in a monocentric cohort of SLE patients: A seven-year follow-up study. *Clin. Exp. Rheumatol.* **2012**, *30*, 856–863. [PubMed]

33. Zen, M.; Iaccarino, L.; Gatto, M.; Bettio, S.; Nalotto, L.; Ghirardello, A.; Punzi, L.; Doria, A. Prolonged remission in Caucasian patients with SLE: Prevalence and outcomes. *Ann. Rheum. Dis.* **2015**, *74*, 2117–2122. [CrossRef] [PubMed]

34. Morand, E.F.; Trasieva, T.; Berglind, A.; Illei, G.G.; Tummala, R. Lupus Low Disease Activity State (LLDAS) attainment discriminates responders in a systemic lupus erythematosus trial: Post-hoc analysis of the Phase IIb MUSE trial of anifrolumab. *Ann. Rheum. Dis.* **2018**, *77*, 706–713. [CrossRef] [PubMed]

35. Ramirez, G.A.; Canti, V.; Moiola, L.; Magnoni, M.; Rovere-Querini, P.; Coletto, L.A.; Dagna, L.; Manfredi, A.A.; Bozzolo, E.P. Performance of SLE responder index and lupus low disease activity state in real life: A prospective cohort study. *Int. J. Rheum. Dis.* **2019**, *22*, 1752–1761. [CrossRef] [PubMed]

36. Nevskaya, T.; Gamble, M.P.; Pope, J.E. A meta-analysis of avascular necrosis in systemic lupus erythematosus: Prevalence and risk factors. *Clin. Exp. Rheumatol.* **2017**, *35*, 700–710.

37. Petri, M.; Purvey, S.; Fang, H.; Magder, L.S. Predictors of organ damage in systemic lupus erythematosus: The Hopkins Lupus Cohort. *Arthritis Rheum.* **2012**, *64*, 4021–4028. [CrossRef]

38. Tektonidou, M.G.; Lewandowski, L.B.; Hu, J.; Dasgupta, A.; Ward, M.M. Survival in adults and children with systemic lupus erythematosus: A systematic review and Bayesian meta-analysis of studies from 1950 to 2016. *Ann. Rheum. Dis.* **2017**, *76*, 2009–2016. [CrossRef]

39. Urowitz, M.B.; Ohsfeldt, R.L.; Wielage, R.C.; Kelton, K.A.; Asukai, Y.; Ramachandran, S. Organ damage in patients treated with belimumab versus standard of care: A propensity score-matched comparative analysis. *Ann. Rheum. Dis.* **2019**, *78*, 372–379. [CrossRef]

40. Iaccarino, L.; Bettio, S.; Reggia, R.; Zen, M.; Frassi, M.; Andreoli, L.; Gatto, M.; Piantoni, S.; Nalotto, L.; Franceschini, F.; et al. Effects of Belimumab on Flare Rate and Expected Damage Progression in Patients with Active Systemic Lupus Erythematosus. *Arthritis Care Res.* **2017**, *69*, 115–123. [CrossRef]
41. Parodis, I.; Sjowall, C.; Jonsen, A.; Ramskold, D.; Zickert, A.; Frodlund, M.; Sohrabian, A.; Arnaud, L.; Rönnelid, J.; Malmström, V.; et al. Smoking and pre-existing organ damage reduce the efficacy of belimumab in systemic lupus erythematosus. *Autoimmun. Rev.* **2017**, *16*, 343–351. [CrossRef] [PubMed]

Journal of
*Clinical Medicine*

*Article*

# The Impacts of the Clinical and Genetic Factors on Chronic Damage in Caucasian Systemic Lupus Erythematosus Patients

Fulvia Ceccarelli [1,†], Giulio Olivieri [1,*,†], Carmelo Pirone [1], Cinzia Ciccacci [2], Licia Picciariello [1], Francesco Natalucci [1], Carlo Perricone [3], Francesca Romana Spinelli [1], Cristiano Alessandri [1], Paola Borgiani [4] and Fabrizio Conti [1]

[1] Reumatologia, Dipartimento di Scienze Cliniche Internistiche, Anestesiologiche e Cardiovascolari, Sapienza University of Rome, Viale del Policlnico 155, 00161 Rome, Italy; fulvia.ceccarelli@uniroma1.it (F.C.); carmelo.pirone@uniroma1.it (C.P.); licia.picciariello@uniroma1.it (L.P.); francesco.natalucci@uniroma1.it (F.N.); francescaromana.spinelli@uniroma1.it (F.R.S.); cristiano.alessandri@uniroma.it (C.A.); fabrizio.conti@uniroma1.it (F.C.)

[2] Università UniCamillus—Saint Camillus International University of Health and Medical Sciences, 00131 Rome, Italy; cinzia.ciccacci@unicamillus.org

[3] Reumatologia, Dipartimento di Medicina e Chirurgia, Università di Perugia, 06123 Perugia, Italy; carlo.perricone@unipg.it

[4] Genetics Section, Department of Biomedicine and Prevention, University of Rome Tor Vergata, 00133 Rome, Italy; borgiani@med.uniroma2.it

* Correspondence: giulio.olivieri@uniroma1.it; Tel.: +39-06-4997-4631

† These authors contributed equally to this work.

**Citation:** Ceccarelli, F.; Olivieri, G.; Pirone, C.; Ciccacci, C.; Picciariello, L.; Natalucci, F.; Perricone, C.; Spinelli, F.R.; Alessandri, C.; Borgiani, P.; et al. The Impacts of the Clinical and Genetic Factors on Chronic Damage in Caucasian Systemic Lupus Erythematosus Patients. *J. Clin. Med.* **2022**, *11*, 3368. https://doi.org/10.3390/jcm11123368

Academic Editors: Christopher Sjöwall and Ioannis Parodis

Received: 23 May 2022
Accepted: 10 June 2022
Published: 12 June 2022

**Publisher's Note:** MDPI stays neutral with regard to jurisdictional claims in published maps and institutional affiliations.



**Abstract:** Objective: The purpose of this study was to determine the distribution of organ damage in a cohort of systemic lupus erythematosus (SLE) patients and to evaluate the roles of clinical and genetic factors in determining the development of chronic damage. Methods: Organ damage was assessed by the SLICC Damage Index (SDI). We analyzed a panel of 17 single-nucleotide polymorphism (SNPs) of genes already associated with SLE, and we performed a phenotype–genotype correlation analysis by evaluating specific domains of the SDI. Results: Among 175 Caucasian SLE patients, 105 (60%) exhibited damage (SDI $\geq$1), with a median value of 1.0 (IQR 3.0). The musculoskeletal (26.2%), neuropsychiatric (24.6%) and ocular domains (20.6%) were involved most frequently. The presence of damage was associated with higher age, longer disease duration, neuropsychiatric (NP) manifestations, anti-phospholipid syndrome and the positivity of anti-dsDNA. Concerning therapies, cyclophosphamide, mycophenolate mofetil and glucocorticoids were associated with the development of damage. The genotype–phenotype correlation analysis showed an association between renal damage, identified in 6.9% of patients, and rs2205960 of TNFSF4 ($p$ = 0.001; OR 17.0). This SNP was significantly associated with end-stage renal disease ($p$ = 0.018, OR 9.68) and estimated GFR < 50% ($p$ = 0.025, OR 1.06). The rs1463335 of MIR1279 gene was associated with the development of NP damage ($p$ = 0.029; OR 2.783). The multivariate logistic regression analysis confirmed the associations between TNFSF4 rs2205960 SNP and renal damage ($p$ = 0.027, B = 2.47) and between NP damage and rs1463335 of MIR1279 gene ($p$ = 0.014, B = 1.29). Conclusions: Our study could provide new insights into the role of genetic background in the development of renal and NP damage.

**Keywords:** systemic lupus erythematosus; genetics; chronic damage; polymorphisms; TNFSF4; MIR1279

## 1. Introduction

Systemic lupus erythematosus (SLE) is a chronic autoimmune disease characterized by multifactorial pathogenesis in which genetic background and environmental factors interplay, determining disease development [1].

In recent decades, there has been a significant improvement in managing patients with SLE in terms of survival rates; however, morbidity due to organ damage remains unresolved. The assessment of accumulated SLE-related damage has been recognized as an important achievement because it is known that specific organ damage and subsequent dysfunction are significant causes of morbidity and mortality in patients with SLE [2].

The Systemic Lupus International Collaborating Clinics/American College of Rheumatology Damage Index (SDI) was developed in 1996 [3] to assess ongoing manifestations of disease activity in SLE patients and to measure irreversible damage resulting from SLE disease activity as well as its treatment and comorbidities [4–6]. Moreover, the presence of specific autoantibodies, such as anti-phospholipid and anti-dsDNA, could be considered predictive factors for the development of chronic damage [7,8].

The SDI is a robust instrument for quantifying damage and has been extensively validated [9]. This tool has prognostic value; in fact, many studies have shown that damage predicts morbidity and mortality [9]. For instance, a prospective study of 230 patients over 10 years of disease duration showed that early damage was associated with a higher mortality rate [10]. High SDI scores were also associated with increased economic costs and reduced health-related quality of life [11]. Risk factors for damage include older age at diagnosis, longer duration of SLE, African-Caribbean or Asian ethnicity, high disease activity at diagnosis and greater overall activity during the disease course [12]. We previously showed that machine learning models could predict the development of chronic damage and the achievement of the Lupus Comprehensive Disease Control (*LupusCDC*) [13,14]. These models have suggested that despite the control of disease activity and the absence of adverse drug events, the chronic damage progresses in some patients, meaning that there may be other risk factors such as genetic background.

During the past two decades, genome-wide association studies have been conducted to screen hundreds of thousands of single nucleotide polymorphisms (SNPs) across the genome [15]. Meta-analyses and large-scale replication/fine-mapping studies have revealed over 100 genomic loci linked to SLE susceptibility, enhancing the understanding of SLE pathogenesis at the molecular level [15,16].

Whereas most studies have looked for an association between susceptibility loci and SLE, only a few have examined the relationships between these markers and selected disease manifestations and clinical subsets or organ damage [17,18]. For example, variants of signal transducer and activator of transcription 4 (STAT4) have been associated with a more severe disease phenotype, including ischemic stroke, nephritis and increased SDI scores [19,20]. Recently, Reid and colleagues reported that high genetic risk scores were associated with a more severe SLE phenotype, renal dysfunction, and organ damage [21]. Moving from these premises, the purpose of this study was to evaluate the contribution on chronic damage development of clinical features and genetic factors in a cohort of SLE patients. Moreover, we analyzed variants of previously identified loci associated with SLE to verify their possible contribution to the development of chronic damage evaluated as specific SDI domains.

## 2. Materials and Methods

### 2.1. Study Design and Population

A cross-sectional study was executed by enrolling Caucasian adult SLE patients attending the Lupus Clinic of the Rheumatology Unit, Sapienza University of Rome (Sapienza Lupus Cohort). SLE diagnosis was performed according to the revised 1997 American College of Rheumatology criteria [22]. We limited the analysis to subjects with a minimum disease duration of five years and at least two visits per year to the Sapienza Lupus Clinic. The Ethical Committee of AOU Policlinico Umberto I, Rome, approved the study protocol. All patients signed the informed consent for the use of their clinical and laboratory data for study purposes.

### 2.2. Clinical and Laboratory Evaluation

The clinical and laboratory data for each SLE patient were collected in a standardized, computerized, electronically filled form, including demographics, past medical history with the date of diagnosis, comorbidities and previous and concomitant treatments. Antinuclear antibodies (ANA) were determined with IIF on HEp-2, anti-dsDNA with IIF on Crithidia luciliae (titer $\geq$ 1:10), ENA (including anti-Ro/SSA, anti-La/SSB, anti-Sm and anti-RNP) analyzed by ELISA considering titers above the cut-off of the reference laboratory, aCL (IgG/IgM isotype) analyzed by ELISA, in serum, at medium or high titers (e.g., >40 GPL or MPL or above the 99th percentile), anti-B2 glycoprotein-I (IgG/IgM isotype) analyzed by ELISA, in serum (above the 99th percentile), and lupus anticoagulant (LA), according to the guidelines of the International Society on Thrombosis and Hemostasis. Finally, C3 and C4 serum levels were determined with nephelometry. All subjects underwent blood drawing (5 mL supplemented with 0.5% EDTA) to perform genetic analysis. We registered clinical and laboratory data referring to the whole patient's disease history.

### 2.3. Disease Activity

We assessed the disease activity according to the SLE Disease Activity Index 2000 [SLEDAI-2K] in all visits available in the three years prior to the SDI assessment [23]. We identified, in our cohort, three different patterns of disease activity according to SLEDAI-2K values, as follows: (1) patients with SLEDAI-2K $\leq$ 2 on all the available visits [minimal disease activity (MDA)]; (2) patients with SLEDAI-2K $\geq$ 4 on at least two consecutive visits [persistent active disease (PAD)]; (3) patients with at least one flare defined as an increase in SLEDAI-2K $\geq$ 4 from the previous visit [relapsing–remitting disease (RRD)].

### 2.4. Chronic Damage

Chronic damage was determined based on SDI at the last available examination in our center. The SDI score was calculated based on organ damage that occurred after diagnosis with SLE [3]. The SDI assesses 41 items across 12 organ systems: ocular (range 0–2), neuropsychiatric (0–6), renal (0–3), pulmonary (0–5), cardiovascular (0–6), peripheral vascular (0–5), gastrointestinal (0–6), musculoskeletal (0–7), skin (0–3), gonadal (0–1), endocrine (0–1) and malignancy (0–2). Most items were assigned 1 point if present, with 2 points possible for recurrent events and 3 points for end-stage renal disease, for a possible total score of 47. To distinguish damage from reversible disease activity, an item must be present for at least six months to be scored, irrespective of the cause. Four items of the SDI focus specifically on glucocorticoid- (GC) related adverse effects (cataracts, osteoporotic fracture, avascular necrosis, diabetes mellitus). From the sum of these items, we generated a single glucocorticoid-related SDI domain (GC-SDI) as previously described [24,25].

### 2.5. DNA Extraction and Genotyping

Genomic DNA was isolated from peripheral blood mononuclear cells using a Qiagen blood DNA mini kit. Based on literature data, we selected a panel of 17 SNPs of genes involved in immune response, autophagy and inflammation that were already described as associated with SLE [15–18]. We analyzed polymorphisms of genes linked to: innate/adaptive immune response [Toll-like receptor and type I interferon signaling: rs7574865 (STAT4), rs3027898 (IRAK1)]; T cell signaling [rs22205960 (TNFSF4)]; T and B cell signaling and interaction [rs1800872 and rs3024505 (IL10), rs4810485 (CD40); self-antigen clearance defects [rs2241880 (ATG16L1)]; autophagy [rs6568431, rs2245214 and rs573775 (ATG5)], rs13361189 and rs4958847 (IRGM)]; genes located in the HLA region [rs9469003 and rs3099844 (HCP5)] and microRNAs [rs1463335 (MIR1279), rs2431697 (MIR146a), rs531564 (MIR124A)].

Genotyping was performed with a TaqMan allelic discrimination assay (Applied Biosystems, Foster City, CA, USA) and real-time PCR. Each assay was run including samples with known genotypes previously confirmed by direct sequencing as genotype controls.

### 2.6. Statistical Analysis

The statistical evaluation was performed using dedicated software: Statistical Package for Social Sciences 13.0 (SPSS, Chicago, IL, USA) and GraphPad 5.0 (GraphPad Software, La

Jolla, CA, USA). Normally distributed variables were summarized using the mean, standard deviation (SD) and nonnormally distributed variables by the median and interquartile range [IQR]. Wilcoxon's matched pairs test and paired t-test were performed. For the univariate analysis, two groups of patients, with and without damage by SDI score, were considered. The differences between categorical variables were calculated using a chi-square test or Fisher's exact test where appropriate. A Spearman correlation analysis was performed for measuring the correlation between variables. Two-tailed P values were reported, and P values less than or equal to 0.05 were considered significant. Odds ratios (ORs) with 95% confidence interval (CI) were calculated. A genotype–phenotype correlation analysis was performed considering the heterozygotes and variant homozygotes together (one degree of freedom). A binary logistic regression analysis (stepwise) was performed to analyze the contributions of specific SNP variants to the development of chronic damage as specific SDI domains.

## 3. Results

We analyzed 175 Caucasian SLE patients (M/F 15/160, median age at disease diagnosis 31 years, IQR 18; median disease duration 227 months, IQR 138). Table 1 summarizes the primary demographic and clinical data, laboratory features and patterns of disease activity in the whole SLE cohort, including ongoing and previous treatments. In our cohort, joint involvement was the most frequent clinical feature (89.1%), followed by skin manifestation (85.7%).

At the time of study enrollment, 105 out of 175 (60%) of SLE patients showed chronic damage in at least one organ/system (SDI $\geq$ 1), with a median value of 1.0 (IQR 3.0).

As expected, a significantly higher median age and median disease duration were observed in patients who had SDI $\geq$ 1 in comparison with patients without chronic damage [age: 54 years (IQR 14) versus 46 years (IQR 16); $p = 0.0001$; disease duration: 267 months (IQR 156) versus 183 months (IQR 108); $p = 0.0001$).

We registered a significant difference in the prevalence of some disease-associated manifestations, serological parameters and drugs prescribed between the two groups (Table 1). In detail, damage accrual was significantly associated with neuropsychiatric manifestations ($p = 0.00001$), anti-phospholipid syndrome ($p = 0.0017$) and the positivity of anti-dsDNA antibodies ($p = 0.0099$) and LA ($p = 0.04$). Concerning therapies, cyclophosphamide (CY) and mycophenolate mofetil (MMF) were more frequently prescribed in patients with chronic damage ($p = 0.00001$, $p = 0.0058$, respectively). Moreover, as is well-known, GC treatment influenced irreversible damage development. All our patients have received GC therapy, but a significantly higher proportion of SLE patients with chronic damage took glucocorticoids for a period longer than ten years ($p = 0.00001$).

Looking at disease activity, we found a similar prevalence of disease activity patterns in patients with and without chronic damage. However, when comparing the median SDI, patients with PAD showed a significant higher value in comparison with MDA patients [2.0 (IQR 4.5) vs. 1.0 (IQR 3.0); $p = 0.04$].

Figure 1 reports the SDIs for our SLE cohort. Most patients had low values [29 patients (16.6%) showed SDI = 1, 26 patients (14.8%) had SDI = 2], whereas only a minority of our patients had SDI scores higher than five points. In detail, five patients (2.8%) had SDI = 6, and five patients showed SDI = 7. Only one patient, who had a disease duration of 508 months, had SDI = 10, which is the highest score registered in our cohort.

In Table 2, we show the distribution of damage according to each SDI domain. The musculoskeletal domain was the most frequently involved organ/system (46/175 patients, 26.2%), followed by the neuropsychiatric and ocular domains, detected in 43 (24.6%) and 53 (20.6%) patients, respectively.

**Table 1.** Demographic and disease activity data, clinical features, serological parameters, and therapies of our SLE cohort and in the two main groups identified.

| | Whole SLE Cohort N = 175 | SLE Patients with SDI = 0 N = 70 | SLE Patients with SDI ≥ 1 N = 105 | *p* Value |
|---|---|---|---|---|
| M/F | 15/160 | 3/67 | 12/93 | n.s. |
| Median age–years [IQR] | 31 (18) | 46 (16) | 54 (14) | *p* = 0.0001 |
| Median disease duration -months [IQR] | 227 (138) | 183 (108) | 267 (156) | *p* = 0.0001 |
| **Disease activity patterns, n (%)** | | | | |
| Minimal Disease Activity | 121 (69.2) | 51 (72.8) | 70 (66.7) | n.s. |
| Persistent Active Disease | 24 (13.7) | 9 (12.) | 15 (14.3) | n.s. |
| Relapsing Remitting | 30 (17.1) | 10 (14.3) | 20 (19.0) | n.s. |
| **Clinical features, n (%)** | | | | |
| Skin manifestation | 150 (85.7) | 56 (80.1) | 94 (89.5) | n.s. |
| *Malar rash* | 119 (68.0) | 46 (65.7) | 73 (69.5) | n.s. |
| *Photosensitivity* | 129 (73.7) | 47 (67.1) | 82 (78.1) | n.s. |
| *Oral ulcers* | 44 (25.1) | 18 (25.7) | 26 (26.7) | n.s. |
| *Alopecia* | 21 (12.0) | 8 (11.4) | 13 (12.4) | n.s. |
| *Discoid rash* | 16 (9.1) | 7 (10.0) | 9 (8.6) | n.s. |
| Joint involvement | 156 (89.1) | 59 (84.3) | 97 (92.4) | n.s. |
| Renal involvement | 67 (38.3) | 22 (31.4) | 45 (42.8) | n.s. |
| *Mesangial nephritis* | 19 (10.8) | 7 (10.0) | 12 (11.4) | n.s. |
| *Proliferative nephritis* | 38 (21.7) | 13 (18.6) | 25 (23.8) | n.s. |
| *Membranous nephritis* | 10 (5.7) | 2 (2.8) | 8 (7.6) | n.s. |
| Hematological manifestation | 101 (57.7) | 39 (55.7) | 62 (59.0) | n.s. |
| *Leukopenia* | 78 (44.6) | 31 (44.3) | 47 (44.7) | n.s. |
| *Thrombocytopenia* | 44 (25.1) | 14 (20.0) | 30 (28.6) | n.s. |
| *Hemolytic anemia* | 10 (5.7) | 5 (7.1) | 5 (4.7) | n.s. |
| Neuropsychiatric involvement | 47 (26.8) | 6 (8.6) | 41 (39.0) | *p* = 0.00001 |
| *Central NPSLE* | 36 (20.6) | 5 (7.1) | 31 (29.5) | *p* = 0.00005 |
| *Peripheral NPSLE* | 11 (6.3) | 1 (1.4) | 10 (9.5) | *p* = 0.009 |
| Serositis | 48 (27.4) | 15 (21.4) | 33 (31.4) | n.s. |
| *Pericarditis* | 38 (21.7) | 12 (17.1) | 26 (24.7) | n.s. |
| *Pleuritis* | 30 (17.1) | 9 (12.8) | 21 (20.0) | n.s. |
| Anti-phospholipid syndrome | 42 (24.0) | 10 (12.3) | 32 (30.5) | *p* = 0.0017 |
| **Laboratory parameters, n (%)** | | | | |
| Anti-dsDNA | 132 (75.4) | 46 (65.7) | 86 (81.9) | *p* = 0.0099 |
| Low C3/C4 serum levels | 107 (61.1) | 40 (57.1) | 67 (63.8) | n.s. |
| Anti-cardiolipin antibodies IgM/IgG | 69 (39.4) | 26 (37.1) | 43 (40.9) | n.s. |
| Anti-B2-glycoprotein I antibodies IgM/IgG | 37 (21.1) | 11 (15.7) | 26 (24.7) | n.s. |
| Lupus anticoagulant | 43 (24.6) | 12 (17.1) | 31 (29.5) | *p* = 0.04 |
| Anti-Ro/SSA | 51 (29.1) | 20 (28.6) | 31 (29.5) | n.s. |
| Anti-La/SSB | 21 (12.0) | 7 (10.0) | 14 (13.3) | n.s. |
| Anti-RNP | 29 (16.6) | 19 (18.1) | 10 (14.3) | n.s. |
| Anti-Sm | 26 (14.8) | 18 (17.1) | 8 (11.4) | n.s. |
| **Treatments, n (%)** | | | | |
| Glucocorticoids [GC] | 175 (100) | 70 (100) | 105 (100) | n.s. |
| GC intake ≥ 10 years | 78 (44.6) | 18 (24.0) | 60 (57.1) | *p* = 0.00001 |
| Hydroxychloroquine | 162 (92.6) | 60 (85.7) | 102 (97.1) | n.s. |
| Azathioprine | 62 (35.4) | 21 (30.0) | 41 (39.0) | n.s. |
| Cyclosporine A | 39 (22.3) | 11 (15.7) | 28 (26.6) | n.s. |
| Methotrexate | 58 (33.1) | 20 (28.6) | 38 (36.2) | n.s. |
| Mycophenolate Mofetil | 69 (39.4) | 12 (17.1) | 36 (34.2) | *p* = 0.0058 |
| Cyclophosphamide | 25 (14.3) | 1 (1.4) | 24 (22.8) | *p* = 0.00001 |
| Belimumab | 28 (16.0) | 8 (11.4) | 19 (18.1) | n.s. |
| Rituximab | 8 (4.6) | 3 (4.2) | 5 (4.7) | n.s. |

Legend: non-significant (n.s).

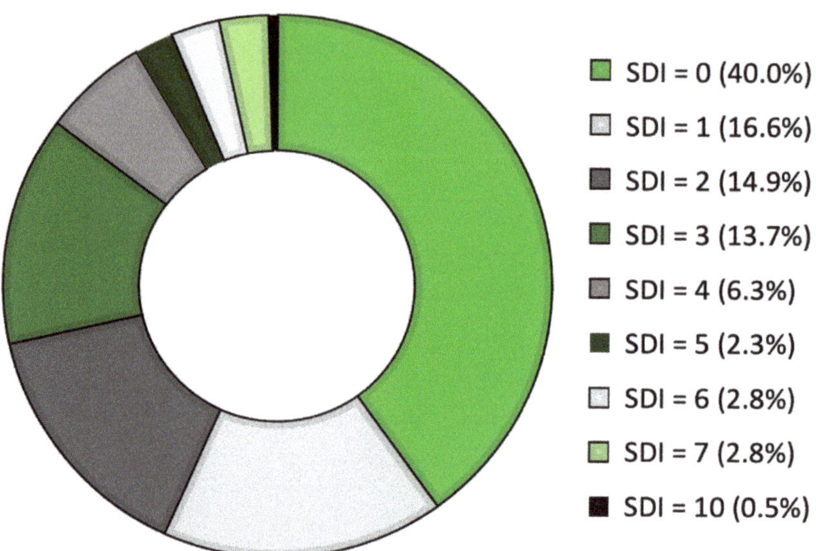


**Figure 1.** The distribution of SDIs in our SLE cohort [please read clockwise starting from the largest group with SDI = 0].

**Table 2.** The distribution of damage according to the involved organ/system.

| Domain N (%) | Item | Patients N (%) |
|---|---|---|
| Ocular 36 (20.6) | *Any cataract ever* | 25 (14.3) |
| | *Retinal change OR optic atrophy* | 15 (8.6) |
| | *Cognitive impairment OR major psychosis* | 19 (10.8) |
| Neuropsychiatric 43 (24.6) | *Seizures requiring therapy for >6 months* | 7 (4.0) |
| | *Cerebral vascular accident ever OR resection not for malignancy* | 10 (5.7) |
| | *Cranial or peripheral neuropathy [excluding optic]* | 18 (10.3) |
| | *Transverse myelitis* | 1 (0.6) |
| Renal 12 (6.9) | *Estimated or measured GFR < 50%* | 4 (2.3) |
| | *Proteinuria >3.5 g/24 h* | 1 (0.6) |
| | *ESRF [regardless of dialysis or transplantation]* | 7 (4.0) |
| Pulmonary 5 (2.8) | *Pulmonary hypertension [right ventricular prominence or loud P2]* | 2 (1.1) |
| | *Pulmonary fibrosis [clinically and/or by X-ray]* | 4 (2.3) |
| | *Shrinking lung [by X-ray] 0 Pleural fibrosis [by X-ray]* | 0 (0) |
| | *Pulmonary infarction [by X-ray] OR resection not for malignancy* | 0 (0) |
| Cardiovascular 15 (8.6) | *Angina OR Coronary artery bypass* | 1 (0.6) |
| | *Myocardial infarction ever* | 2 (1.1) |
| | *Cardiomyopathy [ventricular dysfunction]* | 1 (0.6) |
| | *Valvular disease [diastolic murmur, or systolic murmur >3/6] Pericarditis* | 10 (5.7) |
| | *OR pericardiectomy* | 1 (0.6) |
| | *Claudication* | 0 (0) |
| Peripheral vascular 5 (2.8) | *Minor tissue loss [pulp space]* | 1 (0.6) |
| | *Significant tissue loss ever [at least loss or resection of a digit]* | 2 (1.1) |
| | *Venous thrombosis with swelling, ulceration, OR venous stasis* | 2 (1.1) |
| | *Infarction or resection of bowel below duodenum, spleen, liver, or gall bladder ever* | 24 (13.7) |
| Gastrointestina l24 (13.7) | *Mesenteric insufficiency* | 0 (0) |
| | *Chronic peritonitis* | 0 (0) |
| | *Stricture OR upper gastrointestinal tract surgery ever* | 0 (0) |
| | *Pancreatic insufficiency requiring enzyme replacement OR with pseudocyst* | 0 (0) |

**Table 2.** *Cont.*

| Domain<br>N (%) | Item | Patients<br>N (%) |
|---|---|---|
| Musculoskeletal<br>45 (26.2) | *Muscle atrophy OR weakness* | 8 (4.6) |
| | *Deforming or erosive arthritis* | 27 (15.4) |
| | *Osteoporosis with fracture or vertebral collapse* | 13 (7.4) |
| | *Avascular necrosis* | 5 (2.9) |
| | *Osteomyelitis* | 0 (0) |
| Skin<br>12 (6.9) | *Scarring chronic alopecia* | 2 (1.1) |
| | *Extensive scarring of panniculus other than scalp and pulp space Skin ulceration [excluding* | 2 (1.1) |
| | *thrombosis] of more than 6 months* | 8 (4.6) |
| Gonadal | *Premature gonadal failure* | 12 (6.9) |
| Endocrine | *Diabetes requiring therapy regardless of treatment* | 9 (5.1) |
| Malignancy | *Malignancy [excluded dysplasia]* | 18 (10.3) |

Moving to the assessment of chronic damage related to the side effect of GC treatment, 41 patients (23.4%) developed damage in the GC-SDI domain, of whom 11 had more than one organ/system affected in this peculiar domain.

*Genotype-Phenotype Correlation Analysis*

We further performed a genotype–phenotype correlation analysis to evaluate the possible associations between the above-reported polymorphisms and the development of chronic damage evaluated as specific SDI domains and items. Our study showed a potential role for variants of three different genes.

In detail, we found an association between renal damage, identified in 6.9% of patients, and TNF Superfamily Member 4 (TNFSF4) rs2205960 SNP (G > T) ($p = 0.001$). Only this genetic variant was significantly associated with renal damage, showing that individuals carrying the variant T allele (GT and TT genotypes) had a higher risk of developing this kind of damage in comparison with individuals carrying the wildtype genotype (GG) ($p = 0.001$, OR 17.0, 95% CI 2.122–136.769) (Figure 2).

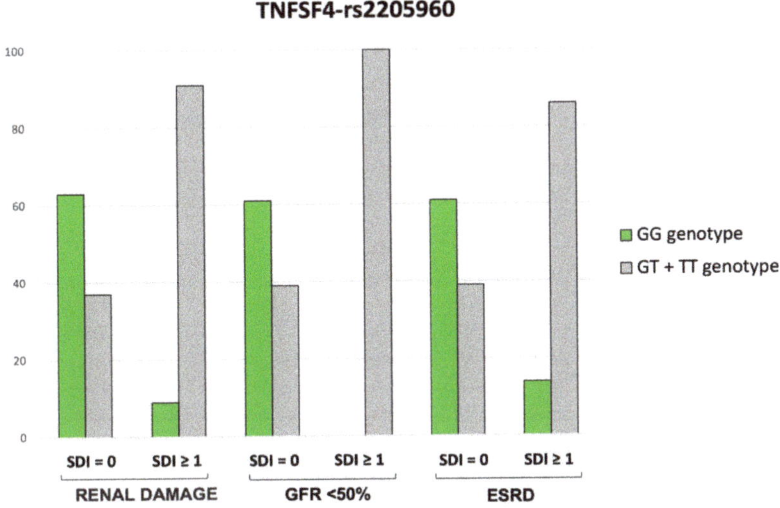


**Figure 2.** The associations between renal damage and rs2205960 of TNFSF4 [$p = 0.001$]. In addition, this SNP was significantly associated with the development of two specific items on the SDI renal domain: estimated glomerular filtration rate (GFR) <50% and end-stage renal disease (ESRD) ($p = 0.025$, $p = 0.018$ respectively).

Moreover, this SNP was significantly associated with the development of two specific items of renal domain: end-stage renal disease (ESDR) ($p$ = 0.018, OR 9.68, 95% CI 1.136–82.527) and estimated glomerular filtration rate (GFR) <50% ($p$ = 0.025) (Figure 2).

Furthermore, we found an association between the rs1463335 SNP (T > A) of microRNA 1279 (MIR1279) gene, mapping to chromosome 12q15, and the development of neuropsychiatric damage (29.1% of patients; $p$ = 0.029; Figure 3A). Patients carrying the variant A allele seem to have an increased risk of developing this type of damage ($p$ = 0.029, OR 2.783. 95% CI 1.081–7.165), but this polymorphism is not associated with the development of specific neuropsychiatric domain items.

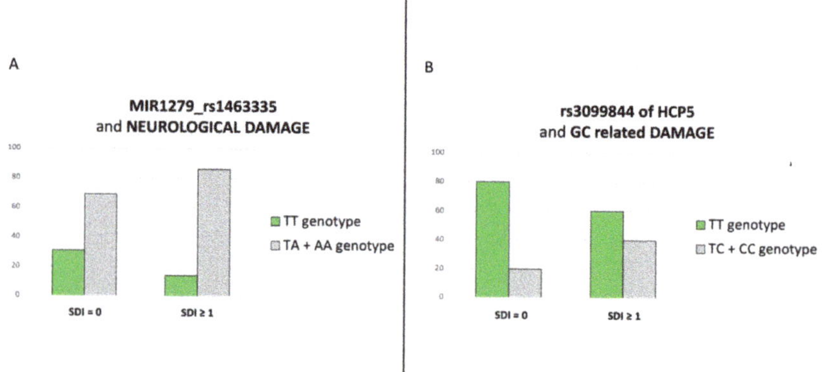


**Figure 3.** (**A**): Associations between rs1463335 of MIR1279 and the development of neuropsychiatric damage ($p$ = 0.029). (**B**): rs9469003 of HCP5 locus was significantly associated with the GC-SDI domain ($p$ = 0.028).

The multivariate logistic regression analysis adjusted for main confounders [sex, age, disease duration, GC treatment duration, MMF and CY treatment and, aPL positivity) confirmed the association between renal damage and rs2205960 of TNFSF4 ($p$ = 0.027, B = 2.47) and between neuropsychiatric damage and rs1463335 of MIR1279 ($p$ = 0.014, B = 1.29).

Finally, we observed a significant association between HLA complex P5 (HCP5) rs9469003 SNP (T > C), on chromosome 6, and the GC-SDI domain ($p$ = 0.028), suggesting that the variant C allele may confer an increased risk of developing damage related to the side effects of GC treatment (OR 2.6; 95% CI 1.091–6.197; Figure 3B).

## 4. Discussion

In the present study, we aimed to evaluate the contribution of genetic background to the development of chronic damage in terms of specific SDI domains.

In our cohort, 60% of SLE patients showed damage after a median disease duration of almost 19 years. In previous reports, more than 60% of patients had irreversible damage within 7 years of diagnosis of SLE [26,27]. Furthermore, our results identified demographic factors, anti-phospholipid antibodies and treatment with GC as predictors of chronic damage [5,6,11,12,24,28]. In fact, we confirm the worse prognostic effect of high age and disease duration, and we acknowledge the main role of GC treatment in determining chronic damage. Thus, most patients with SDI $\geq$ 1 receive GC for a cumulative period of more than 10 years and, 23.4% of them had an impairment in 1 or more items of the GC-SDI domain.

Moreover, we found a significant association between CY and MMF administration and the presence of irreversible damage. Certainly, the involvement of major organs could be a confounding factor because these drugs are generally administered to patients with more severe disease manifestations, such as proliferative nephritis or central nervous system vasculitis. Of note, treatment with CY was reported associated with higher SDIs and remained a predictor of damage [29,30].

Moving to the SDI domains in our cohort, damage more frequently involved the musculoskeletal system (26.2%), followed by neuropsychiatric and ocular involvement (24.6% and 20.6%, respectively). These results agree with several studies, conducted in patients with different ethnic backgrounds, reporting that the musculoskeletal system was the most frequently damaged in SLE patients [6,31–33]. Data from the Hopkins Lupus Cohort and the Toronto Lupus Cohort also showed that musculoskeletal damage accrued linearly, with osteonecrosis being the most frequent subtype, followed by deforming arthritis [31,34].

It should be considered that most of our patients had erosive arthritis, and this could be related to the fact that our research group has consistently and thoroughly focused on the presence of bone erosions, as previously reported [35].

On the other hand, renal damage was uncommon in our cohort [6.9%], in contrast to most other studies (14–32.4%) [7,28,29,31–33], which may be because the expression of renal disease is more aggressive in some ethnic groups (11). Although renal damage was infrequent, it was significantly associated with rs2205960 of TNFSF4 gene, which was previously associated with SLE susceptibility and lupus nephritis (LN) [17,30,36].

Investigating the genetics contribution to the development of chronic damage, we found a significant association between rs2205960 SNP of TNFSF4, the development of irreversible renal damage and two specific items of this domain [end-stage renal disease and estimated GFR < 50%]. Moreover, in our analysis, we described the correlation between neuropsychiatric damage and the rs1463335 SNP in MIR1279 gene, while rs9469003 SNP in the HCP5 gene showed an association with GC-related damage.

The TNFSF4 gene is located on human chromosome 1 and encodes the TNFSF4 protein, also known as OX40 ligand (OX40L), a cytokine of the TNF ligand family. The TNFSF4 molecule is a type II transmembrane protein, which is mainly expressed on several activated immune cells. It plays an important role in effector T-cell survival, B-cell differentiation and proliferation, cytokine production and memory cell formation [36].

In 2011, Sanchez et al., analyzing the SLE susceptibility loci in a large cohort with different ethnicities, found a significant association between renal involvement and TNFSF4 gene [17]. In agreement, Aten and colleagues detected an increased TNFSF4 expression in renal biopsies of patients with LN [37]. In 2017, two different groups verified the role of TNFSF4 in SLE and LN pathogenesis using a conditional knockout mouse system and in vivo agonist and antagonist approaches in an SLE mouse model [38,39]. Both studies suggested that the OX40/OX40L pathway contributes to lupus pathogenesis by promoting the T follicular helper cell response.

Here, we described for the first time the association between rs2205960 of TNFSF4 gene and the development of irreversible renal damage, as estimated GFR < 50% and ESDR. This result could reinforce the role of this gene in SLE-related renal involvement, suggesting different pathogenic pathways for the specific disease manifestations.

The second interesting finding of our study is the association between neuropsychiatric damage and rs1463335 of the MIR1279 gene. MicroRNAs (miRNAs) are small non-coding RNA that play important functions in cell differentiation and development, cell cycle regulation and apoptosis. Emerging evidence suggests that miRNAs have various essential roles in the normal brain and that abnormal miRNA expression contributes to neurological and psychiatric diseases such as fronto-temporal dementia, Alzheimer's disease, Parkinson's disease, multiple sclerosis, major depression and stroke [40,41].

Interestingly, it was reported that MIR1279 presents target sites among paralogous genes of the human tyrosine family and recognizes five target miRNAs, including PTPN12 miRNA [42]. Notably, the PTPN22 gene, which is associated with SLE and multiple sclerosis susceptibility, belongs to the same family of PTPN12 [43,44]. Thus, we can speculate that this gene could be involved in neuroinflammation.

Finally, our study also describes an association between GC-related damage and the rs9469003 of HCP5 gene. HCP5 gene (major histocompatibility complex P5), located in HLA region, is expressed primarily in immune cells; thus, it could potentially play a role in autoimmune response [45]. Polymorphisms in HCP5 gene were previously described as associated with different types of severe drug reactions, such as Steven–Johnson Syndrome and toxic epidermal necrolysis [46,47]. According to this association with drug toxicity, we evaluated the two HCP5 SNPs in relation to the development of chronic GC side effects: we found only a weak association with the rs9469003.

Certainly, our study shows some limitations. The inclusion of participants of different ethnicity is needed to further investigate the role of these genetic polymorphisms in the development of chronic damage. Another limitation of our study is the relatively small number of subjects with chronic damage in a specific SDI domain, such as renal, and we lacked data regarding cumulative prednisolone dose, which is an important risk factor for the development of organ damage. Finally, although we have evaluated several SNPs, the contribution of other genetic variants should be addressed.

However, it should be underlined that this is a monocentric cohort of patients of the same ethnicity that was strictly followed and thus was well-characterized by clinical/laboratory findings and treated according to the same therapeutic approach.

In conclusion, our study showed the association of TNFSF4 with MIR1279 polymorphisms with, respectively, irreversible renal damage and the development of neuropsychiatric damage.

Our results appear promising and possibly useful in identifying patients more prone to developing specific chronic damage. These data should be considered preliminary, and a replication study in larger cohorts is strongly recommended.

**Author Contributions:** All authors made substantial contributions to the conception or design of the work and acquisition of the data. All authors contributed to the critical review and revision of the manuscript and approved the final version. F.C. (Fulvia Ceccarelli), G.O., F.N., L.P., C.P. (Carlo Perricone), C.P. (Carmelo Pirone), C.C., F.R.S., P.B. and C.A. acquired and analyzed the data, analyzed the results, and prepared the manuscript; F.C. (Fulvia Ceccarelli), C.P. (Carmelo Pirone), G.O. and C.C. performed the final data analyses; G.O. and F.C. (Fulvia Ceccarelli) wrote the manuscript; F.C. (Fabrizio Conti) supervised the study. All authors have read and agreed to the published version of the manuscript.

**Funding:** This research received no external funding.

**Institutional Review Board Statement:** The Policlinico Umberto I Ethical Committee approved the study. Number: 3908.

**Informed Consent Statement:** Informed written consent was obtained from all subjects involved in the study.

**Data Availability Statement:** The datasets generated during and analyzed during the current study are available from the corresponding author on reasonable request.

**Acknowledgments:** This work has been supported by Fondazione Umberto Di Mario ONLUS.

**Conflicts of Interest:** The authors declare no conflict of interest.

## References

1. Ruiz-Irastorza, G.; Khamashta, M.A.; Castellino, G.; Hughes, G.R. Systemic lupus erythematosus. *Lancet* **2001**, *357*, 1027–1032. [CrossRef]
2. Nived, O.; Jönsen, A.; Bengtsson, A.A.; Bengtsson, C.; Sturfelt, G. High predictive value of the Systemic Lupus International Collaborating Clinics/American College of Rheumatology damage index for survival in systemic lupus erythematosus. *J. Rheumatol.* **2002**, *29*, 1398–1400. [PubMed]
3. Gladman, D.; Ginzler, E.; Goldsmith, C.; Fortin, P.; Liang, M.; Urowitz, M.; Bacon, P.; Bombardieri, S.; Hanly, J.; Hay, E.; et al. The development and initial validation of the Systemic Lupus International Collaborating Clinics/American College of Rheumatology damage index for systemic lupus erythematosus. *Arthritis Rheum.* **1996**, *39*, 363–369. [CrossRef] [PubMed]

4.   Ghazali, W.S.W.; Daud, S.M.M.; Mohammad, N.; Wong, K.K. Slicc damage index score in systemic lupus erythematosus patients and its associated factors. *Medicine* **2018**, *97*, e12787. [CrossRef]

5.   Apostolopoulos, D.; Kandane-Rathnayake, R.; Louthrenoo, W.; Luo, S.F.; Wu, Y.J.; Lateef, A.; Golder, V.; Sockalingam, S.; Navarra, S.T.V.; Zamora, L.; et al. Factors associated with damage accrual in patients with systemic lupus erythematosus with no clinical or serological disease activity: A multicentre cohort study. *Lancet Rheumatol.* **2020**, *2*, e24–e30. [CrossRef]

6.   Conti, F.; Ceccarelli, F.; Perricone, C.; Leccese, I.; Massaro, L.; Pacucci, V.A.; Truglia, S.; Miranda, F.; Spinelli, F.R.; Alessandri, C.; et al. The chronic damage in systemic lupus erythematosus is driven by flares, glucocorticoids and antiphospholipid antibodies: Results from a monocentric cohort. *Lupus* **2016**, *25*, 719–726, Epub 27 January 2016. Erratum in *Lupus* **2017**, *26*, 1012. [CrossRef]

7.   Stoll, T.; Seifert, B.; Isenberg, D.A. SLICC/ACR Damage Index is valid, and renal and pulmonary organ scores are predictors of severe outcome in patients with systemic lupus erythematosus. *Br. J. Rheumatol.* **1996**, *35*, 248–254. [CrossRef]

8.   Ruiz-Irastorza, G.; Egurbide, M.V.; Martinez-Berriotxoa, A.; Ugalde, J.; Aguirre, C. Antiphospholipid antibodies predict early damage in patients with systemic lupus erythematosus. *Lupus* **2004**, *13*, 900–905. [CrossRef] [PubMed]

9.   Barber, M.R.W.; Johnson, S.R.; Gladman, D.D.; Clarke, A.E.; Bruce, I.N. Evolving concepts in systemic lupus erythematosus damage assessment. *Nat. Rev. Rheumatol.* **2021**, *17*, 307–308. [CrossRef]

10.  Rahman, P.; Gladman, D.D.; Urowitz, M.B.; Hallett, D.; Tam, L.S. Early damage as measured by the SLICC/ACR damage index is a predictor of mortality in systemic lupus erythematosus. *Lupus* **2001**, *10*, 93–96. [CrossRef]

11.  Sutton, E.J.; Davidson, J.E.; Bruce, I.N. The systemic lupus international collaborating clinics [SLICC] damage index: A systematic literature review. *Semin. Arthritis Rheum.* **2013**, *43*, 352–361. [CrossRef] [PubMed]

12.  Petri, M.; Purvey, S.; Fang, H.; Magder, L.S. Predictors of organ damage in systemic lupus erythematosus: The Hopkins Lupus Cohort. *Arthritis Rheum.* **2012**, *64*, 4021–4028. [CrossRef] [PubMed]

13.  Ceccarelli, F.; Sciandrone, M.; Perricone, C.; Galvan, G.; Morelli, F.; Vicente, L.N.; Leccese, I.; Massaro, L.; Cipriano, E.; Spinelli, F.R.; et al. Prediction of chronic damage in systemic lupus erythematosus by using machine-learning models. *PLoS ONE* **2017**, *12*, e0174200. [CrossRef] [PubMed]

14.  Ceccarelli, F.; Olivieri, G.; Sortino, A.; Dominici, L.; Arefayne, F.; Celia, A.I.; Cipriano, E.; Garufi, C.; Lapucci, M.; Mancuso, S.; et al. Comprehensive disease control in systemic lupus erythematosus. *Semin. Arthritis Rheum.* **2021**, *51*, 404–408. [CrossRef]

15.  Kwon, Y.C.; Chun, S.; Kim, K.; Mak, A. Update on the Genetics of Systemic Lupus Erythematosus: Genome-Wide Association Studies and Beyond. *Cells* **2019**, *8*, 1180. [CrossRef] [PubMed]

16.  Dai, C.; Deng, Y.; Quinlan, A.; Gaskin, F.; Tsao, B.P.; Fu, S.M. Genetics of systemic lupus erythematosus: Immune responses and end organ resistance to damage. *Curr. Opin. Immunol.* **2014**, *31*, 87–96. [CrossRef]

17.  Sanchez, E.; Nadig, A.; Richardson, B.C.; Freedman, B.I.; Kaufman, K.M.; Kelly, J.A.; Niewold, T.B.; Kamen, D.L.; Gilkeson, G.S.; Ziegler, J.T.; et al. Phenotypic associations of genetic susceptibility loci in systemic lupus erythematosus. *Ann. Rheum. Dis.* **2011**, *70*, 1752–1757. [CrossRef]

18.  Ceccarelli, F.; Perricone, C.; Borgiani, P.; Ciccacci, C.; Rufini, S.; Cipriano, E.; Alessandri, C.; Spinelli, F.R.; Sili Scavalli, A.; Novelli, G.; et al. Genetic Factors in Systemic Lupus Erythematosus: Contribution to Disease Phenotype. *J. Immunol. Res.* **2015**, *2015*, 745647. [CrossRef]

19.  Svenungsson, E.; Gustafsson, J.; Leonard, D.; Sandling, J.; Gunnarsson, I.; Nordmark, G.; Jönsen, A.; Bengtsson, A.A.; Sturfelt, G.; Rantapää-Dahlqvist, S.; et al. A STAT4 risk allele is associated with ischaemic cerebrovascular events and anti-phospholipid antibodies in systemic lupus erythematosus. *Ann. Rheum. Dis.* **2010**, *69*, 834–840. [CrossRef]

20.  Bolin, K.; Sandling, J.K.; Zickert, A.; Jönsen, A.; Sjöwall, C.; Svenungsson, E.; Bengtsson, A.A.; Eloranta, M.L.; Rönnblom, L.; Syvänen, A.C.; et al. Association of STAT4 polymorphism with severe renal insufficiency in lupus nephritis. *PLoS ONE* **2013**, *8*, e84450. [CrossRef]

21.  Reid, S.; Alexsson, A.; Frodlund, M.; Morris, D.; Sandling, J.K.; Bolin, K.; Svenungsson, E.; Jönsen, A.; Bengtsson, C.; Gunnarsson, I.; et al. High genetic risk score is associated with early disease onset, damage accrual and decreased survival in systemic lupus erythematosus. *Ann. Rheum. Dis.* **2020**, *79*, 363–369. [CrossRef] [PubMed]

22.  Hochberg, M.C. Updating the American College of Rheumatology revised criteria for the classification of systemic lupus erythematosus. *Arthritis Rheum.* **1997**, *40*, 1725. [CrossRef] [PubMed]

23.  Gladman, D.D.; Ibañez, D.; Urowitz, M.B. Systemic lupus erythematosus disease activity index 2000. *J. Rheumatol.* **2002**, *29*, 288–291. [PubMed]

24.  Ruiz-Arruza, I.; Ugarte, A.; Cabezas-Rodriguez, I.; Medina, J.A.; Moran, M.A.; Ruiz-Irastorza, G. Glucocorticoids and irreversible damage in patients with systemic lupus erythematosus. *Rheumatology* **2014**, *53*, 1470–1476. [CrossRef] [PubMed]

25.  Apostolopoulos, D.; Kandane-Rathnayake, R.; Raghunath, S.; Hoi, A.; Nikpour, M.; Morand, E.F. Independent association of glucocorticoids with damage accrual in SLE. *Lupus Sci. Med.* **2016**, *3*, e000157. [CrossRef]

26.  Rivest, C.; Lew, R.A.; Welsing, P.M.; Sangha, O.; Wright, E.A.; Roberts, W.N.; Liang, M.H.; Karlson, E.W. Association between clinical factors, socioeconomic status, and organ damage in recent onset systemic lupus erythematosus. *J. Rheumatol.* **2000**, *27*, 680–684.

27.  Bruce, I.N.; O'Keeffe, A.G.; Farewell, V.; Hanly, J.G.; Manzi, S.; Su, L.; Gladman, D.D.; Bae, S.C.; Sanchez-Guerrero, J.; Romero-Diaz, J.; et al. Factors associated with damage accrual in patients with systemic lupus erythematosus: Results from the Systemic Lupus International Collaborating Clinics [SLICC] Inception Cohort. *Ann. Rheum. Dis.* **2015**, *74*, 1706–1713. [CrossRef]

28. Alarcón, G.S.; Roseman, J.M.; McGwin, G., Jr.; Uribe, A.; Bastian, H.M.; Fessler, B.J.; Baethge, B.A.; Friedman, A.W.; Reveille, J.D.; LUMINA Study Group. Systemic lupus erythematosus in three ethnic groups. XX. Damage as a predictor of further damage. *Rheumatology* **2004**, *43*, 202–205. [CrossRef]

29. Gladman, D.D.; Urowitz, M.B.; Rahman, P.; Ibañez, D.; Tam, L.S. Accrual of organ damage over time in patients with systemic lupus erythematosus. *J. Rheumatol.* **2003**, *30*, 1955–1959.

30. Zhou, X.J.; Cheng, F.J.; Qi, Y.Y.; Zhao, M.H.; Zhang, H. A replication study from Chinese supports association between lupus-risk allele in TNFSF4 and renal disorder. *Biomed. Res. Int.* **2013**, *2013*, 597921. [CrossRef]

31. Petri, M. Musculoskeletal complications of systemic lupus erythematosus in the Hopkins Lupus Cohort: An update. *Arthritis Care Res.* **1995**, *8*, 137–145. [CrossRef] [PubMed]

32. Nossent, J.C. SLICC/ACR Damage Index in Afro-Caribbean patients with systemic lupus erythematosus: Changes in and relationship to disease activity, corticosteroid therapy, and prognosis. *J. Rheumatol.* **1998**, *25*, 654–659. [PubMed]

33. Zonana-Nacach, A.; Camargo-Coronel, A.; Yáñez, P.; de Lourdes Sánchez, M.; Jímenez-Balderas, F.J.; Aceves-Avila, J.; Martínez-Osuna, P.; Fuentes, J.; Medina, F. Measurement of damage in 210 Mexican patients with systemic lupus erythematosus: Relationship with disease duration. *Lupus* **1998**, *7*, 119–123. [CrossRef]

34. Ceccarelli, F.; Perricone, C.; Colasanti, T.; Massaro, L.; Cipriano, E.; Pendolino, M.; Natalucci, F.; Mancini, R.; Spinelli, F.R.; Valesini, G.; et al. Anti-carbamylated protein antibodies as a new biomarker of erosive joint damage in systemic lupus erythematosus. *Arthritis Res. Ther.* **2018**, *20*, 126. [CrossRef]

35. Song, K.; Liu, L.; Zhang, X.; Chen, X. An update on genetic susceptibility in lupus nephritis. *Clin. Immunol.* **2020**, *210*, 108272. [CrossRef] [PubMed]

36. Wang, J.M.; Yuan, Z.C.; Huang, A.F.; Xu, W.D. Association of TNFSF4 rs1234315, rs2205960 polymorphisms and systemic lupus erythematosus susceptibility: A meta-analysis. *Lupus* **2019**, *28*, 1197–1204. [CrossRef]

37. Aten, J.; Roos, A.; Claessen, N.; Schilder-Tol, E.J.M.; Ten Berge, I.J.M.; Weening, J.J. Strong and selective glomerular localization of CD134 ligand and TNF receptor-1 in proliferative lupus nephritis. *J. Am. Soc. Nephrol.* **2000**, *11*, 1426–1438. [CrossRef]

38. Cortini, A.; Ellinghaus, U.; Malik, T.H.; Cunninghame Graham, D.S.; Botto, M.; Vyse, T.J. B cell OX40L supports T follicular helper cell development and contributes to SLE pathogenesis. *Ann. Rheum. Dis.* **2017**, *76*, 2095–2103. [CrossRef]

39. Sitrin, J.; Suto, E.; Wuster, A.; Eastham-Anderson, J.; Kim, J.M.; Austin, C.D.; Lee, W.P.; Behrens, T.W. The Ox40/Ox40 Ligand Pathway Promotes Pathogenic Th Cell Responses, Plasmablast Accumulation, and Lupus Nephritis in NZB/W F1 Mice. *J. Immunol.* **2017**, *199*, 1238–1249. [CrossRef]

40. Bhalala, O.G.; Srikanth, M.; Kessler, J.A. The emerging roles of microRNAs in CNS injuries. *Nat. Rev. Neurol.* **2013**, *9*, 328–339. [CrossRef]

41. Wang, H. MicroRNAs, Multiple Sclerosis, and Depression. *Int. J. Mol. Sci.* **2021**, *22*, 7802. [CrossRef] [PubMed]

42. Ivashchenko, A.T.; Issabekova, A.S.; Berillo, O.A. miR-1279, miR-548j, miR-548m, and miR-548d-5p binding sites in CDSs of paralogous and orthologous PTPN12, MSH6, and ZEB1 Genes. *Biomed. Res. Int.* **2013**, *2013*, 902467. [CrossRef] [PubMed]

43. Rawlings, D.J.; Dai, X.; Buckner, J.H. The role of PTPN22 risk variant in the development of autoimmunity: Finding common ground between mouse and human. *J. Immunol.* **2015**, *194*, 2977–2984. [CrossRef] [PubMed]

44. Zheng, J.; Ibrahim, S.; Petersen, F.; Yu, X. Meta-analysis reveals an association of PTPN22 C1858T with autoimmune diseases, which depends on the localization of the affected tissue. *Genes Immun.* **2012**, *13*, 641–652. [CrossRef] [PubMed]

45. Kulski, J.K. Long Noncoding RNA HCP5, a Hybrid HLA Class I Endogenous Retroviral Gene: Structure, Expression, and Disease Associations. *Cells* **2019**, *8*, 480. [CrossRef]

46. Génin, E.; Schumacher, M.; Roujeau, J.C.; Naldi, L.; Liss, Y.; Kazma, R.; Sekula, P.; Hovnanian, A.; Mockenhaupt, M. Genome-wide association study of Stevens-Johnson Syndrome and Toxic Epidermal Necrolysis in Europe. *Orphanet J. Rare Dis.* **2011**, *6*, 52. [CrossRef] [PubMed]

47. Tohkin, M.; Kaniwa, N.; Saito, Y.; Sugiyama, E.; Kurose, K.; Nishikawa, J.; Hasegawa, R.; Aihara, M.; Matsunaga, K.; Abe, M.; et al. A whole-genome association study of major determinants for allopurinol-related Stevens-Johnson syndrome and toxic epidermal necrolysis in Japanese patients. *Pharm. J.* **2013**, *13*, 60–69. [CrossRef]

Journal of
*Clinical Medicine*

*Review*

# Long-Term Outcome in Systemic Lupus Erythematosus; Knowledge from Population-Based Cohorts

Sigrid Reppe Moe [1,*], Hilde Haukeland [1,2], Øyvind Molberg [1,3] and Karoline Lerang [1]

[1] Department of Rheumatology, Oslo University Hospital, 0424 Oslo, Norway; hihauk@ous-hf.no (H.H.); oyvind.molberg@medisin.uio.no (Ø.M.); klerang@ous-hf.no (K.L.)
[2] Department of Rheumatology, Martina Hansens Hospital, 1346 Gjettum, Norway
[3] Institute of Clinical Medicine, University of Oslo, 0424 Oslo, Norway
[*] Correspondence: sigmoe@ous-hf.no

**Abstract:** Background: Accurate knowledge of outcomes in Systemic Lupus Erythematosus (SLE) is crucial to understanding the true burden of the disease. The main objective of this systematic review was to gather all population-based studies on mortality, end-stage renal disease (ESRD) and cancer in SLE. Method: We performed a systematic literature search in two electronic databases (MEDLINE and Embase) to identify all population-based articles on SLE and survival, mortality, ESRD and cancer. The SLE diagnosis had to be verified. We used the Preferred Reporting Items for Systematic Reviews and Meta-Analyses guidelines (PRISMA). Results: We included 40/1041 articles on mortality (27), ESRD (11) and cancer (3), of which six were defined as inception studies. In the total SLE cohort, the standardized mortality ratio ranged from 1.9 to 4.6. Cardiovascular disease was the most frequent cause of death in studies with follow-up times over 15 years. SLE progressed to ESRD in 5–11% of all SLE patients. There are no data supporting increased cancer incidence from population-based inception cohorts. Conclusion: There is a need for more population-based studies on outcomes of SLE, especially inception studies, with the use of control groups and follow-up times over 15 years.

**Keywords:** epidemiology; Systemic Lupus Erythematosus; outcome; mortality; survival; end-stage renal disease; cancer

**Citation:** Reppe Moe, S.; Haukeland, H.; Molberg, Ø.; Lerang, K. Long-Term Outcome in Systemic Lupus Erythematosus; Knowledge from Population-Based Cohorts. *J. Clin. Med.* **2021**, *10*, 4306. https://doi.org/10.3390/jcm10194306

Academic Editors: Christopher Sjöwall, Ioannis Parodis and Angelo Valerio Marzano

Received: 2 August 2021
Accepted: 13 September 2021
Published: 22 September 2021

**Publisher's Note:** MDPI stays neutral with regard to jurisdictional claims in published maps and institutional affiliations.

## 1. Introduction

Systemic Lupus Erythematosus (SLE) is a rare systemic and chronic disease often referred to as the prototype of autoimmune rheumatic diseases because of the varied spectrum of clinical manifestations and diversity of phenotypes. The etiology of SLE is believed to be multifactorial, and both genetic predisposition and environmental triggers are most likely involved [1]. The incidence, severity and phenotypic expression of the disease differ between ethnic groups, gender and age at disease onset. The annual incidence of SLE varies from 0.3 to 23.3/100,000, and the prevalence varies from 0 to 241/100,000 [1]. The variations are highly dependent on the method of retrieval and the definition of SLE diagnosis.

Several aspects of SLE make it one of the most challenging conditions to study at the population level. First, no diagnostic criteria for SLE exist and the diagnosis is based on the judgement of an experienced clinician. Diagnosing SLE can be challenging since SLE is a great imitator of other diseases. The symptoms of SLE overlap many other diseases that can easily be mistaken for SLE in as much as 40% [2–5] of cases.

Secondly, in many countries, SLE patients are not treated in the same hospital and/or specialization since different organs may be affected and the severity of the disease varies. Selected patient populations from tertiary hospitals tend to miss milder cases, and therefore underestimate the incidence and overestimate the severity of SLE. Thus, a closer estimate of the true frequency of clinical and laboratory SLE manifestations and outcomes is more likely from a geographically complete cohort of patients. All these aspects of the disease make it difficult and labour-intensive to collect epidemiological data. In Georgia, Lim et al. found 45,000 potential patients, screened 3142 records and found 1320 patients with a verified SLE diagnosis. In Sweden, Ingvarsson et al. screened 2461 cases and found 55 patients with a verified diagnosis, and Voss et al. in Denmark screened 980 cases to find 95 patients with a verified SLE diagnosis [4–6].

Earlier publications on SLE and epidemiology differ greatly in study-design. A good epidemiological study is highly dependent on valid data to obtain reliable results that are indicative of the total size of the problem and thus, a reliable assessment of outcome. Truly population-based research, with a verified and ascertained SLE diagnosis by chart review, is the best way to achieve the most accurate knowledge possible on this disease and its outcome measures. The use of standardized methods gives the best basis for comparison of epidemiological data across different studies and countries.

The objective of this study was to conduct a review of literature on population-based epidemiologic data on SLE and well-defined and hard outcomes; mortality, end-stage renal disease (ESRD) and cancer. The elected publications were thoroughly reviewed to ensure that they were from population-based cohorts and that the SLE diagnose was verified.

## 2. Materials and Methods

A senior medical librarian searched two electronic databases: MEDLINE (Ovid) and Embase (Ovid), from their inception to 25 June 2021, with language restricted to English. The systematic search used both controlled vocabulary (MeSH terms or EMTREE terms) and text word search in title, abstract or author keywords. The search consisted of two searches with different approaches. Search 1: Concepts for systemic lupus, SLE criteria, mortality or cancer, were combined with the Boolean operator AND. Search 2: Concepts for lupus nephritis, end stage renal disease or kidney transplantation were combined with the Boolean operator AND (Supplementary Materials S2). Both searches were restricted to population-based cohorts.

Two investigators (HH and KL or SRM and KL) independently evaluated all abstracts and titles to determine eligibility for inclusion. When necessary, the articles were reviewed in full, and, if in conflict, discussed in plenum (HH, SRM, KL). The authors also searched the reference list of included articles to find additional relevant studies.

For inclusion in this systematic review, the SLE diagnosis had to be verified by chart review. Studies on SLE were included on the relevant outcomes: mortality, overall and renal survival and risk of malignancy.

We excluded: (1) Studies that failed to validate the SLE diagnosis by chart review; (2) Studies based on administrative data; (3) Studies from tertiary centers only, if it was not specified that it was the only hospital serving the region; (4) Animal studies; (5) Meta-analysis; (6) Case reports; (7) Studies on unrelated outcomes; (8) Studies of selected SLE subsets (paediatric SLE, biopsy-proven lupus nephritis (LN), hospital inpatients); (9) Studies with fewer than 30 patients; (10) Studies on subset of relevant outcome (cardiovascular mortality).

Causes of death analyses were excluded from this review if the study reported only multiple causes of death. We defined the study period as years from start of inclusion to end of follow-up. The total SLE population was defined as all SLE patients in the given study-period. Incident SLE were defined as patients diagnosed within the study-period. Inception SLE was defined as patients diagnosed within the study-period and captured within one year of the diagnosis.

This review was carried out in accordance with the Preferred Reporting Items for Systematic Reviews and Meta-Analyses (PRISMA) guidelines [7].

## 3. Results

We screened 1041 titles/abstracts. Through the screening process, we identified 40 studies that met the criteria for inclusion, whereof 27 were for survival and mortality,

11 were for ESRD and three were for cancer (Figure 1). We found seven articles through manual search of the reference list of included articles. The case finding methodology and SLE ascertainment in all included cohorts is described in Supplementary Materials Table S1. All but three study locations included only SLE patients who fulfilled four or more of the American College of Rheumatology SLE classification criteria [8–11].

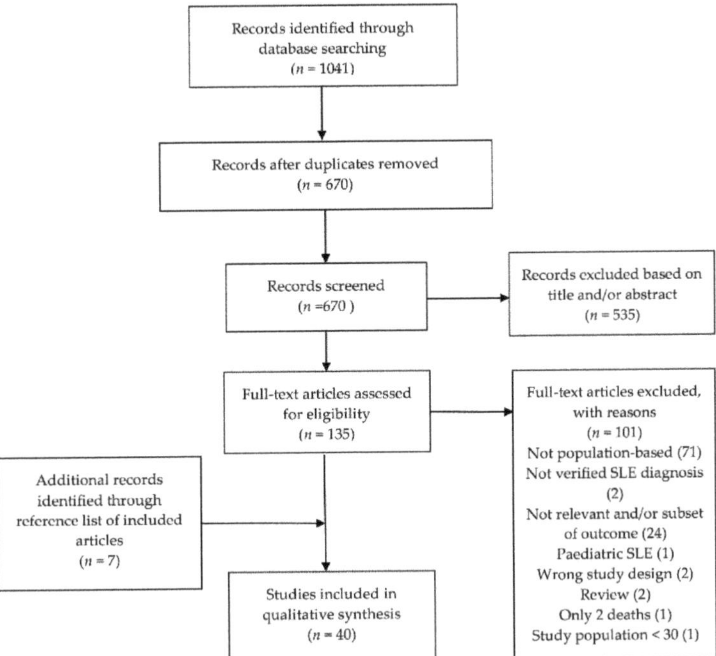


**Figure 1.** Flowchart of literature search and study inclusion. Studies identified through MEDLINE (Ovid) and Embase (Ovid) through 25 June 2021.

### 3.1. Standardized Mortality Rate and Survival

Twenty-three population-based studies reported survival with SLE, while a standardized mortality rate (SMR) was reported in 13 studies. Eighteen studies used incident patients for survival analysis, while five included all SLE patients (total). Six studies used only incident patients and seven used the total SLE population for SMR analysis (Table 1).

**Table 1.** Survival and standardized mortality rate (SMR) in Systemic Lupus Erythematosus, from population-based cohorts.

| Author; Year (Ref.) | Study Location | Study Period * | Ethnicity | Follow-Up Time | SLE Cases, $n$ | | Deaths, $n$ | | SMR, 95% CI | | | Survival (Controls)% | | | |
| --- | --- | --- | --- | --- | --- | --- | --- | --- | --- | --- | --- | --- | --- | --- | --- |
| | | | | | Total | Incident | Total | Incident | Total | Female | Male | 5 Years | 10 Years | 15 Years | 20 Years |
| **NORTH AMERICA** | | | | | | | | | | | | | | | |
| Peschken et al.; 2000 [13] | Canada | 1980–1997 | 177 Caucasian / 49 NAI | NA | 257 | 257 | NA | NA | | | | 97 / 95 | 95 / 83 | 91 / 75 | |
| Uramoto et al.; 1999 [26] | MN, USA | 1950–1979 / 1980–1992 | Mainly Caucasian | 7.2 years (μ) | 79 | 79 | NA | NA | 2.7 (1.7–4.2) | | | 75 (95) / 90 (90) | 50 (92) / 71 (90) | | |
| Naleway et al.; 2005 [24] | WI, USA | 1991–2001 | Mainly Caucasian | 5.8 years (μ) | 117 | 44 | NA | 8 | | | | 88 | 76 | | |
| Bartels et al.; 2014 [25] | WI, USA | 1991–2009 | NA | 7.7 years (μ) | 70 | 70 | | | | | | 87 (90) | 74 (81) | 59 (73) | |
| Jarukitsopa, S et al.; 2015 [27] | MN, USA | 1993–2005 | 80% white | 7.8 years (μ) | 117 | 45 | NA | 6 | 2.6 (1.0–5.6) | | | 93 | 89 | 64 | |
| Lim et al.; 2019 [32] | GA, USA | 2002–2016 | 76% black | NA | 1689 | 336 | 401 | 97 | 3.1 (2.8–3.4) | 3.1 (2.8–3.5) | 3.0 (2.3–3.9) | | | | |
| Flower et al.; 2012 [12] | Barbados | 2000–2009 | 98% African Caribbean | NA | 183 | 183 | 24 | 24 | | | | 88 | 80 | | |
| **SOUTH AMERICA** | | | | | | | | | | | | | | | |
| Lucero et al.; 2020 [29] | Argentina | 2005–2012 | 83% Mestizos | NA | 353 | NA | 32 | NA | | | | 96 | 93 | | |
| Nossent; 1992 [15] | Curaçao | 1980–1990 | All of African descent | NA | 94 | 68 | 25 | NA | | | | 60 | 46 | | |
| **ASIA** | | | | | | | | | | | | | | | |
| Iseki et al.;, 1994 [35] | Japan | 1972–1993 | NA | 4877 PY | 566 | NA | 104 | NA | | | | 89 | 78 | 72 | 69 |
| Mok et al.; 2005 [14] | Hong Kong, China | 1991–2003 | All ethnic Chinese | NA | 258 | 258 | 29 | 29 | | | | 92 | 83 | 80 | |
| Mok et al.; 2008 [33] | Hong Kong, China | 2000–2006 | Mainly Asian | NA | 442 | NA | 30 | NA | 3.9 ** | 4.0 ** | 9.6 ** | | | | |
| Yeh et al.; 2013 [34] | Taiwan | 2003–2008 | NA | NA | 6675 | 6675 | 1611 | 1611 | 11.1 (NA) | | | | | | |
| Al-Adhoubi et al.; 2021 [11] | Oman | 2006–2020 | NA | NA | 1160 | NA | 54 | NA | | | | 100 | 100 | | 99 |

**Table 1.** *Cont.*

| Author; Year (Ref.) | Study Location | Study Period * | Ethnicity | Follow-Up Time | SLE Cases, *n* | | Deaths, *n* | | SMR, 95% CI | | | Survival (Controls)% | | | |
|---|---|---|---|---|---|---|---|---|---|---|---|---|---|---|---|
| | | | | | Total | Incident | Total | Incident | Total | Female | Male | 5 Years | 10 Years | 15 Years | 20 Years |
| **EUROPE** | | | | | | | | | | | | | | | |
| Gudmundsson et al.; 1990 [17] | Iceland | 1975–1988 | NA | NA | 76 | 76 | 17 | 17 | 3.4 (2.0–5.4) | | | 84 | 78 | | |
| Jacobsen et al., 1999 [30] | Denmark | 1975–1995 | NA | 4185 PY | 513 | NA | 122 | NA | 4.6 (3.8–5.5) | 4.7 (3.9–5.8) | 4.0 (3.8–5.5) | 91 | 76 | | 53 |
| Nossent et al.; 2001 [18] | Norway, north | 1978–1999 | >96% Caucasian | NA | 105 | 83 | 18 | 11 | | | | 92 | 75 | | |
| Eilertsen et al.; 2009 [19] | Norway, north | 1978–1995 | 98.8% Caucasian | NA | 81 | 81 | 25 | 25 | 2.0 (1.4–2.8) | 2.1 (1.5–3.1) | 1.5 (0.6–3.5) | 91(98) | 81 (96) | | |
| | | 1996–2007 | 98.3% Caucasian | NA | 58 | 58 | 5 | 5 | | | | 96 (98) | 92 (96) | | |
| Lerang et al.; 2014 [20] | Norway, Oslo | 1999–2009 | 84% of European descent | 2665/812 PY | 325 | 129 | 50 | 7 | 3.0 (2–3.8) | 2,7 (2,0–3.7) | 4.6 (2.3–8.1) | 95 (99) | 90 (96) | | |
| Jonsson et al.; 1989 [28] | Sweden, Lund | 1981–1986 | NS | 342 PY | 86 | 38 | 9 | NA | | | | 97 (97) | | | |
| Ståhl-Hallengren et al.; 2000 [22] | Sweden, Lund | 1981–1991 | NA | NA | 162 | 162 | 17 | 17 | | | | 93 (98) | 83 (96) | | |
| Ingvarsson et al.; 2019 [10] | Sweden, Lund | 1981–2014 | 98.3% Caucasian | 3053 PY | 174 | 174 | 60 | 60 | 2.5 (1.9–3.3) | 2.7 (2.0–3.6) | 1.9 (1.0–3.4) | 91 (97) | 85 (91) | 73 (86) | 62 (77) |
| Alamanos et al.; 2003 [16] | Greece | 1982–2001 | NA | NA | 178 | 178 | 12 | 12 | 1.3 (NA) | | | 97 | 90 | | |
| Alonso et al.; 2011 [21] | Spain | 1987–2006 | NA | 7.8 years (μ) | 150 | 150 | 19 | 19 | | | | 94 (97) | 87 (94) | 80 (89) | |
| Laustrup et al.; 2009 [31] | Denmark, Funen | 1995–2003 | 94% Caucasian | 767 PY | 138 | NA | 15 | NA | 1.9 (1.0–3.0) | 1.8 (0.9–3.2) | 2.1 (0.4–6.2) | | | | |
| Voss et al.; 2013 [9] | Denmark, Funen | 1995–2010 | 94% Caucasian | 2052 PY | 215 | NA | 38 | NA | 2.3 (1.6–3.2) | 1.9 (1.3–2.9) | 3.2 (1.5–6.3) | 94 | 73 | | |
| Pamuk et al.; 2015 [23] | Turkey | 2003–2014 | NA | 48 months (mdn) | 331 | 331 | 17 | 17 | | | | 95 | 90 | | |

SLE: Systemic Lupus Erythematosus, SMR: Standardized Mortality Rate, PY: Patient years, μ: Mean, mdn: Median, CI: Confidence Interval, NA: Not avaliable, GA: Georgia, WI: Wisconsin, MN: Minnesota, NAI: Native American Indians. * Years from start of inclusion to end of follow-up, ** Calculated. Inception SLE : Incident SLE cases captured within one year from diagnosis. Total SLE : All SLE cases within the given study-period. Incident SLE : SLE cases diagnosed within the study-period.

The ten-year survival in incident cohorts ranges from 46% in Curacao to 92% in northern Norway, and from 90 to 92% in Europe and 76 to 89% in North America [10,12–27]. Five and ten-year survival differed in incident cohorts with patient inclusion before and after 1990 (five-year survival 80% versus 92% and ten-year survival 63% versus 88%) [10,12–28]. For all studies with patient inclusion starting after 1990, the five-year survival was 90% or more, except for Barbados and Wisconsin [9,11,14,20,23,27,29]. In studies on total SLE cohorts, the SMR ranges from 1.9 to 4.6 [9,19,20,26,27,30–32]. For female SLE patients, the SMR ranges from 1.8 to 4.7, while in male patients the SMR ranges from 1.5 to 4.6 [9,10,19,20,30,31,33]. There was no significant difference between the two groups. Among the incident SLE patients the SMR varied from 1.3 to 11.1, depending on follow-up time (one to 33 years) [10,16,17,33,34]. Only one incident study reported 25-year survival with SLE (60% survival versus 73% in the general population) [10].

### 3.2. The Main Causes of Death in Systemic Lupus Erythematosus

An average of 41% of patients in the studies from Asia died of infections, compared to an average of 12% in studies from Europe (Table 2) [9,10,14,16,17,20,21,23,30,33,35]. Renal failure was the underlying cause of death in about 17% (median) of SLE patients, except for a much higher frequency in Barbados (46%) [12,16,21,23,25,27,30,33,35]. From the article with the shortest follow-up time versus the longest, the causes of death varied from 60% infections and 6% cardio- and cerebrovascular disease (CVD) in Hong Kong [33] to 15% infections and 59% CVD in Sweden [10]. CVD was the most frequent cause of death in the two study locations with population-based cohorts over time [9,10].

### 3.3. End Stage Renal Disase

Within the primary studies reviewed, ESRD developed in 5–11% of the total SLE patients [35–37], of which 5–6% were in a Scandinavian population (Table 4) [36,38]. The incidence rate of ESRD varied from 2.3 to 11.1/1000 patient years in incident patient populations, depending on the population studied (Table 4) [38–40].

### 3.4. Cancer

We found only three studies on cancer in population-based cohorts, from three different countries. Only the study from Sweden was an inception study (Table 3) [41].

**Table 2.** The main causes of death in Systemic Lupus Erythematosus, from population-based studies.

| Author; Year (Ref.) | Study Location | Study Period * | Follow-Up Time | Deaths/SLE Cases; *n*/*N* | Cause of Death, % | | | | | |
|---|---|---|---|---|---|---|---|---|---|---|
| | | | | | Active SLE | CVD | Infections | PD | Malignancy | Renal Failure |
| | **NORTH AMERICA** | | | | | | | | | |
| Bartels et al.; 2014 [25] | WI, USA | 1991–2009 | 540 patient years | 19/70 | | 32% | 16% | | 13% | 13% |
| Jarukitsopa et al.; 2015 [27] | MN, USA | 1993–2005 | 7.8 years (mean) | 6/45 | | | 67% | | | 33% |
| Flower et al.; 2012 [12] | Barbados | 2000–2009 | NA | 24/181 | | | 42% [d] | 8% | | 46% [b] |
| | **SOUTH AMERICA** | | | | | | | | | |
| Lucero et al.; 2020 [29] | Argentina | 2005–2012 | NA | 32/353 | | | 44% | | | |
| | **ASIA** | | | | | | | | | |
| Mok et al.; 2005 [14] | China | 1991–2003 | NA | 29/258 | | 31% [c] | 55% | | 3% | |
| Iseki et al.; 1994 [35] | Japan | 1972–1993 | 4877 patient years | 104/566 | | 15% | 24% | | | 12% |
| Mok et al.; 2008 [33] | China | 2000–2006 | NA | 30/422 | | 6% | 60% | 3% | | 7% |
| | **EUROPE** | | | | | | | | | |
| Jacobsen et al.; 1999 [30] | Denmark | 1975–1995 | 4185 patient years | 122/513 | 19% | 24% | 20% | | 7% | 10% |
| Voss et al.; 2013 [a] [9] | Denmark | 1995–2010 | 2052 patient years | 38/214 | 8% | 32% | 8% | 16% | 13% | |
| Gudmundsson et al.; 1990 [17] | Iceland | 1975–1988 | NA | 17/76 | 35% [b] | 29% | 6% | | | |
| Ingvarsson et al.; 2019 [a] [10] | Sweden | 1981–2014 | 3053 patient years | 60/174 | 7% | 59% | 15% | 5% | 13% | |
| Alamanos et al.; 2003 [16] | Greece | 1982–2001 | NA | 12/178 | | | 17% | | | 17% |
| Alonso et al.; 2011 [21] | Spain | 1987–2006 | 7.8 years (mean) | 19/150 | | 21% | 21% | | 26% | 11% |
| Lerang et al.; 2014 [20] | Norway | 1999–2009 | 2665 patient years | 50/325 | 12% | 16% | 6% | | 20% | |
| Pamuk et al.; 2015 [23] | Turkey | 2003–2014 | 48 months (mdn) | 17/331 | | 24% | 23% [d] | | 12% | 12% |

CVD: Cardio- and cerebrovascular Disease, PD: Pulmonary Disease, mdn: Median, NA: Not available, WI: Wisconsin, MN: Minnesota. * Years from start of inclusion to end of follow-up [a] Last articles of multiple over time, [b] Death attributed to Lupus Nephritis, [c] Including hemorrhagic stroke, [d] Sepsis. Total SLE : All SLE cases within the given study-period. Incident SLE : SLE cases diagnosed within the study-period.

**Table 3.** Cancer risk in Systemic Lupus Erythematosus relative to the general population, from population-based studies.

| Author; Year (Ref.) | Study Location | Study Period * | Follow-Up Time, Mean | Age, Years | SLE Cases, $n$ | | SLE Cases with Malignancies, $n$ | Malignancies O/E Ratio (95% CI) | Subgroups of Malignancy **, O/E Ratio (95% CI)/($p$) *** |
|---|---|---|---|---|---|---|---|---|---|
| | | | | | Total | Incident | | | |
| **EUROPE** | | | | | | | | | |
| Ragnarsson et al.; 2003 [45] | Iceland | 1957–2001 | 12.8 years | All | 238 | NA | 27 | O/E 1.4 (0.9–1.9) | Skin SCC 6.4 (1.3–18.5)<br>Lymphoma 5.5 (0.6–19.6)<br>Lung 1.7 (0.4–5.0)<br>Breast 1.6 (0.7–3.2)<br>Prostate 1.2 (0.0–6.2) |
| Nived et al.; 2001 [41] | Sweden | 1981–1998 | 9.4 years | >15 | 116 | 116 | 11 | SMR 1.2 ****<br>Male 2.2 (0.6–5.7)<br>Female 1.0 (0.4–2.1) | NHL 11.6 (1.4–42)<br>Prostate 6.4 (1.3–18.7)<br>Lung 5.6 (0.7–20.1) |
| **ASIA** | | | | | | | | | |
| Chen et al.; 2010 [46] [a] | Taiwan | 1996–2007 | 6.1 years | All | 11,763 | 11,763 | 259 | SIR 1.8 (1.7–1.8) | NHL 7.3 (7.0–7.6)<br>Vagina/vulva 4.8 (4.2–5.3)<br>Nasopharynx, siunus, ears 4.2 (3.9–4.5)<br>Leukemia 2.6 (2.5–2.8)<br>Skin 1.7 (1.6–1.8)<br>Breast 1.6 (1.5–1.6)<br>Cervix 1.4 (1.3–1.5)<br>Lung/mediastinum 1.2 (1.2–1.3)<br>Prostate 0.8 (0.7–0.9) |

SLE: Systemic Lupus Erythematosus, O/E ratio: observed/expected events, SMR: Standardized Morbidity Rate, SIR: Standardized Incidence Ratio, RR: Relative Risk, NA: Not available, NHL: Non-Hodgkin Lymphoma, SCC: Squamous Cell Carcinoma. * Years from start of inclusion to end of follow-up, ** Not all results included, *** $p < 0.001$ indicates statistical significance, **** Calculated. [a] Main article on cohort, sub-analysis not included. Inception SLE : Incident SLE cases captured within one year from diagnosis. Total SLE : All SLE cases within the given study-period. Incident SLE : SLE cases diagnosed within the study-period.

**Table 4.** Risk of End Stage Renal Disease in Systemic Lupus Erythematosus, from population-based studies.

| Author, Year (Ref.) | Study Location | Study Period * | Follow-Up Time | SLE Cases, *n* | | LN,% | Age, Years | Ethnicity | ESRD Development | | |
|---|---|---|---|---|---|---|---|---|---|---|---|
| | | | | Total | Incident | | | | Total SLE | Incident SLE | LN |
| **NORTH AMERICA** | | | | | | | | | | | |
| Somers et al., 2014 [37] | MI, USA | 2002–2004 | NA | 2129 | 399 | 32 | All | 56% black patients | Total 10.8%; black 15.3%, white 4.5% | | |
| Plantinga et al., 2016 [40] | GA, USA | 2002–2004 | 2603 patient years | 344 | 344 | NA | All | 76.1% black patients | | Total 11.1; black 13.8, white 3.3/1000 patient years | |
| **ASIA** | | | | | | | | | | | |
| Iseki et al., 1994 [35] | Japan | 1972–1991 | 4788 patient years | 566 | NA | 49 | All | NA | 9% | | |
| Yu et al., 2016 [39] | Taiwan | 2000–2008 | NA | 1196 | 1196 | NA | All | NA | | 6.1/1000 patient years | |
| Lin et al., 2017 [42] | Taiwan | 2000–2011 | 8.1 years (mean) | 7326 | 7326 | NA | All | NA | | 4.3% | |
| Lin et al., 2013 [43] | Taiwan | 2003–2008 | NA | 4130 | 4130 | NA | All | NA | | 2.5% | |
| **EUROPE** | | | | | | | | | | | |
| Jacobsen et al., 1998 [36] | Denmark | 1975–1995 | 8.2 years (mean) | 513 | NA | 42 | All | NA | 5% | | |
| Eilertsen et al., 2011 [44] | Norway | 1978–1995 | NA | 62 | 62 | 32 | ≥16 | 98% Caucasian | | 10 years renal survival: 100% | |
| | | 1996–2007 | NA | 87 | 87 | 18 | ≥16 | 99% Caucasian | | 10 years renal survival 88.5% | |
| Jonsson et al., 1989 [28] | Sweden | 1981–1986 | NA | 86 | 38 | 30 | ≥15 | NA | | | 3.8% |
| Gergianaki et al., 2017 [3] | Greece | 1999–2013 | 7.2 years (mean) | 750 | NA | 13 | ≥15 | 97% Greek | | | 4.4% |
| Reppe Moe et al., 2019 [38] | Norway | 1999–2017 | 18.4/10.6 years (mean) | 325 | 129 | 30 | ≥16 | 84% of European descent | 6% | 2,3/1000 patient years | |

SLE: Systemic Lupus Erythematosus, LN: Lupus Nephritis, ESRD: End Stage Renal Disease, GA: Georgia, MI: Minnesota, NA: Not available. * years from start of inclusion to end of follow-up. Inception SLE : Incident SLE cases captured within one year from diagnosis. Total SLE : All SLE cases within the given study-period. Incident SLE : SLE cases diagnosed within the study-period.

## 4. Discussion

The literature search on outcomes in SLE and mortality, ESRD and cancer revealed population-based studies from 22 different locations around the world. The main discovery is that from 1990 there is a higher survival rate during the first five to ten years of the disease. A cardiovascular cause of death is common later in the disease's course, and improvement in survival is less clear. Death caused by infections differs between geographical area and the death rate due to infections is lowest in Europe. Development of ESRD occurs in 5–10% of SLE patients in cohorts of European and Asian ethnic population. ESRD is, however, more common in the African ethnic population. We only discovered one study on cancer from a population-based cohort with inclusion at the time of the SLE diagnosis.

It is well established that the change in treatment of SLE after the 1950s and 1960s caused the survival rate to improve tremendously, from less than a 50% survival rate over five years in the 1950s [47,48]. There are, however, some aspects of selected patient populations that may influence the reported outcome; a tertiary center may overestimate the severity of SLE by missing the diagnosis of milder SLE cases, for example.

Our search on survival with SLE revealed a ten-year survival rate varying with time and location, from 46% in Curacao in the 1980s to 93% in a more recent study from northern Norway [15,19]. The overall trend in survival indicates an improvement in five- and ten-year survival rates after 1990, with a five-year survival similar to the control population. This discovery is in accordance with the conclusion in a recent meta-analysis that survival with SLE improved up to the 1990s, but since appears to have stabilized [48].

A control group is necessary to enhance the quality of survival estimates in SLE. As survival from SLE improves, it may become similar to the survival rate in the general population. The reported survival rate from studies depends on the age composition of the SLE cohort and hence, the time since inception. In this systematic review, nine of the studies included made use of a control group in their survival analysis. They all included only incident cases and five studies were also defined as inception studies. From the inception studies with control groups conducted after 1990, the ten-year survival is only slightly lower in the SLE groups versus the control groups (91% vs. 96%) [19,20]. However, the gap seems to increase with time from diagnosis [10,21].

Findings from this review also indicate that the main causes of death from SLE differ with the length of follow-up time of the studies; CVD is more frequent in studies with the longest follow-up time [10,21]. It is well known from earlier studies that death due to CVD is more frequent later in the course of the disease [49,50]. Urowitch et al. identified this bimodal pattern of mortality in 1976 [51]. In the included studies, European SLE patients died less often of infections compared to Asians. It appears we still do not manage to prevent CVD over time, as up to 59% of SLE patients die of CVD. This might indicate better treatment for the acute phase of SLE, but not for damage accrual due to SLE. However, death from infections remains prominent in certain parts of the world.

In this review, SMR in total SLE cohorts ranges from 1.9 to 4.6, similar, but with a slightly lower range of variation, compared to two previous meta-analyses [52,53]. Studying SMR in incident populations makes comparison difficult as the inclusion periods differ, the highest SMR being from Taiwan within the first year after diagnosis [34]. Several studies have identified ethnicity as a modifier of outcome in SLE, with lower survival in patients of African descent [47,54]. This corresponds with our findings of the lowest SMR in a predominantly white Scandinavian population. The discrepancy in prognosis might be due to both genetic and socioeconomic factors. A possible gender disparity in SLE prognosis has been proposed; however, the results have been inconsistent and contradictory [55]. In this review, we found no significant sex differences in SMR.

Many studies have reported the risk of ESRD development in SLE, and, as registries of biopsy-proven LN are quite common, outcomes in this particular patient subset have been widely investigated. However, as many as 44% of all LN patients are not biopsy-proven [56]. Thus, we excluded studies of this selected SLE patient subset, as they might differ from other LN patients. In this review, we found that only 11 population-based studies estimated the frequency of ESRD in SLE populations. An estimated 5–11% of SLE patients progressed to ESRD, fewer than in a recent meta-analysis [57]. A lower frequency of ESRD in the white population is in line with previous reports [54,57]. The trend in ESRD development seems to be stable over time, despite improvements in therapy. This corresponds to findings from a recent meta-analysis where the risk of ESRD development remained unchanged during the last decade [57].

We identified only three studies on cancer development in SLE patients. Only one was an inception study [41]. In these studies, the cancer risk was increased by 1.2–1.8 times. By comparison, a prior review, which also included non-population-based studies, found an increased risk of cancer ranging from 1.1 to 3.6 times in the SLE population [58]. The lowest cancer risk (SMR 1.2) found in our review was from an old Swedish study with 116 SLE patients. The study from the National Health Insurance Research Database from Taiwan is on the other end of the scale, with a SIR of 1.8 [46].

Earlier studies, mostly non-population-based or without a verified diagnosis, have found that hematological cancers appear more often in the SLE population compared to the general population [58]. All three studies in this review found significantly higher numbers of lymphomas, and especially non-Hodgkin lymphomas, with a reported SMR of 11.6 from Sweden and SIR of 7.3 from Taiwan [41,45,46]. In addition, all three studies found an increased incidence of lung cancer [41,45,46]. Taiwan reports a significant increase for lung/mediastinum (SIR 1.2) [46], yet data from Sweden (SMR 5.6) and Iceland (O/E ratio 1.7) are not significant [41,45].

Cancer development in SLE patients is particularly difficult to study for two reasons. First, cancer sometimes leads to death; subsequently, patients who get cancer early in the course of the disease may not be captured. Secondly, some people with cancer might have paraneoplastic symptoms that may mimic SLE and then be mistakenly diagnosed with SLE. This emphasizes both the importance of a verified SLE diagnosis in studies on cancer and SLE, and the need for further population-based, and preferably inception-based (early capture), studies on cancer.

Considerable differences in the methods for case finding, verification of diagnosis, and study design can make comparing the results of the SLE outcomes difficult. To overcome some of these problems, all studies in this systematic review have employed comprehensive case-finding and case ascertainment methods, or it has been indicated in the article that all patients in a defined geographic region were included. However, the geographic area and its location for care of SLE patients is not always described in detail, and it is likely that we have missed some population-based studies.

The composition of the cohorts used for analysis of outcomes differs as some studies include all patients and some include only incident patients, making comparisons more difficult. Only seven studies of incident SLE patients had a follow-up period over 15 years [10,13,14,21,25,27,41]. The reason for this may be that hospital data registries going back before the year 2000 are rare and not so easily accessible. They may also not contain the entire volume of ICD-codes on outpatients [59].

Most of the population-based studies, except for Taiwan, are small due to the work effort necessary to identify all patients and verify their diagnoses. Taiwan has a good health system, and 96%–99% of its population is included in the National Health Insurance Database. All SLE patients must fulfill the ACR criteria to receive their benefit claim checks as in- and outpatients [34,39,42,43,46]. However, this may also give the patients and their doctors an additional motive towards approving the SLE diagnosis. In addition, verification of the SLE diagnosis is processed earlier on in the course of the disease in Taiwan compared to the other studies. Hence, an early misdiagnosis of SLE would not be reclassified retrospectively.

We found that six locations (Iceland, Lund in Sweden, Funen in Denmark, northern Norway, Rochester in the USA and New Territories in Hong Kong) have repeated the retrieval of patients at several time points [10,19,27,31,33]. Scandinavia is highly represented in publishing from population-based studies, probably due to the health care system being mostly public, making it easier to identify the patients. Despite small study populations, these are valuable contributions to population-based knowledge of outcomes for SLE. Lund in Sweden already published the very first data on survival from a population-based cohort in 1989 and has, to date, the longest follow-up time on an inception cohort reporting on 25-year survival (60%) [10]. However, four locations from the USA have made a tremendous effort collecting larger population-based cohorts that were published in the last decade [25,27,32,37,40].

## 5. Conclusions

Population-based studies on SLE patients with a verified diagnosis is considered the gold standard in the pursuit of finding the true outcomes of suffering from SLE. Studies using the 1997 ACR criteria are easier to compare over time, as most studies included only SLE patients with four or more ACR criteria. There is a special need for cancer studies and studies with longer follow-up time on survival in population-based inception cohorts.

**Supplementary Materials:** The following are available online at https://www.mdpi.com/article/10.3390/jcm10194306/s1, Table S1: Case finding methodology and SLE ascertainment in population-based Systemic Lupus Erythematosus cohorts; Supplementary Materials S2: Search strings.

**Author Contributions:** K.L. received the invitation to write the article. Ø.M. and K.L. concepted and designed the article. H.H. and S.R.M. performed the literature search. H.H., S.R.M. and K.L. performed the screening and eligibility phase and study identification. H.H. and S.R.M. wrote the first draft of the article; they contributed equally to this paper and share first authorship. All authors contributed to the writing of subsequent drafts. All authors have read and agreed to the published version of the manuscript.

**Funding:** This research received funding from The Dam Foundation (Grant number 296618), The Norwegian Women's Public Health Association (Grant number 40108) and Vivi Irene Hansens fond (Grant number 190613).

**Institutional Review Board Statement:** Not applicable.

**Informed Consent Statement:** Not applicable.

**Data Availability Statement:** Not applicable.

**Acknowledgments:** The authors would like to thank Liberian Hilde Iren Flaatten, University of Oslo, for performing the literature search, and Torild Garen for valuable technical assistance.

**Conflicts of Interest:** The authors declare no conflict of interest.

## References

1. Rees, F.; Doherty, M.; Grainge, M.; Lanyon, P.; Zhang, W. The worldwide incidence and prevalence of systemic lupus erythematosus: A systematic review of epidemiological studies. *Rheumatology* **2017**, *56*, 1945–1961. [CrossRef]
2. Lerang, K.; Gilboe, I.-M.; Gran, J.T. Differences between rheumatologists and other internists regarding diagnosis and treatment of systemic lupus erythematosus. *Rheumatology* **2011**, *51*, 663–669. [CrossRef]
3. Gergianaki, I.; Fanouriakis, A.; Repa, A.; Tzanakakis, M.; Adamichou, C.; Pompieri, A.; Spirou, G.; Bertsias, A.; Kabouraki, E.; Tzanakis, I.; et al. Epidemiology and burden of systemic lupus erythematosus in a Southern European population: Data from the community-based lupus registry of Crete, Greece. *Ann. Rheum. Dis.* **2017**, *76*, 1992–2000. [CrossRef]
4. Voss, A.; Green, A.; Junker, P. Systemic lupus erythematosus in Denmark: Clinical and epidemiological characterization of a county-based cohort. *Scand. J. Rheumatol.* **1998**, *27*, 98–105.
5. Ingvarsson, R.F.; Bengtsson, A.A.; Jönsen, A. Variations in the epidemiology of systemic lupus erythematosus in southern Sweden. *Lupus* **2016**, *25*, 772–780. [CrossRef] [PubMed]
6. Lim, S.S.; Bayakly, A.R.; Helmick, C.G.; Gordon, C.; Easley, K.; Drenkard, C. The Incidence and Prevalence of Systemic Lupus Erythematosus, 2002–2004: The Georgia Lupus Registry. *Arthritis Rheumatol.* **2013**, *66*, 357–368. [CrossRef] [PubMed]
7. Moher, D.; Liberati, A.; Tetzlaff, J.; Altman, D.G.; The PRISMA Group. Preferred Reporting Items for Systematic Reviews and Meta-Analyses: The PRISMA Statement. *PLoS Med.* **2009**, *6*, e1000097. [CrossRef]
8. Hochberg, M.C. Updating the American college of rheumatology revised criteria for the classification of systemic lupus erythematosus. *Arthritis Rheum.* **1997**, *40*, 1725. [CrossRef] [PubMed]
9. Voss, A.; Laustrup, H.; Hjelmborg, J.; Junker, P. Survival in systemic lupus erythematosus, 1995–2010. A prospective study in a Danish community. *Lupus* **2013**, *22*, 1185–1191. [CrossRef]

10. Ingvarsson, R.F.; Landgren, A.J.; Bengtsson, A.A.; Jönsen, A. Good survival rates in systemic lupus erythematosus in southern Sweden, while the mortality rate remains increased compared with the population. *Lupus* **2019**, *28*, 1488–1494. [CrossRef]

11. Al-Adhoubi, N.K.; Al-Balushi, F.; Al Salmi, I.; Ali, M.; Al Lawati, T.; Al Lawati, B.S.H.; Abdwani, R.; Al Shamsi, A.; Al Kaabi, J.; Al Mashaani, M.; et al. A multicenter longitudinal study of the prevalence and mortality rate of systemic lupus erythematosus patients in Oman: Oman Lupus Study. *Int. J. Rheum. Dis.* **2021**, *24*, 847–854. [CrossRef]

12. Flower, C.; Hennis, A.J.M.; Hambleton, I.R.; Nicholson, G.D.; Liang, M.H.; the The Barbados National Lupus Registry Group. Systemic lupus erythematosus in an Afro-Caribbean population: Incidence, clinical manifestations and survival in the Barbados national lupus registry. *Arthritis Rheum.* **2012**, *64*, 1151–1158. [CrossRef]

13. Peschken, C.A.; Esdaile, J.M. Systemic lupus erythematosus in North American Indians: A population based study. *J. Rheumatol.* **2000**, *27*.

14. Mok, C.C.; Mak, A.; Chu, W.P.; To, C.H.; Nin Wong, S. Long-term Survival of Southern Chinese Patients With Systemic Lupus Erythematosus. *Medicine* **2005**, *84*, 218–224. [CrossRef] [PubMed]

15. Nossent, J. Systemic lupus erythematosus on the Caribbean island of Curacao: An epidemiological investigation. *Ann. Rheum. Dis.* **1992**, *51*, 1197–1201. [CrossRef] [PubMed]

16. Alamanos, Y.; Voulgari, P.V.; Siozos, C.; Katsimpri, P.; Tsintzos, S.; Dimou, G.; Politi, E.N.; Rapti, A.; Laina, G.; Drosos, A.A. Epidemiology of systemic lupus erythematosus in northwest Greece 1982–2001. *J. Rheumatol.* **2003**, *30*, 731–735.

17. Gudmundsson, S.; Steinsson, K. Systemic lupus erythematosus in Iceland 1975 through 1984. A nationwide epidemiological study in an unselected population. *J. Rheumatol.* **1990**, *17*, 1162–1167.

18. Nossent, H.C. Systemic lupus erythematosus in the Arctic region of Norway. *J. Rheumatol.* **2001**, *28*, 539–546.

19. Eilertsen, G.; Becker-Merok, A.; Nossent, J.C. The Influence of the 1997 Updated Classification Criteria for Systemic Lupus Erythematosus: Epidemiology, Disease Presentation, and Patient Management. *J. Rheumatol.* **2009**, *36*, 552–559. [CrossRef]

20. Lerang, K.; Gilboe, I.-M.; Thelle, D.S.; Gran, J.T. Mortality and years of potential life loss in systemic lupus erythematosus: A population-based cohort study. *Lupus* **2014**, *23*, 1546–1552. [CrossRef]

21. Alonso, M.D.; Llorca, J.; Martinez-Vazquez, F.; Miranda-Filloy, J.A.; de Teran, T.D.; Dierssen, T.; Rodriguez, T.R.V.; Gomez-Acebo, I.; Blanco, R.; Gonzalez-Gay, M.A. Systemic Lupus Erythematosus in Northwestern Spain. *Medicine* **2011**, *90*, 350–358. [CrossRef]

22. Ståhl-Hallengren, C.; Jönsen, A.; Nived, O.; Sturfelt, G. Incidence studies of systemic lupus erythematosus in Southern Sweden: Increasing age, decreasing frequency of renal manifestations and good prognosis. *J. Rheumatol.* **2000**, *27*, 685–691. [PubMed]

23. Pamuk, O.N.; Balci, M.A.; Donmez, S.; Tsokos, G.C. The incidence and prevalence of systemic lupus erythematosus in Thrace, 2003–2014: A 12-year epidemiological study. *Lupus* **2015**, *25*, 102–109. [CrossRef] [PubMed]

24. Naleway, A.L.; Davis, M.E.; Greenlee, R.T.; Wilson, D.A.; Mccarty, D.J. Epidemiology of systemic lupus erythematosus in rural Wisconsin. *Lupus* **2005**, *14*, 862–866. [CrossRef] [PubMed]

25. Bartels, C.M.; Buhr, K.A.; Goldberg, J.W.; Bell, C.L.; Visekruna, M.; Nekkanti, S.; Greenlee, R.T. Mortality and Cardiovascular Burden of Systemic Lupus Erythematosus in a US Population-based Cohort. *J. Rheumatol.* **2014**, *41*, 680–687. [CrossRef] [PubMed]

26. Uramoto, K.M.; Michet, C.J., Jr.; Thumboo, J.; Sunku, J.; O'Fallon, W.M.; Gabriel, S.E. Trends in the incidence and mortality of systemic lupus erythematosus, 1950-1992. *Arthritis Rheum.* **1999**, *42*, 46–50. [CrossRef]

27. Jarukitsopa, S.; Hoganson, D.D.; Crowson, C.S.; Sokumbi, O.; Davis, M.D.; Michet, C.J.; Matteson, E.L.; Kremers, H.M.; Chowdhary, V.R. Epidemiology of Systemic Lupus Erythematosus and Cutaneous Lupus Erythematosus in a Predominantly White Population in the United States. *Arthritis Rheum.* **2014**, *67*, 817–828. [CrossRef] [PubMed]

28. Jonsson, H.; Nived, O.; Sturfelt, G. Outcome in systemic lupus erythematosus: A prospective study of patients from a defined population. *Medicine* **1989**, *68*, 141–150. [CrossRef] [PubMed]

29. Lucero, L.G.; Barbaglia, A.L.; Bellomio, V.I.; Bertolaccini, M.C.; Escobar, M.A.M.; Sueldo, H.R.; Yacuzzi, M.S.; Carrizo, G.A.; Robles, N.; Rengel, S.; et al. Prevalence and incidence of systemic lupus erythematosus in Tucumán, Argentina. *Lupus* **2020**, *29*, 1815–1820. [CrossRef]

30. Jacobsen, S. Mortality and Causes of Death of 513 Danish Patients with Systemic Lupus Erythematosus. *Scand. J. Rheumatol.* **1999**, *28*, 75–80. [CrossRef]

31. Laustrup, H.; Voss, A.; Green, A.; Junker, P. Occurrence of systemic lupus erythematosus in a Danish community: An 8-year prospective study. *Scand. J. Rheumatol.* **2009**, *38*, 128–132. [CrossRef] [PubMed]

32. Lim, S.S.; Helmick, C.G.; Bao, G.; Hootman, J.; Bayakly, R.; Gordon, C.; Drenkard, C. Racial Disparities in Mortality Associated with Systemic Lupus Erythematosus—Fulton and DeKalb Counties, Georgia, 2002–2016. *MMWR. Morb. Mortal. Wkly. Rep.* **2019**, *68*, 419–422. [CrossRef] [PubMed]

33. Mok, C.C.; To, C.H.; Ho, L.Y.; Yu, K.L. Incidence and mortality of systemic lupus erythematosus in a southern Chinese population, 2000-2006. *J. Rheumatol.* **2008**, *35*, 1978–1982.

34. Yeh, K.-W.; Yu, C.-H.; Chan, P.-C.; Horng, J.-T.; Huang, J.-L. Burden of systemic lupus erythematosus in Taiwan: A population-based survey. *Rheumatol. Int.* **2013**, *33*, 1805–1811. [CrossRef]

35. Iseki, K.; Miyasato, F.; Oura, T.; Uehara, H.; Nishime, K.; Fukiyama, K. An Epidemiologic Analysis of End-stage Lupus Nephritis. *Am. J. Kidney Dis.* **1994**, *23*, 547–554. [CrossRef]

36. Jacobsen, S.; Petersen, J.; Ullman, S.; Junker, P.; Voss, A.; Rasmussen, J.M.; Tarp, U.; Poulsen, L.H.; Hansen, G.V.O.; Skaarup, B.; et al. A multicentre study of 513 Danish patients with systemic lupus erythematosus. I. Disease manifestations and analyses of clinical subsets. *Clin. Rheumatol.* **1998**, *17*, 468–477. [CrossRef]

37. Somers, E.C.; Marder, W.; Cagnoli, P.; Lewis, E.E.; Deguire, P.; Gordon, C.; Helmick, C.G.; Wang, L.; Wing, J.J.; Dhar, J.P.; et al. Population-Based Incidence and Prevalence of Systemic Lupus Erythematosus: The Michigan Lupus Epidemiology and Surveillance Program. *Arthritis Rheumatol.* **2013**, *66*, 369–378. [CrossRef]

38. Moe, S.E.R.; Molberg, Ø.; Strøm, E.H.; Lerang, K. Assessing the relative impact of lupus nephritis on mortality in a population-based systemic lupus erythematosus cohort. *Lupus* **2019**, *28*, 818–825. [CrossRef]

39. Yu, K.-H.; Kuo, C.-F.; Chou, I.-J.; Chiou, M.-J.; See, L.-C. Risk of end-stage renal disease in systemic lupus erythematosus patients: A nationwide population-based study. *Int. J. Rheum. Dis.* **2016**, *19*, 1175–1182. [CrossRef] [PubMed]

40. Plantinga, L.C.; Lim, S.S.; Patzer, R.E.; McClellan, W.M.; Kramer, M.; Klein, M.; Pastan, S.; Gordon, C.; Helmick, C.G.; Drenkard, C. Incidence of End-Stage Renal Disease Among Newly Diagnosed Systemic Lupus Erythematosus Patients: The Georgia Lupus Registry. *Arthritis Rheum.* **2015**, *68*, 357–365. [CrossRef]

41. Nived, O.; Bengtsson, A.; Jönsen, A.; Sturfelt, G.; Olsson, H. Malignancies during follow-up in an epidemiologically defined systemic lupus erythematosus inception cohort in southern Sweden. *Lupus* **2001**, *10*, 500–504. [CrossRef]

42. Lin, C.-H.; Hung, P.-H.; Hu, H.-Y.; Chen, Y.-J.; Guo, H.-R.; Hung, K.-Y. Infection-related hospitalization and risk of end-stage renal disease in patients with systemic lupus erythematosus: A nationwide population-based study. *Nephrol. Dial. Transplant.* **2016**, *32*, 1683–1690. [CrossRef]

43. Lin, W.-H.; Guo, C.-Y.; Wang, W.-M.; Yang, D.-C.; Kuo, T.-H.; Liu, M.-F.; Wang, M.-C. Incidence of progression from newly diagnosed systemic lupus erythematosus to end stage renal disease and all-cause mortality: A nationwide cohort study in Taiwan. *Int. J. Rheum. Dis.* **2013**, *16*, 747–753. [CrossRef]

44. Eilertsen, G.O.; Fismen, S.; Hanssen, T.-A.; Nossent, J.C. Decreased incidence of lupus nephritis in northern Norway is linked to increased use of antihypertensive and anticoagulant therapy. *Nephrol. Dial. Transplant.* **2010**, *26*, 620–627. [CrossRef]

45. Ragnarsson, O.; Gröndal, G.; Steinsson, K. Risk of malignancy in an unselected cohort of Icelandic patients with systemic lupus erythematosus. *Lupus* **2003**, *12*, 687–691. [CrossRef]

46. Chen, Y.-J.; Chang, Y.-T.; Wang, C.-B.; Wu, C.-Y. Malignancy in Systemic Lupus Erythematosus: A Nationwide Cohort Study in Taiwan. *Am. J. Med.* **2010**, *123*, 1150.e1–1150.e6. [CrossRef]

47. Borchers, A.T.; Keen, C.L.; Shoenfeld, Y.; Gershwin, M. Surviving the butterfly and the wolf: Mortality trends in systemic lupus erythematosus. *Autoimmun. Rev.* **2004**, *3*, 423–453. [CrossRef]

48. Tektonidou, M.; Lewandowski, L.B.; Hu, J.; Dasgupta, A.; Ward, M.M. Survival in adults and children with systemic lupus erythematosus: A systematic review and Bayesian meta-analysis of studies from 1950 to 2016. *Ann. Rheum. Dis.* **2017**, *76*, 2009–2016. [CrossRef]

49. Schoenfeld, S.R.; Kasturi, S.; Costenbader, K.H. The epidemiology of atherosclerotic cardiovascular disease among patients with SLE: A systematic review. *Semin. Arthritis Rheum.* **2013**, *43*, 77–95. [CrossRef]

50. Lu, X.; Wang, Y.; Zhang, J.; Pu, D.; Hu, N.; Luo, J.; An, Q.; He, L. Patients with systemic lupus erythematosus face a high risk of cardiovascular disease: A systematic review and Meta-analysis. *Int. Immunopharmacol.* **2021**, *94*, 107466. [CrossRef]

51. Urowitz, M.B.; Bookman, A.A.; Koehler, B.E.; Gordon, D.A.; Smythe, H.A.; Ogryzlo, M.A. The bimodal mortality pattern of systemic lupus erythematosus. *Am. J. Med.* **1976**, *60*, 221–225. [CrossRef]

52. Yurkovich, M.; Vostretsova, K.; Chen, W.; Aviña-Zubieta, J.A. Overall and Cause-Specific Mortality in Patients With Systemic Lupus Erythematosus: A Meta-Analysis of Observational Studies. *Arthritis Rheum.* **2013**, *66*, 608–616. [CrossRef]

53. Lee, Y.H.; Choi, S.J.; Ji, J.D.; Song, G.G. Overall and cause-specific mortality in systemic lupus erythematosus: An updated meta-analysis. *Lupus* **2016**, *25*, 727–734. [CrossRef] [PubMed]

54. Lewis, M.J.; Jawad, A.S. The effect of ethnicity and genetic ancestry on the epidemiology, clinical features and outcome of systemic lupus erythematosus. *Rheumatology* **2016**, *56*, i67–i77. [CrossRef] [PubMed]

55. Lu, L.-J.; Wallace, D.; Ishimori, M.; Scofield, R.; Weisman, M. Review: Male systemic lupus erythematosus: A review of sex disparities in this disease. *Lupus* **2009**, *19*, 119–129. [CrossRef] [PubMed]

56. Hanly, J.G.; O'Keeffe, A.G.; Su, L.; Urowitz, M.B.; Romero-Diaz, J.; Gordon, C.; Bae, S.-C.; Bernatsky, S.; Clarke, A.E.; Wallace, D.J.; et al. The frequency and outcome of lupus nephritis: Results from an international inception cohort study. *Rheumatology* **2015**, *55*, 252–262. [CrossRef]

57. Tektonidou, M.; Dasgupta, A.; Ward, M.M. Risk of End-Stage Renal Disease in Patients With Lupus Nephritis, 1971-2015: A Systematic Review and Bayesian Meta-Analysis. *Arthritis Rheumatol.* **2016**, *68*, 1432–1441. [CrossRef]

58. Choi, M.Y.; Flood, K.; Bernatsky, S.; Ramsey-Goldman, R.; Clarke, A.E. A review on SLE and malignancy. *Best Pr. Res. Clin. Rheumatol.* **2017**, *31*, 373–396. [CrossRef]

59. Jørgensen, K.T.; Pedersen, B.V.; Nielsen, N.M.; Jacobsen, S.; Frisch, M. Childbirths and risk of female predominant and other autoimmune diseases in a population-based Danish cohort. *J. Autoimmun.* **2012**, *38*, J81–J87. [CrossRef]

Journal of
*Clinical Medicine*

*Article*

# Good Long-Term Prognosis of Lupus Nephritis in the High-Income Afro-Caribbean Population of Martinique with Free Access to Healthcare

Benoit Suzon [1],* , Fabienne Louis-Sidney [2], Cédric Aglaé [3], Kim Henry [1], Cécile Bagoée [1], Sophie Wolff [1], Florence Moinet [1], Violaine Emal-Aglaé [3], Katlyne Polomat [1], Michel DeBandt [2], Christophe Deligny [1] and Aymeric Couturier [1]

[1] Department of Internal Medicine, Martinique University Hospital, CEDEX CS, 90632 Fort-de-France, Martinique, France
[2] Department of Rheumatology, Martinique University Hospital, CEDEX CS, 90632 Fort-de-France, Martinique, France
[3] Department of Nephrology, Martinique University Hospital, CEDEX CS, 90632 Fort-de-France, Martinique, France
* Correspondence: benoit.suzon@chu-martinique.fr

**Abstract:** Lupus nephritis (LN) has been described as having worse survival and renal outcomes in African-descent patients than Caucasians. We aimed to provide long-term population-based data in an Afro-descendant cohort of LN with high income and easy and free access to specialized healthcare. Study design: We performed a retrospective population-based analysis using data from 2002–2015 of 1140 renal biopsies at the University Hospital of Martinique (French West Indies). All systemic lupus erythematosus patients with a diagnosis of LN followed for at least 12 months in Martinique or who died during this period were included. Results: A total of 89 patients were included, of whom 68 (76.4%) had proliferative (class III or IV), 17 (19.1%) had membranous (class V), and 4 (4.5%) had class I or II lupus nephritis according to the ISN/RPS classification. At a mean follow-up of 118.3 months, 51.7% of patients were still in remission. The rates of end-stage renal disease were 13.5%, 19.1%, and 21.3% at 10, 15, and 20 years of follow-up, respectively, and mortality rates were 4.5%, 5.6%, and 7.9% at 10, 15, and 20 years of follow-up, respectively. Conclusions: The good survival of our Afro-descendant LN patients, similar to that observed in Caucasians, shades the burden of ethnicity but rather emphasizes and reinforces the importance of optimizing all modifiable factors associated with poor outcome, especially socioeconomics.

**Keywords:** Afro-Caribbean; systemic lupus; lupus nephritis; long-term prognosis; end-stage renal disease; mortality

**Citation:** Suzon, B.; Louis-Sidney, F.; Aglaé, C.; Henry, K.; Bagoée, C.; Wolff, S.; Moinet, F.; Emal-Aglaé, V.; Polomat, K.; DeBandt, M.; et al. Good Long-Term Prognosis of Lupus Nephritis in the High-Income Afro-Caribbean Population of Martinique with Free Access to Healthcare. *J. Clin. Med.* **2022**, *11*, 4860. https://doi.org/10.3390/jcm11164860

Academic Editors: Christopher Sjöwall and Ioannis Parodis

Received: 28 March 2022
Accepted: 18 May 2022
Published: 19 August 2022

**Publisher's Note:** MDPI stays neutral with regard to jurisdictional claims in published maps and institutional affiliations.



## 1. Introduction

Systemic lupus erythematosus (SLE) is an autoimmune disease with severe renal involvement. SLE and lupus nephritis (LN) are described as more prevalent and severe in Afro-Caribbean and Afro-American populations [1,2] with a five-year survival of less than 70% [2–4] compared with more than 88% in the Caucasian population [5–7]. However, there are discrepancies about whether race is an independent risk factor, with some studies emphasizing the critical role of healthcare access for minorities [8–10]. Martinique is a French Caribbean island whose population is more than 90% Afro-descendant and has full and free access to medical care [1,11,12]. A previous study in Martinique estimated the incidence of SLE to be 4.7/100,000 population and showed ten-year overall survival rates similar to those observed in the Caucasian population, suggesting that prognosis in the Afro-Caribbean population is not related to ethnicity [1]. In the present study, we sought to assess the long-term prognosis of lupus nephritis in a population with free and easy access to specialized healthcare.

## 2. Materials and Methods

We practiced a monocentric hospital-based retrospective study of renal biopsies analyzed from 1 January 2002 to 16 September 2015 in the pathology unit of Fort de France University Hospital. Inclusion criterion was histological LN diagnosis according to ISN/RPS classification in SLE patients fulfilling ACR 1997 criteria, with at least 12 months of follow-up. Samples first analyzed with WHO classification were converted to International Society of Nephrology/Renal Pathology Society (ISN/RPS) classification [13]. Patients with less than 12 months of follow-up were excluded, excepted for those who died during this period.

At baseline (first LN) and during follow-up, we collected data such as age, sex, date of SLE diagnosis, SLE treatment, date and histopathological class of LN occurrence, and number of LN flares; biological data such as serum creatinine, urine protein/creatinine ratio hematuria, and detection of anti-nuclear, anti-extractable nuclear antigen (ENA) and anti-phospholipid (aPL) auto-antibodies. Medical history and treatment of high blood pressure (defined as uncontrolled if >130/80 mmHg during 2 consecutive visits) were also collected.

### 2.1. Definitions

In SLE patients fulfilling ACR 1997 criteria, LN was suspected with proteinuria > 0.5 g/24 h or a urine protein/creatinine ratio > 0.5 g/g on two consecutive samples, with or without hematuria (>10,000 red blood cells (RBCs)/mL). LN suspicion was confirmed and classified by kidney histopathological analysis according to ISN/RPS classification. Time to remission was defined by the time from LN diagnosis to complete or partial remission, with or without treatment. Immunosuppressive therapy initiated for LN purposes was referred to as "induction" or "maintenance therapy". Non-adherence to treatment was not systematically evaluated but was considered when stopping of steroids or immunosuppressant drugs was self-reported. Complete response (CR) was defined if proteinuria was $\leq$0.5 g/24 h or urine protein/creatinine ratio < 50 mg/mmol, and creatinine no greater than 15% above baseline. Partial response (PR) was defined if proteinuria was $\leq$3 g/24 h or urine protein/creatinine ratio <300 mg/mmol, with a reduction > 50% from baseline, and creatinine no greater than 15% above baseline [14]. Relapse was defined by the serum level of creatinine increasing >25% above nadir, persistent for more than a month and/or proteinuria increasing on 2 consecutive urinary samples: >1 g/24 h if the patient was in CR, and doubling of proteinuria or >2 g/24 h in case of PR [3]. Relapse duration was considered until re-remission. End-stage renal disease (ESRD) was defined when renal replacement therapy was necessary more than 3 consecutive months.

In case of patients lost to follow-up or missing values, we considered the patient's status unchanged until a testified new condition. Serum level creatinine, ESRD, and vital status were reviewed for each patient lost to follow-up. If the patient could not be reached, the family physician or other treating physician was contacted by telephone or mail.

Finally, we separated a poor prognosis group for LN, defined by ESRD or death, and compared them to other patients.

### 2.2. Statistics and Ethics

Quantitative data are expressed as mean with standard deviation and qualitative data in unit and percentage. Comparison of the data between groups was performed by Fisher's exact test or student's t-test as appropriate with a significant value if $p < 0.05$. Survival curves were computed using the Kaplan Meier method. Analyses were conducted using Prism-GraphPad software. This study was declared to the Commission Nationale Informatique et Libertés (CNIL) with the registration number 1899602v0.

## 3. Results

A total of 89 Afro-Caribbean patients with LN fulfilling ACR 1997 criteria were included, with a mean follow-up of 119.3 $\pm$ 73.3 months. Baseline characteristics are given in Table 1.

**Table 1.** Clinical, biological, and histological characteristics at lupus nephritis onset.

| | Total (*n* = 89) |
|---|---|
| **Age,** mean (years) | 32.5 ± 13 |
| **Women,** % (*n*) | 93.2 (83) |
| **Duration of SLE before LN (months)** | 39.7 ± 60.8 |
| **LN follow-up (months),** mean ± SD | 119 ± 72.9 |
| **Clinical features** % (*n*) | |
| Fever | 20.22 (18) |
| Neurolupus | 8.99 (8) |
| Arthritis | 44.94 (40) |
| Myositis | 4.49 (4) |
| Cutaneous rash | 20.22 (18) |
| Alopecia | 10.11 (9) |
| Mucosal ulcer | 2.24 (2) |
| Serositis | 29.21 (26) |
| SLEDAI, mean | 17.11 ± 5.73 |
| **Biological features** % (*n*) | |
| Anti-ds-DNA Ab | 97.75 (87) |
| Anti-ds-DNA Ab title, mean ± SD | 241.34 ± 235.40 |
| Anti-Sm | 58.42 (52) |
| Anti-SSA | 56.18 (50) |
| Anti-SSB | 23.59 (21) |
| Anti-RNP | 60.67 (54) |
| aPL positivity | 58.42 (52) |
| APS | 17.98 (16) |
| Hematuria | 73 (65) |
| Leucocyturia | 58.42 (52) |
| Proteinuria, mean, g/24 h | 3.55 ± 3.72 |
| Serum level albumine, mean, g/L | 25.19 ± 8.17 |
| Serum level creatinine, mean, μmol/L | 118.94 ± 93.21 |
| Low C3 | 57.3 (51) |
| Low C4 | 60.67 (54) |
| Thrombopenia | 5.61 (5) |
| Leucopenia | 12.35 (11) |
| **Histologic features at first renal biopsy (ISN/RPS),** % (*n*) | |
| Class I | 3.37 (3) |
| Class II | 1.12 (1) |
| Class III | 19.10 (17) |
| Class IV | 24.72 (22) |
| Class V | 19.10 (17) |
| Class III + V | 20.22 (18) |
| Class IV + V | 12.36 (11) |
| Proliferative LN (III, III + V, IV or IV + V), % (*n*) | 76.4 (68) |
| Activity index, mean, % | 35.93 ± 28.8 (36) |
| Chronicity index, mean, % | 23.1 ± 20.6 (36) |

Abbreviations: Ab: antibody; aPL: anti-phospholipid auto-antibody positivity; APS: anti-phospholipid syndrome; ISN/RPS: International Society of Nephrology/Renal Pathology Society; LN: lupus nephritis; anti-RNP: anti-ribonucleoprotein auto-antibody; SLE: systemic lupus erythematosus; anti-Sm: anti-Smith auto-antibody; anti-SSA: anti-Sjögren's-syndrome-related antigen A auto-antibody; and anti-SSB: anti-Sjögren's-syndrome-related antigen B auto-antibodies.

Among all LN (*n* = 89), 68 (76.4%) were proliferative. Mean SLEDAI score was 17.11 ± 5.73 (*n* = 89). Although the overall time to loss to follow-up was 4.2% of the total follow-up time, the renal and vital status of every single patient were known at the end of the study. Mean time from SLE diagnosis to first LN was 39.7 ± 60.8 months. Hydroxychloroquine (HCQ) was taken for at least 24 months by 74% of patients. Induction therapy consisted of intravenous cyclophosphamide (CYC) in 54 (60.7%) and mycophenolate in 20 (22.5%) cases. Twelve patients received inadequate induction therapy (azathioprine or steroids alone). Three patients did not receive induction therapy: one because of concomitant treatment for lymphoma and the other two for no apparent reason. Regarding maintenance therapy, 64 patients (70.8%) received mycophenolate, 4 (4.5%) received CYC, and 4 (4.5%) received AZA. Other patients received either steroids, ciclosporine, methotrexate, or no treatment. Furthermore, 13 out of 17 (76.47%) initial isolated membranous LN secondarily evolved to a proliferative LN (class IV + V; $n = 10$ and class III + V; $n = 3$) at a mean time of 95 ± 41.2 months. Finally, one of three class I LN progressed to class V and the single class II had a proliferative recurrence (IV + V).

*Evolution and Long-Term Prognosis*

Seventy patients (78.7%) went into remission in a mean of 18.7 months [0.9–115]. CR and PR were achieved in 49.5% and 29.2%, respectively. Among responders, 33 (47.1%) never relapsed. Among relapsers, 21 patients (23.6%) presented one, 10 patients presented two (11.2%), and 6 presented patients (6.7%) three renal flares. Evolution features at 1, 5, 10, 15, and 20 years are detailed in Figures 1 and 2. Renal and global survival analysis are given in Figure 3.

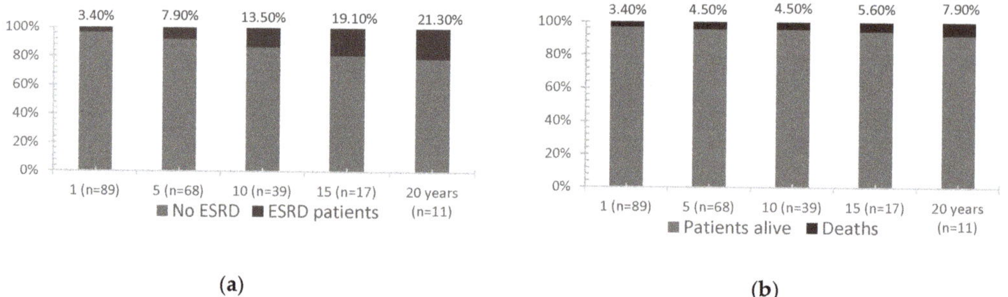

(a)

(b)

**Figure 1.** Cumulative rates of ESRD (**a**) and mortality (**b**) during 1, 5, 10, 15, and 20 years in Martinican patients with lupus nephritis.

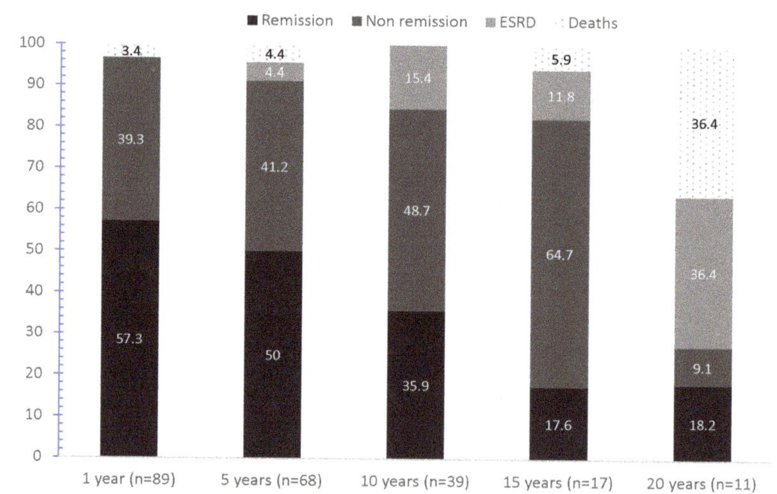

**Figure 2.** Remission, ESRD, and death rates in Martinican patients with lupus nephritis followed for 1, 5, 10, 15, and 20 years. "$(n = x)$" indicates the remaining patients at the end points. After 1, 5, 10, 15, and 20 years of follow-up, the number of patients lost to follow-up was 2, 4, 7, 4, 2, and 0, respectively. At 1, 5, 10, 15, and 20 years, the number of patients in CR and PR were 38 and 18, 26 and 8, 10 and 4, 1 and 3, and 1 and 1, respectively. Eight patients died: three of infectious origin, one of hemorrhage following abdominal surgery, one of probable massive pulmonary embolism, one of heart failure, and two of unknown cause.

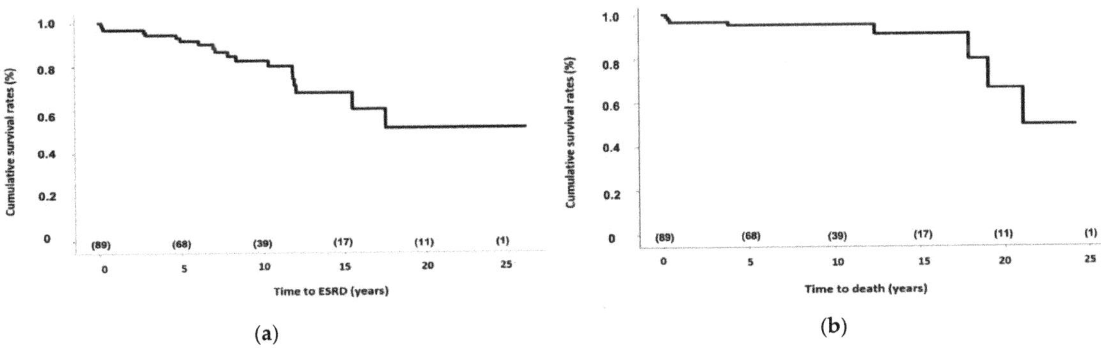

**Figure 3.** Kaplan Meier analysis of the probability of end-stage renal disease (ESRD) (**a**) or death (**b**) in Martinican patients with lupus nephritis. The numbers in brackets refer to the number of remaining patients. Estimated renal survival at 5, 10, 15, and 20 years were 93.2%, 82.3%, 68%, and 51.7%, respectively. Estimated vital survival at 5, 10, 15, and 20 years were 94.9%, 94.9%, 91%, and 66.7%, respectively.

Mean estimated renal and global survival were $18.67 \pm 1.42$ and $20.61 \pm 1.06$ years, respectively. At the end of follow-up, 19 (21.3%) patients were in ESRD: 8 had a kidney transplant and 11 were treated with hemodialysis, with a mean time from LN to hemodialysis initiation of $92 \pm 60$ months. Eight (9%) patients died: three of infectious origin (severe chikungunya, bilateral pneumonia, acute pyelonephritis), one of hemorrhage following abdominal surgery, one of probable massive pulmonary embolism, one of heart failure, and two of unknown cause. Poor outcomes were associated with time to remission ($27.5 \pm 33.4$ versus $87.1 \pm 58.9$ months, $p < 0.0001$) and uncontrolled hypertension (16% vs. 53.8%, $p = 0.0002$).

## 4. Discussion

Afro-descendant or African race or ethnicity is cited as an independent risk factor for adverse outcomes in LN. This finding is not shared by most authors who have highlighted the unfavorable socioeconomic factors of Afro-descendants compared with Caucasians, and their consequences [15–19]. Most populations of African descent worldwide, even in developed countries, face barriers to care, poor socioeconomic conditions, and poverty, which have a negative impact on the prognosis of SLE and LN [20–22]. On the contrary, our population cohort of African-descent is unique because of the easy and completely free access to healthcare and specialized care providers (national reference center dedicated to SLE). Despite a high activity score and high proportion of proliferative LN, we reported good long-term outcomes of our Afro-Caribbean population, similar to that observed in Caucasian patients [5–7].

To compare survival and ESRD in studies coming from tertiary centers or population based can be hazardous but all available data on African-descent patients go on the same way. To date, survival in observational studies concerning LN African-descent patients remains consistently low: 69% at 5 years in an Afro-Caribbean population from the UK [23], 59% at 10 years for an African-American population [4]; and 91 and 59%, 93 and 68%, 90.9 and 60.7% at 1 and 5 years in Curacao [2], Barbados [24], and Jamaica [25], respectively. Mortality was also superior in African-American LN patients than Caucasians with Medicaid or Medicare insurance [26]. In the same way, ESRD has been consistently found more frequently in studies including African-descent patients in [27–29] and outside the United States [23] compared to Caucasian populations. Some studies have compared the long-term outcomes of lupus nephritis between patients of African origin and Caucasians. For example, one-, five-, and ten-year survival of African-descent versus Caucasian LN patients was 95 vs. 94%, 71 vs. 85%, and 59 vs. 81%, respectively. At the same end points, renal survival was 85 vs. 91%, 50 vs. 74%, and 38 vs. 68%, respectively [4]. Another study reported a five-year renal survival in African-descendants of 57% compared to 94.5% in Caucasians [27]. In a largely Caucasian population (76 Caucasians, 8 Afro-descendants, and 6 Asians), who had received recent immunosuppressive treatments (Eurolupus protocol), the ten-year survival was 92% [5]. In our predominantly Afro-descendant population, we reported five- and ten-year renal survival of 93.2 and 82.3%, respectively; overall survival was 94.9% at 10 years, similar to that observed in Caucasians.

Many initial risk factors influencing the outcomes of SLE and LN have been described and highlighted: non-modifiable risk factors such as sex [30–32], age at LN onset [3,33], early LN [34], and LN recurrence, clinical and biological parameters such as hypertension [35], elevated initial serum creatinine [5,33,36–38], elevated initial proteinuria [33,36,39], pathological information [33,40], chronicity index [41–43], and ISN/RPS class [44,45], and genetic factors such as Apol1 [46].

The role of some of these non-modifiable risk factors is still debated, and race or ethnicity is not considered by some authors to be a risk factor for poor outcomes in SLE, in contrast to many modifiable factors including socioeconomic conditions [16,47–50]. Thus, these non-modifiable parameters at presentation can likely be less unfavorable with better management of modifiable factors.

Some of these modifiable factors are of particular importance among minorities: suboptimal access to care related to distance from the rheumatologist and nephrologist, unaffordable consultation due to insufficient medical insurance [51], access to specialized SLE and LN care [52], ideally organized in a network [40] and without delay in LN diagnosis [33,41], and access to renal biopsy [42]. Additional factors include the choice and dose of induction and maintenance therapies [34,53,54], the availability, prescription, and affordability of immunosuppressants [9,55], the timeliness of medication prescription after diagnosis of LN [33], the use of antiproteinuric and anti-malarial medicine [47,56], non-adherence [22,57], smoking [58], and socioeconomic factors [9,10,19,23,47,57,59].

In Martinique, numerous non-modifiable factors were present in SLE and LN patients, and they shared high-disease activity with other Afro-descendant groups [1]. Nevertheless, many modifiable factors are favorable, and poverty is not an obstacle to care. Transportation is covered by public insurance if necessary to the University Hospital with a Lupus Clinic, labeled by the French Ministry of Health since 2008, included in the national network of rare diseases. All medications are free of charge and there is easy and free access to renal biopsies (which are read twice, locally and in a second specialized center in Paris) and hospitalizations. It should be noted that in Curaçao, less than 10% of LNs were biopsied [60] and that in Barbados, renal biopsy was performed in only 28% of LNs due to lack of resources [61]. Thus, in Martinique, access to SLE specialists is easy and the combination with the proximity of their subspecialists facilitates early diagnosis and optimal treatment of LN patients.

Persistent negative factors can be corrected: adequate induction and maintenance immunosuppressive drugs and generalization of anti-hypertensive drugs and RAAS for better control of hypertension and proteinuria [5,34,35]. It should be balanced that some data transcribed obsolete management practices of different departments (internal medicine, rheumatology, and nephrology departments), before the establishment of the current common strategies. As in numerous studies, our patients with worse outcomes had higher serum creatinine levels at the onset of LN. We did not find any risk factors of severe evolution among the chronicity index or ISN-RPS class but this was probably related to a lack of power.

Regarding treatment, mycophenolate seemed to be associated with better remission rates in Afro-descendants, but no definitive proof has been provided [62]. A low-dose IV CYC regimen but with a high initial dose of steroids has been considered to have the same efficacy in African-American patients than European patients in the ACCESS trial [5,63,64]. This regimen has been systematically proposed. Anti-malarial drugs, associated with better survival, was received by three-quarter of our patients [56]. Adherence to treatment, which remains a concern in the treatment of LN, especially in minorities, is largely underestimated in studies [55,65–67]. In our work, all patients were included, treatment compliant or non-compliant, without prior selection as in prospective interventional studies. Although assessed in an unsystematic, self-reported manner, poor adherence to treatment affected one-quarter of our patients. Prescribed to nearly 100% of patients, HCQ was taken for at least 2 consecutive years by 74% of them, mainly due to poor adherence. While data on non-compliance from real-world studies such as ours may be less conclusive than those from controlled studies, none of the latter provide a systematic survey of adherence after treatment initiation. Some limitations of the study should be noted. The number of patients was small, and the retrospective nature of the study did expose it to the risk of missing data. Although vital status or ESRD diagnosis were known for all patients at the end of the study, some patients were lost to follow-up at certain times. Finally, the data collection was old but reflects the reality of the sometimes obsolete or inadequate practices of the different departments before the constitution of a national reference center.

Nevertheless, our data shade the burden of race or ethnicity and highlighted the impact of modifiable factors and access to care in minorities with lupus nephritis. After three decades of improvement, LN survival has not improved since the 2000s, and this may be related to less effective management in minorities living in low-income areas. There is a need for further studies to confirm these results.

## 5. Conclusions

We reported good long-term global and renal survival in a population-based cohort of Afro-Caribbeans with lupus nephritis, similar to those observed in Caucasian patients, arguing for the weight of socioeconomic and modifiable factors.

**Author Contributions:** Conceptualization, B.S., F.L.-S., C.D. and A.C.; methodology, B.S., F.L.-S., C.D. and A.C.; validation, C.A., C.D. and A.C.; formal analysis, B.S. and A.C.; investigation, B.S. and A.C.; data curation, A.C.; writing—original draft preparation, B.S. and A.C.; writing—review and editing, B.S., F.L.-S., C.A., K.H., C.B., S.W., V.E.-A., K.P., F.M., M.D., C.D., and A.C.; visualization, B.S.; supervision, C.D.; project administration, C.D. All authors have read and agreed to the published version of the manuscript.

**Funding:** This research received no external funding.

**Institutional Review Board Statement:** The study was conducted in accordance with the Declaration of Helsinki, and approved by the Institutional Review Board of Martinique University Hospital (protocol code 2022/175; 26 April 2022).

**Informed Consent Statement:** In accordance with French law, no patient included in this study expressed opposition.

**Data Availability Statement:** The data presented in this study are available on request from the corresponding author.

**Conflicts of Interest:** The authors declare no conflict of interest.

## References

1. Deligny, C.; Thomas, L.; Dubreuil, F.; Théodose, C.; Garsaud, A.M.; Numéric, P.; Ranlin, A.; Jean-Baptiste, G.; Arfi, S. Systemic lupus erythematosus in Martinique: An epidemiologic study. *Rev. Med. Interne* **2002**, *23*, 21–29. [CrossRef]
2. Nossent, J.C. Clinical Renal Involvement in Afro-Caribbean Lupus Patients. *Lupus* **1993**, *2*, 173–176. [CrossRef] [PubMed]
3. Parikh, S.V.; Nagaraja, H.N.; Hebert, L.; Rovin, B.H. Renal Flare as a Predictor of Incident and Progressive CKD in Patients with Lupus Nephritis. *Clin. J. Am. Soc. Nephrol. CJASN* **2014**, *9*, 279–284. [CrossRef] [PubMed]
4. Korbet, S.M.; Schwartz, M.M.; Evans, J.; Lewis, E.J. Collaborative Study Group Severe Lupus Nephritis: Racial Differences in Presentation and Outcome. *J. Am. Soc. Nephrol. JASN* **2007**, *18*, 244–254. [CrossRef]
5. Houssiau, F.A.; Vasconcelos, C.; D'Cruz, D.; Sebastiani, G.D.; de Ramon Garrido, E.; Danieli, M.G.; Abramovicz, D.; Blockmans, D.; Cauli, A.; Direskeneli, H.; et al. The 10-Year Follow-up Data of the Euro-Lupus Nephritis Trial Comparing Low-Dose and High-Dose Intravenous Cyclophosphamide. *Ann. Rheum. Dis.* **2010**, *69*, 61–64. [CrossRef]
6. Huong, D.L.; Papo, T.; Beaufils, H.; Wechsler, B.; Blétry, O.; Baumelou, A.; Godeau, P.; Piette, J.C. Renal Involvement in Systemic Lupus Erythematosus. A Study of 180 Patients from a Single Center. *Medicine* **1999**, *78*, 148–166. [CrossRef]

7. Cervera, R.; Khamashta, M.A.; Font, J.; Sebastiani, G.D.; Gil, A.; Lavilla, P.; Mejía, J.C.; Aydintug, A.O.; Chwalinska-Sadowska, H.; de Ramón, E.; et al. Morbidity and Mortality in Systemic Lupus Erythematosus during a 10-Year Period: A Comparison of Early and Late Manifestations in a Cohort of 1000 Patients. *Medicine* **2003**, *82*, 299–308. [CrossRef]

8. Hoover, P.J.; Costenbader, K.H. Insights into the Epidemiology and Management of Lupus Nephritis from the US Rheumatologist's Perspective. *Kidney Int.* **2016**, *90*, 487–492. [CrossRef]

9. Hopkinson, N.D.; Jenkinson, C.; Muir, K.R.; Doherty, M.; Powell, R.J. Racial Group, Socioeconomic Status, and the Development of Persistent Proteinuria in Systemic Lupus Erythematosus. *Ann. Rheum. Dis.* **2000**, *59*, 116–119. [CrossRef]

10. Barr, R.G.; Seliger, S.; Appel, G.B.; Zuniga, R.; D'Agati, V.; Salmon, J.; Radhakrishnan, J. Prognosis in Proliferative Lupus Nephritis: The Role of Socio-Economic Status and Race/Ethnicity. *Nephrol. Dial. Transplant. Off. Publ. Eur. Dial. Transpl. Assoc. Eur. Ren. Assoc.* **2003**, *18*, 2039–2046. [CrossRef]

11. INSEE: Statistiques. Available online: https://www.cleiss.fr/docs/regimes/regime_france/an_1.html (accessed on 1 March 2022).

12. Levinson, D. *Ethnic Groups Worldwide: A Ready Handbook*; Oryx Press: Phoenix, AZ, USA, 1998.

13. Weening, J.J.; D'agati, V.D.; Schwartz, M.M.; Seshan, S.V.; Alpers, C.E.; Appel, G.B.; Balow, J.E.; Bruijn, J.A.N.A.; Cook, T.; Ferrario, F.; et al. The Classification of Glomerulonephritis in Systemic Lupus Erythematosus Revisited. *Kidney Int.* **2004**, *65*, 521–530. [CrossRef] [PubMed]

14. Condon, M.B.; Ashby, D.; Pepper, R.J.; Cook, H.T.; Levy, J.B.; Griffith, M.; Cairns, T.D.; Lightstone, L. Prospective Observational Single-Centre Cohort Study to Evaluate the Effectiveness of Treating Lupus Nephritis with Rituximab and Mycophenolate Mofetil but No Oral Steroids. *Ann. Rheum. Dis.* **2013**, *72*, 1280–1286. [CrossRef] [PubMed]

15. Karlson, E.W.; Daltroy, L.H.; Lew, R.A.; Wright, E.A.; Partridge, A.J.; Fossel, A.H.; Roberts, W.N.; Stern, S.H.; Straaton, K.V.; Wacholtz, M.C.; et al. The Relationship of Socioeconomic Status, Race, and Modifiable Risk Factors to Outcomes in Patients with Systemic Lupus Erythematosus. *Arthritis Rheum.* **1997**, *40*, 47–56. [CrossRef] [PubMed]

16. Yee, C.-S.; Su, L.; Toescu, V.; Hickman, R.; Situnayake, D.; Bowman, S.; Farewell, V.; Gordon, C. Birmingham SLE Cohort: Outcomes of a Large Inception Cohort Followed for up to 21 Years. *Rheumatology* **2015**, *54*, 836–843. [CrossRef]

17. Drenkard, C.; Lim, S.S. Update on Lupus Epidemiology: Advancing Health Disparities Research through the Study of Minority Populations. *Curr. Opin. Rheumatol.* **2019**, *31*, 689–696. [CrossRef]

18. Cooper, G.S.; Treadwell, E.L.; William St. Clair, E.; Gilkeson, G.S.; Dooley, M.A. Sociodemographic Associations with Early Disease Damage in Patients with Systemic Lupus Erythematosus. *Arthritis Care Res.* **2007**, *57*, 993–999. [CrossRef]

19. Bae, S.C.; Hashimoto, H.; Karlson, E.W.; Liang, M.H.; Daltroy, L.H. Variable Effects of Social Support by Race, Economic Status, and Disease Activity in Systemic Lupus Erythematosus. *J. Rheumatol.* **2001**, *28*, 1245–1251.

20. Yelin, E.; Trupin, L.; Yazdany, J. A Prospective Study of the Impact of Current Poverty, History of Poverty, and Exiting Poverty on Accumulation of Disease Damage in Systemic Lupus Erythematosus. *Arthritis Rheumatol.* **2017**, *69*, 1612–1622. [CrossRef]

21. Walsh, S.J.; Gilchrist, A. Geographical Clustering of Mortality from Systemic Lupus Erythematosus in the United States: Contributions of Poverty, Hispanic Ethnicity and Solar Radiation. *Lupus* **2006**, *15*, 662–670. [CrossRef]

22. Ward, M.M. Access to Care and the Incidence of Endstage Renal Disease Due to Systemic Lupus Erythematosus. *J. Rheumatol.* **2010**, *37*, 1158–1163. [CrossRef]

23. Alarcón, G.S.; Bastian, H.M.; Beasley, T.M.; Roseman, J.M.; Tan, F.K.; Fessler, B.J.; Vilá, L.M.; McGwin, G. LUMINA Study Group Systemic Lupus Erythematosus in a Multi-Ethnic Cohort (LUMINA) XXXII: [Corrected] Contributions of Admixture and Socioeconomic Status to Renal Involvement. *Lupus* **2006**, *15*, 26–31. [CrossRef] [PubMed]

24. Flower, C.; Hennis, A.J.M.; Hambleton, I.R.; Nicholson, G.D.; Liang, M.H. Barbados National Lupus Registry Group Systemic Lupus Erythematosus in an Afro-Caribbean Population: Incidence, Clinical Manifestations and Survival in the Barbados National Lupus Registry. *Arthritis Care Res.* **2012**, *64*, 1151–1158. [CrossRef]

25. Williams, W.; Sargeant, L.A.; Smikle, M.; Smith, R.; Edwards, H.; Shah, D. The Outcome of Lupus Nephritis in Jamaican Patients. *Am. J. Med. Sci.* **2007**, *334*, 426–430. [CrossRef]

26. Gómez-Puerta, J.A.; Barbhaiya, M.; Guan, H.; Feldman, C.H.; Alarcón, G.S.; Costenbader, K.H. Racial/Ethnic Variation in All-Cause Mortality Among United States Medicaid Recipients With Systemic Lupus Erythematosus: A Hispanic and Asian Paradox: Racial and Ethnic Mortality Variation Among SLE Patients. *Arthritis Rheumatol.* **2015**, *67*, 752–760. [CrossRef] [PubMed]

27. Dooley, M.A.; Hogan, S.; Jennette, C.; Falk, R. Cyclophosphamide Therapy for Lupus Nephritis: Poor Renal Survival in Black Americans. Glomerular Disease Collaborative Network. *Kidney Int.* **1997**, *51*, 1188–1195. [CrossRef]

28. Cooper, G.S.; Parks, C.G.; Treadwell, E.L.; St Clair, E.W.; Gilkeson, G.S.; Cohen, P.L.; Roubey, R.A.S.; Dooley, M.A. Differences by Race, Sex and Age in the Clinical and Immunologic Features of Recently Diagnosed Systemic Lupus Erythematosus Patients in the Southeastern United States. *Lupus* **2002**, *11*, 161–167. [CrossRef] [PubMed]

29. Costenbader, K.H.; Desai, A.; Alarcón, G.S.; Hiraki, L.T.; Shaykevich, T.; Brookhart, M.A.; Massarotti, E.; Lu, B.; Solomon, D.H.; Winkelmayer, W.C. Trends in the Incidence, Demographics, and Outcomes of End-Stage Renal Disease Due to Lupus Nephritis in the US from 1995 to 2006. *Arthritis Rheum.* **2011**, *63*, 1681–1688. [CrossRef]

30. de Carvalho, J.F.; do Nascimento, A.P.; Testagrossa, L.A.; Barros, R.T.; Bonfá, E. Male Gender Results in More Severe Lupus Nephritis. *Rheumatol. Int.* **2010**, *30*, 1311–1315. [CrossRef]

31. Soni, S.S.; Gowrishankar, S.; Adikey, G.K.; Raman, A. Sex-Based Differences in Lupus Nephritis: A Study of 235 Indian Patients. *J. Nephrol.* **2008**, *21*, 570–575.

32. Hsu, C.-Y.; Chiu, W.-C.; Yang, T.-S.; Chen, C.-J.; Chen, Y.-C.; Lai, H.-M.; Yu, S.-F.; Su, Y.-J.; Cheng, T.-T. Age- and Gender-Related Long-Term Renal Outcome in Patients with Lupus Nephritis. *Lupus* **2011**, *20*, 1135–1141. [CrossRef]
33. Esdaile, J.M.; Joseph, L.; MacKenzie, T.; Kashgarian, M.; Hayslett, J.P. The Benefit of Early Treatment with Immunosuppressive Agents in Lupus Nephritis. *J. Rheumatol.* **1994**, *21*, 2046–2051. [PubMed]
34. Yazdany, J.; Feldman, C.H.; Liu, J.; Ward, M.M.; Fischer, M.A.; Costenbader, K.H. Quality of Care for Incident Lupus Nephritis among Medicaid Beneficiaries in the United States. *Arthritis Care Res.* **2014**, *66*, 617–624. [CrossRef] [PubMed]
35. Naiker, I.P.; Chrystal, V.; Randeree, I.G.; Seedat, Y.K. The Significance of Arterial Hypertension at the Onset of Clinical Lupus Nephritis. *Postgrad. Med. J.* **1997**, *73*, 230–233. [CrossRef] [PubMed]
36. Fiehn, C.; Hajjar, Y.; Mueller, K.; Waldherr, R.; Ho, A.D.; Andrassy, K. Improved Clinical Outcome of Lupus Nephritis during the Past Decade: Importance of Early Diagnosis and Treatment. *Ann. Rheum. Dis.* **2003**, *62*, 435–439. [CrossRef]
37. Faurschou, M.; Starklint, H.; Halberg, P.; Jacobsen, S. Prognostic Factors in Lupus Nephritis: Diagnostic and Therapeutic Delay Increases the Risk of Terminal Renal Failure. *J. Rheumatol.* **2006**, *33*, 1563–1569.
38. Walsh, M.; Solomons, N.; Lisk, L.; Jayne, D.R.W. Mycophenolate Mofetil or Intravenous Cyclophosphamide for Lupus Nephritis with Poor Kidney Function: A Subgroup Analysis of the Aspreva Lupus Management Study. *Am. J. Kidney Dis.* **2013**, *61*, 710–715. [CrossRef]
39. Faurschou, M.; Dreyer, L.; Kamper, A.-L.; Starklint, H.; Jacobsen, S. Long-Term Mortality and Renal Outcome in a Cohort of 100 Patients with Lupus Nephritis. *Arthritis Care Res.* **2010**, *62*, 873–880. [CrossRef]
40. Najafi, C.C.; Korbet, S.M.; Lewis, E.J.; Schwartz, M.M.; Reichlin, M.; Evans, J. Significance of Histologic Patterns of Glomerular Injury upon Long-Term Prognosis in Severe Lupus Glomerulonephritis. *Kidney Int.* **2001**, *59*, 2156–2163. [CrossRef]
41. Mahmoud, G.A.; Zayed, H.S.; Ghoniem, S.A. Renal Outcomes among Egyptian Lupus Nephritis Patients: A Retrospective Analysis of 135 Cases from a Single Centre. *Lupus* **2015**, *24*, 331–338. [CrossRef]
42. Austin, H.A.; Boumpas, D.T.; Vaughan, E.M.; Balow, J.E. High-Risk Features of Lupus Nephritis: Importance of Race and Clinical and Histological Factors in 166 Patients. *Nephrol. Dial. Transplant. Off. Publ. Eur. Dial. Transpl. Assoc. Eur. Ren. Assoc.* **1995**, *10*, 1620–1628.
43. Schwartz, M.M.; Lan, S.P.; Bernstein, J.; Hill, G.S.; Holley, K.; Lewis, E.J. Role of Pathology Indices in the Management of Severe Lupus Glomerulonephritis. Lupus Nephritis Collaborative Study Group. *Kidney Int.* **1992**, *42*, 743–748. [PubMed]
44. Gómez-Puerta, J.A.; Feldman, C.H.; Alarcón, G.S.; Guan, H.; Winkelmayer, W.C.; Costenbader, K.H. Racial and Ethnic Differences in Mortality and Cardiovascular Events Among Patients With End-Stage Renal Disease Due to Lupus Nephritis. *Arthritis Care Res.* **2015**, *67*, 1453–1462. [CrossRef] [PubMed]
45. Donadio, J.V.; Hart, G.M.; Bergstralh, E.J.; Holley, K.E. Prognostic Determinants in Lupus Nephritis: A Long-Term Clinicopathologic Study. *Lupus* **1995**, *4*, 109–115. [CrossRef] [PubMed]
46. Freedman, B.I.; Langefeld, C.D.; Andringa, K.K.; Croker, J.A.; Williams, A.H.; Garner, N.E.; Birmingham, D.J.; Hebert, L.A.; Hicks, P.J.; Segal, M.S.; et al. End-Stage Renal Disease in African Americans with Lupus Nephritis Is Associated with APOL1. *Arthritis Rheumatol.* **2014**, *66*, 390–396. [CrossRef]
47. Ward, M.M.; Pyun, E.; Studenski, S. Long-Term Survival in Systemic Lupus Erythematosus. Patient Characteristics Associated with Poorer Outcomes. *Arthritis Rheum.* **1995**, *38*, 274–283. [CrossRef]
48. Ginzler, E.M.; Diamond, H.S.; Weiner, M.; Schlesinger, M.; Fries, J.F.; Wasner, C.; Medsger, T.A.; Ziegler, G.; Klippel, J.H.; Hadler, N.M.; et al. A Multicenter Study of Outcome in Systemic Lupus Erythematosus. I. Entry Variables as Predictors of Prognosis. *Arthritis Rheum.* **1982**, *25*, 601–611. [CrossRef]
49. Studenski, S.; Allen, N.B.; Caldwell, D.S.; Rice, J.R.; Polisson, R.P. Survival in Systemic Lupus Erythematosus. A Multivariate Analysis of Demographic Factors. *Arthritis Rheum.* **1987**, *30*, 1326–1332. [CrossRef]
50. Kasitanon, N.; Magder, L.S.; Petri, M. Predictors of Survival in Systemic Lupus Erythematosus. *Medicine* **2006**, *85*, 147–156. [CrossRef]
51. Yazdany, J.; Trupin, L.; Tonner, C.; Dudley, R.A.; Zell, J.; Panopalis, P.; Schmajuk, G.; Julian, L.; Katz, P.; Criswell, L.A.; et al. Quality of Care in Systemic Lupus Erythematosus: Application of Quality Measures to Understand Gaps in Care. *J. Gen. Intern. Med.* **2012**, *27*, 1326–1333. [CrossRef]
52. Hanly, J.G.; O'Keeffe, A.G.; Su, L.; Urowitz, M.B.; Romero-Diaz, J.; Gordon, C.; Bae, S.-C.; Bernatsky, S.; Clarke, A.E.; Wallace, D.J.; et al. The Frequency and Outcome of Lupus Nephritis: Results from an International Inception Cohort Study. *Rheumatology* **2016**, *55*, 252–262. [CrossRef]
53. Dooley, M.A.; Jayne, D.; Ginzler, E.M.; Isenberg, D.; Olsen, N.J.; Wofsy, D.; Eitner, F.; Appel, G.B.; Contreras, G.; Lisk, L.; et al. Mycophenolate versus Azathioprine as Maintenance Therapy for Lupus Nephritis. *N. Engl. J. Med.* **2011**, *365*, 1886–1895. [CrossRef] [PubMed]
54. Tamirou, F.; D'Cruz, D.; Sangle, S.; Remy, P.; Vasconcelos, C.; Fiehn, C.; del Mar Ayala Guttierez, M.; Gilboe, I.-M.; Tektonidou, M.; Blockmans, D.; et al. Long-Term Follow-up of the MAINTAIN Nephritis Trial, Comparing Azathioprine and Mycophenolate Mofetil as Maintenance Therapy of Lupus Nephritis. *Ann. Rheum. Dis.* **2016**, *75*, 526–531. [CrossRef] [PubMed]
55. Feldman, C.H.; Collins, J.; Zhang, Z.; Subramanian, S.V.; Solomon, D.H.; Kawachi, I.; Costenbader, K.H. Dynamic Patterns and Predictors of Hydroxychloroquine Nonadherence among Medicaid Beneficiaries with Systemic Lupus Erythematosus. *Semin. Arthritis Rheum.* **2018**, *48*, 205–213. [CrossRef] [PubMed]

56. Alarcón, G.S.; Mc Gwin, G.; Bertoli, A.M.; Fessler, B.J.; Calvo-Alén, J.; Bastian, H.M.; Vilá, L.M.; Reveille, J.D. LUMINA Study Group Effect of Hydroxychloroquine on the Survival of Patients with Systemic Lupus Erythematosus: Data from LUMINA, a Multiethnic US Cohort (LUMINA L). *Ann. Rheum. Dis.* **2007**, *66*, 1168–1172. [CrossRef]
57. Petri, M.; Perez-Gutthann, S.; Longenecker, J.C.; Hochberg, M. Morbidity of Systemic Lupus Erythematosus: Role of Race and Socioeconomic Status. *Am. J. Med.* **1991**, *91*, 345–353. [CrossRef]
58. Montes, R.A.; Mocarzel, L.O.; Lanzieri, P.G.; Lopes, L.M.; Carvalho, A.; Almeida, J.R. Smoking and Its Association With Morbidity in Systemic Lupus Erythematosus Evaluated by the Systemic Lupus International Collaborating Clinics/American College of Rheumatology Damage Index: Preliminary Data and Systematic Review. *Arthritis Rheumatol.* **2016**, *68*, 441–448. [CrossRef]
59. Bastian, H.M.; Roseman, J.M.; McGwin, G.; Alarcón, G.S.; Friedman, A.W.; Fessler, B.J.; Baethge, B.A.; Reveille, J.D.; LUMINA Study Group. LUpus in MInority populations: NAture vs nurture Systemic Lupus Erythematosus in Three Ethnic Groups. XII. Risk Factors for Lupus Nephritis after Diagnosis. *Lupus* **2002**, *11*, 152–160. [CrossRef]
60. Nossent, J.C. Systemic Lupus Erythematosus on the Caribbean Island of Curaçao: An Epidemiological Investigation. *Ann. Rheum. Dis.* **1992**, *51*, 1197–1201. [CrossRef]
61. Flower, C.; Hennis, A.; Hambleton, I.R.; Nicholson, G. Lupus Nephritis in an Afro-Caribbean Population: Renal Indices and Clinical Outcomes. *Lupus* **2006**, *15*, 689–694. [CrossRef]
62. Isenberg, D.; Appel, G.B.; Contreras, G.; Dooley, M.A.; Ginzler, E.M.; Jayne, D.; Sánchez-Guerrero, J.; Wofsy, D.; Yu, X.; Solomons, N. Influence of Race/Ethnicity on Response to Lupus Nephritis Treatment: The ALMS Study. *Rheumatology* **2010**, *49*, 128–140. [CrossRef]
63. ACCESS Trial Group. Treatment of Lupus Nephritis with Abatacept: The Abatacept and Cyclophosphamide Combination Efficacy and Safety Study. *Arthritis Rheumatol.* **2014**, *66*, 3096–3104. [CrossRef] [PubMed]
64. Houssiau, F.A.; Vasconcelos, C.; D'Cruz, D.; Sebastiani, G.D.; de Ramon Garrido, E.; Danieli, M.G.; Abramovicz, D.; Blockmans, D.; Mathieu, A.; Direskeneli, H.; et al. Immunosuppressive Therapy in Lupus Nephritis: The Euro-Lupus Nephritis Trial, a Randomized Trial of Low-Dose versus High-Dose Intravenous Cyclophosphamide: Immunosuppressive Therapy in Lupus Nephritis. *Arthritis Rheum.* **2002**, *46*, 2121–2131. [CrossRef] [PubMed]
65. Bruce, I.N.; Gladman, D.D.; Urowitz, M.B. Factors Associated with Refractory Renal Disease in Patients with Systemic Lupus Erythematosus: The Role of Patient Nonadherence. *Arthritis Rheum.* **2000**, *13*, 406–408. [CrossRef]
66. Garcia Popa-Lisseanu, M.G.; Greisinger, A.; Richardson, M.; O'Malley, K.J.; Janssen, N.M.; Marcus, D.M.; Tagore, J.; Suarez-Almazor, M.E. Determinants of Treatment Adherence in Ethnically Diverse, Economically Disadvantaged Patients with Rheumatic Disease. *J. Rheumatol.* **2005**, *32*, 913–919.
67. Koneru, S.; Kocharla, L.; Higgins, G.C.; Ware, A.; Passo, M.H.; Farhey, Y.D.; Mongey, A.-B.; Graham, T.B.; Houk, J.L.; Brunner, H.I. Adherence To Medications In Systemic Lupus Erythematosus. *JCR J. Clin. Rheumatol.* **2008**, *14*, 195–201. [CrossRef]

MDPI

St. Alban-Anlage 66

4052 Basel

Switzerland

Tel. +41 61 683 77 34

Fax +41 61 302 89 18

www.mdpi.com

*Journal of Clinical Medicine* Editorial Office

E-mail: jcm@mdpi.com

www.mdpi.com/journal/jcm

www.ingramcontent.com/pod-product-compliance
Lightning Source LLC
LaVergne TN
LVHW071014170726
843515LV00006B/1130